THE
ADRENAL CORTEX

THE
ADRENAL CORTEX

BY

I. CHESTER JONES
Department of Zoology, University of Liverpool

CAMBRIDGE
AT THE UNIVERSITY PRESS
1957

CAMBRIDGE
UNIVERSITY PRESS

University Printing House, Cambridge CB2 8BS, United Kingdom

Cambridge University Press is part of the University of Cambridge.

It furthers the University's mission by disseminating knowledge in the pursuit of education, learning and research at the highest international levels of excellence.

www.cambridge.org
Information on this title: www.cambridge.org/9781107502413

First published 1957
First paperback edition 2015

A catalogue record for this publication is available from the British Library

ISBN 978-1-107-50241-3 Paperback

Additional resources for this publication at www.cambridge.org/9781107502413

Cambridge University Press has no responsibility for the persistence or accuracy of URLs for external or third-party internet websites referred to in this publication, and does not guarantee that any content on such websites is, or will remain, accurate or appropriate.

....

Every effort has been made in preparing this book to provide accurate and up-to-date information which is in accord with accepted standards and practice at the time of publication. Although case histories are drawn from actual cases, every effort has been made to disguise the identities of the individuals involved. Nevertheless, the authors, editors and publishers can make no warranties that the information contained herein is totally free from error, not least because clinical standards are constantly changing through research and regulation. The authors, editors and publishers therefore disclaim all liability for direct or consequential damages resulting from the use of material contained in this book. Readers are strongly advised to pay careful attention to information provided by the manufacturer of any drugs or equipment that they plan to use.

CONTENTS

Contents

PREFACE

I WROTE this book, at the suggestion of Professor P. B. Medawar, for the series of 'Cambridge Monographs in Experimental Biology'. It proved to be too long for inclusion there and now appears as a separate volume.

The Eutheria apart, the book contains a complete survey of what is known about the adrenal cortex of animals from marsupials to fish, covering the literature up to about the spring of 1955. It includes many data obtained by my research group at Liverpool. In particular, Dr A. Wright contributed to the chapter on Reptilia and, with Mr J. G. Phillips, to the section on Prototheria; Dr M. A. Fowler to that on Amphibia; Miss M. H. Spalding (Mrs Lawson) to that on Teleostei and Mr Phillips to that on steroids. To these and to Dr N. W. Nowell, Dr M. Christianson and Mr W. N. Holmes who have also worked with me in the field of comparative endocrinology, I am very grateful, especially for the lively and stimulating atmosphere they created.

The literature on the adrenal gland of the eutherian mammals is enormous and growing constantly. The chapter on the Eutheria can therefore, of necessity, be only an incomplete account. In it I have tried to give general background information and to provide a survey of current problems. As much of our information has been obtained from the rat and a few other species, it would be greatly to the advantage of the subject if a wide range of mammalian species were investigated. Moreover, it is my hope that workers will be stimulated to study also the lower vertebrates with an assiduity comparable to that bestowed on the common eutherian laboratory animals.

I am in debt to many people for their generous help in the making of this book, which, of course, remains my responsibility. I gained a great deal by being able to restart endocrinological work after the war under Professor R. O. Greep's skilled and sympathetic guidance at Harvard. More recently a travel grant from the Commonwealth Fund, generously arranged by Mr Lansing V. Hammond, allowed me to visit research centres

Preface

in the United States and to talk over problems with many workers. I am grateful to Dr W. Holmes and Dr P. Gérard for Protopterus material (Plate III); to Dr M. Olivereau and Mademoiselle Fromentin for microphotographs of the adrenal of the eel (Plate IV); to Professor Benoit and Dr Leroy for preparations of the duck adrenal (Plate VII); to Professor H. Waring and Dr I. G. Jarrett for freshly fixed adrenals of monotremes and marsupials (Plates VIII and IX). Professor F. G. Young was most kind in clarifying for me the present status of our knowledge of ACTH. I was helped a great deal by Professor Sir Solly Zuckerman, Professor R. J. Pumphrey, Dr C. L. Smith and Dr D. F. Cole, all of whom read either the manuscript or proofs at various stages and made many valuable suggestions. In addition Dr Cole contributed to the discussion on the action of hormones at the kidney level. Mrs Lawson assisted me throughout the preparation of the book and in the compilation of the bibliography and the index. The Cambridge University Press was at all times helpful, and tolerant of my errors.

I. C. J.

Department of Zoology
UNIVERSITY OF LIVERPOOL

ACKNOWLEDGEMENTS

I AM grateful to the following publishers and publications for material used in compiling the tables and text-figures noted: *Johns Hopkins Hospital Bulletin* (Johns Hopkins Press, Baltimore), Text-figures 1 and 4; *Journal of Pathology and Bacteriology* (Oliver and Boyd, London), Table 1; *American Journal of Anatomy*, *Anatomical Record* (Wistar Institute of Anatomy and Biology, Philadelphia, Pennsylvania), Text-figures 2, 5, 26, 28, 32 and 33 and Tables 13, 14, 15, 21, 27, 29, 30; *Handbuch der microskopischen Anatomie des Menschen*, volume 6 (Springer-Verlag, Berlin, Göttingen, Heidelberg), Text-fig. 3; *British Medical Bulletin* (British Council, London), Table 2 and Text-figures 15, 16; *Ciba Foundation Colloquia on Endocrinology* (J. and A. Churchill, London), Table 5 and Text-figure 9; *Proceedings of the Society for Experimental Biology and Medicine* (New York), Text-figure 8; *The Hormones* (Academic Press, New York), Tables 6 and 7; *Biochemical Journal* (Cambridge University Press), Tables 10 and 11; *Transactions of the Zoological Society of London*, Text-figures 17 and 18; *Pflüger's Archiv für die gesamte Physiologie* (Springer-Verlag, Berlin), Text-figure 17; *Zeitschrift für wissenschaftliche Zoologie* (Geest und Portig, Leipzig), Text-figures 17 and 34; *Zeitschrift für mikroskopisch-anatomische Forschung* (Leipzig), Text-figures 19 and 34; *Memorie della reale Accademia delle scienze dell'Istituto di Bologna*, Text-figure 20; *Proceedings of the Royal Society of London* (Cambridge University Press), Text-figure 21; *Monitore zoologico italiano*, Text-figures 22 and 32; *Archives internationales de Physiologie* (Hermann et Cie, Paris), Tables 17 and 18; *Revista Brasileira de Biologia* (Rio de Janeiro, Brazil), Table 21; *Revista de la Sociedad Argentina de Biología* (Buenos Aires, Argentina), Table 20; Universita de Buenos Aires, Table 21; *Growth* (Menasha, Wisconsin), Tables 21 and 25; *Anatomische Anzeiger* (Fischer, Jena), Table 22; *Endocrinology* (Banta, Menasha, Wisconsin), Table 26; *The Auk* (Cambridge, Massachusetts), Table 24; *Ohio Journal of Science* (Ohio State University and Academy of Science, Columbus, Ohio), Table 24; *Journal of Agricultural Research* (U.S. Government Printing

Acknowledgements

Office, Washington D.C.), Text-figure 27; *Journal of Pharmacology* (Williams and Wilkins, Baltimore, Maryland), Text-figure 29; *American Journal of Physiology* (American Physiology Society, Waverly Press, Baltimore, Maryland), Tables 28 and 30; *Rendiconti delle sessioni della Reale Accademia delle scienze dell' Istituto di Bologna*, Text-figure 32; *Journal of the College of Science, Tokyo* (Faculty of Science, Tokyo, Japan), Text-figure 32; *Quarterly Journal of Microscopical Science* (Clarendon Press, Oxford), Text-figure 32; *Hertwig's Handbuch der vergleichenden und experimentellen Entwicklungslehre der Wirbeltiere* (Fischer, Jena), Text-figure 32; *Journal of Clinical Endocrinology* (C. C. Thomas, Springfield, Illinois), Text-figure 32; *Journal of Anatomy*, London (Cambridge University Press), Text-figure 33; *The Mammalian Adrenal Gland* (Clarendon Press, Oxford), Text-figure 34; *The Life of Vertebrates* (Clarendon Press, Oxford), Text-figure 34.

NOTES

Unless otherwise stated, body weights are given in grammes and organ weights in milligrammes.

Wherever variance of a mean figure is shown, this is the standard error.

INTRODUCTION

MOST of the endocrine glands were seen by the very early anatomists, but the adrenal glands were not described until 1563 when Bartholomaeus Eustachius (1520–74), in his account of the kidneys, noted them as 'Glandulae renibus incumbentes'. Eustachius's drawings of human anatomy were published for the first time in 1714 by Lancisius (1655–1720) who indexed the adrenals as 'Renes succenturiati ab Eustachio primum detecti'. The adrenals have gone by many other names 'capsulae renales' (Spigelius, 1578–1625), 'glandulae renales' (Wharton, 1614–73), 'capsulae atrabilariae' (Bartholinus, 1585–1629). The more familiar name of 'suprarenal capsules' was introduced in 1629 by Jean Riolan the younger (1580–1657) (Rolleston, 1936). The idea that certain organs without excretory ducts elaborated material for discharge into the blood was of slow growth, and, although it was formulated by Haller in 1766, Bordeu in 1775, Schmidt in 1785, Legallois in 1801 and others, it was only by the middle of the nineteenth century that the concept became general. Gulliver in 1840 considered that the adrenals 'pour a peculiar matter into the blood, which has doubtless a special use...'. Carpenter (1852) wrote that 'the vascular glands (spleen, thymus, thyroid and adrenals)... exactly correspond with ordinary glands...and they differ only in being unprovided with excretory ducts for the discharge of the products of their operation'. The work of Berthold (1849), showing that the comb atrophy in castrated birds was prevented by testis transplantation, might have stimulated his contemporaries to further experimental work had it not been overlooked at the time (Biedl, 1913).

The undoubted founders of the conception of internal secretion were Claude Bernard and Brown-Séquard. The term 'internal secretion' was first used by Bernard in 1855 and in 1859 he gave as examples of the organs concerned, the spleen, adrenals, thyroid and lymphatic glands. Meanwhile Addison at Guy's Hospital was observing the effects of disease of the adrenal glands in man and in 1855 he published his classical

The Adrenal Cortex

monograph, *The Constitutional and Local Effects of Disease of the Suprarenal Capsules*, giving the characteristics of adrenal insufficiency, subsequently named Addison's disease. Following this, Brown-Séquard (1856, 1857, 1858) removed the adrenal glands of various animals and, finding that this operation proved fatal, considered that the adrenals were essential for life. This conclusion was not received with favour and it was only towards the end of the nineteenth century that it began to receive support and experimental proof (Tizzoni, 1889; Abelous and Langlois, 1891a, b; 1892; and see Biedl, 1913). In this epoch Brown-Séquard, then a septuagenarian, aroused interest in the study of internal secretions beyond the confines of medicine by purporting to show 'la puissance dynamogénique chez l'homme d'un liquide extrait de testicules d'animaux' (1889). This paper was hardly the foundation stone of modern endocrinology, as some have claimed; rather it was a hopeful side-step in the century's steady march of scientific investigation.

The division of the adrenal into a central and a peripheral part was recognized probably from the early seventeenth century onwards, and Cuvier in 1805 gave an account of their differences. The terms 'medulla' and 'cortex' were introduced, however, by Huschke in 1845, Henry Gray in 1852 and Kölliker in 1854. The foundations of our histological knowledge of the adrenal were laid by Ecker (1846), Kölliker (1854), Harley (1858), Arnold (1866), among others (Biedl, 1913; Bachmann, 1954). Knowledge of function came much later. The conjecture of Riolan the younger (1629) that the adrenals were functional only in foetal life was very persistent; equally persistent was the view that the cortex exerted an antitoxic influence by neutralizing poisons produced elsewhere in the body.

The extraction of a pressor substance and the demonstration of its medullary origin by Oliver and Schafer in 1894 led the way to the isolation of adrenaline and the establishment of its chemical structure and its synthesis (Takamine, 1901; Aldrich, 1901; Stolz, 1904). The fitting of the cortex into its physiological role was begun only in recent times and is as yet incomplete.

The study of the adrenal cortex is a part of the whole branch of science called endocrinology. The term itself was made up

2

Introduction

from ἔνδον ('within') and κρίνω ('I separate') and was employed in the first place, according to Rolleston, by Pende in 1909, and used in England originally by Crookshank in 1914. The principal endocrine glands are the adrenal, the anterior, intermediate and posterior lobes of the pituitary, the thyroid, the islets of Langerhans in the pancreas, the gonads and the parathyroids. These glands have the common property of releasing their secretions into the blood stream without anatomical pathways such as excretory ducts. Some organs such as the thymus and pineal are of uncertain status, and at various times endocrine functions have been attributed to other structures. The secretions of the endocrine glands are nowadays referred to as hormones (from ὁρμάω, 'I excite or arouse'). This name was coined after the discovery of secretin by Bayliss and Starling (1902), in response to their request for a better term than 'internal secretions', and it was first used by Starling in 1905 in his Croonian Lecture. Since that time the subject has grown so tremendously that in this book I am dealing with only a part of one gland, at least as it is constituted in some vertebrates, namely the adrenal cortex. The medulla or chromaffin tissue is mentioned only as it affects cortical function.

CHAPTER I

EUTHERIA

(I) GROSS ANATOMY

THE adrenal glands of the Eutheria are paired structures lying in the region of the anterior pole of the kidneys; though they vary in the closeness of their juxtaposition to them and to the main abdominal blood vessels. The glands are surrounded by both white and brown fat (Cowdry, 1950; Fawcett and Chester Jones, 1949) and are contained within the fascia of the kidneys. The glands take various shapes, spheroid, oval, elliptical, cylindrical or rod-like (Hartman and Brownell, 1949). In the dog, the right adrenal has the shape of a 'conch harpoon', the left, that of a dumbbell (Baker, 1937); in the rat the adrenals are bean-shaped; in primates they are wedge-shaped. In man the gland has a concave lower surface which is applied closely to the cephalad end of the kidney.

The general plan of the blood supply to the adrenals is similar throughout the Eutheria. The glands may receive branches from all the main arteries which pass near them and drain into the local major veins. In the dog, Flint (1900) found that the dorsal and ventral surfaces of the anterior part were supplied by three to five branches of the arteria phrenica, and of the posterior part by two to four branches of the arteria abdominalis; in addition, the posterior dorsal surface received two to six branches of the arteria lumbalis and, more caudally, four to six branches of the arteria renalis (text-fig. 1). These numerous adrenal arteries run, without anastomosing, from the main vascular trunks to the surface of the gland where they lie in the loose external connective tissue whence their twigs anastomose to form an ill-defined plexus. The venous drainage in the dog is not typical of the Eutheria for as many as four large trunks empty into the lumbar vein which passes through the hilus on the ventral surface of the gland, and into the renal and phrenic veins. In man (Gérard, 1913; Pick and Anson, 1940; Anson,

4

Cauldwell, Pick and Beaton, 1947), the adrenal arteries arise from phrenic, aortic and renal sources with varying pre-dominance in different individuals. The adrenal arteries, as they run towards the gland, frequently divide so that as many as fifty may reach its periphery.

There is only a single vein leaving each gland and this is typical of most eutherian glands. The vein in man on the right

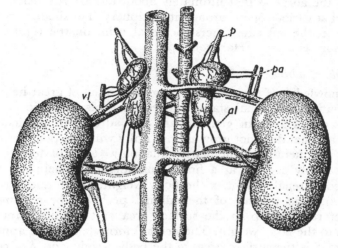

Text-fig. 1. General anatomical relations and gross blood supply of the adrenals in an adult dog (from Flint, 1900). *al*, arteria lumbalis; *vl*, vena lumbalis; *p*, arteria phrenica; *pa*, arteria phrenica access.

side is short and enters the inferior vena cava; on the left it descends to the renal vein receiving, on its way, the inferior phrenic vein and capsular tributaries. In the rat (Gersh and Grollman, 1941; Harrison, 1951), the anterior part of the gland on each side is supplied by an artery which, arising from the aorta, divides into two branches, and after further sub-division ramifies on the surface of the gland without apparent anastomoses. The medial part of the gland is vascularized by a separate artery arising from the ventral aspect of the aorta. A single adrenal vein on each side runs either directly to the vena cava or to the renal vein. In the cat the blood supply of the adrenal comprises numerous small arteries arising from the aorta, both renal arteries, the adrenolumbar arteries, one ileo-lumbar artery, the phrenic arteries and the coeliac axis. The main venous channels open directly through several small apertures into the adrenolumbar vein which lies in a depression on the face of the gland (Bennett and Kilham, 1940). The rabbit adrenal is chiefly supplied by subdivisions of a branch of the adrenolumbar artery, together with subsidiary vessels from the aorta so that ultimately about ten arteries reach the gland at various points around its periphery. The single adrenal vein on the left side enters the renal vein, on the right, the inferior vena cava (Harrison, 1951).

Weight of adrenal glands

Knowledge of the weights of organs is often of great help in estimating functional status and especially is this true of the endocrine glands in general, and the adrenal in particular. The adrenal does not grow isometrically with the body weight and, in general, the allometric equation $y = bx^a$ (Huxley, 1932) must apply at least as a first approximation (where y is the paired adrenal weight, x the body weight, and b and a constants). The weights of the adrenal probably bear a more direct relationship to the surface area of the animal rather than to the body weight. The general formula $S = Kw^{\frac{2}{3}}$ applies where S is the surface area, w the body weight and K a constant with a value of about 10 for many animals (Benedict, 1938). Houssay and Molinelli (1926) thought that, while the relationship of adrenal weight to body weight was exponential, the weight of adrenal per given surface area remained constant. In many endocrinological experiments, adrenal and body weights do not change in the same direction and for this reason gland weights are frequently given in mg./100 g. of body weight (often written briefly mg. %). Such figures may well be misleading and both absolute and relative figures should be used to obviate false comparisons. Furthermore, body fat may be laid down in different amounts in otherwise comparable animals. For this reason Korenchevsky (1942) recommended taking fat-free body weight. He found that the absolute weight of the adrenal in the rat from the age of seven to nine months onwards (in strains which are senile about

6

fourteen to eighteen months of age) increases only very slowly and may remain stationary. At the same time the fat-free body weight continues to increase and adrenal weight expressed in terms of mg./100 g. body weight therefore shows a decrease (table 1) (cf. Yeakel, 1946).

TABLE 1. *Relative weight (mg.) of adrenal glands per 100 g. body weight (minus weight of adipose tissue) in normal male rats of different ages (extracted from Korenchevsky, 1942)*

	Age groups in days				
	21–30	31–40	41–50	51–60	61–70
No. of rats	84	64	41	47	36
Adrenal weight (mg. %)	42	33	27	24	21
	Age group in days				
	81–90	101–125	151–200	300–400	401–500
No. of rats	20	45	14	8	6
Adrenal weight (mg. %)	22	19	16	14	14

A vast amount of information has been collected about organ weights of the laboratory rat, a fair amount for man, the guinea-pig and dog, some for the rabbit, cat, mouse and hamster, and for the rest of the Eutheria only random and scattered data (see Bachmann, 1954). Donaldson (1924) produced very useful reference tables for the albino and the Norway rat (text-fig. 2), which show the weight of the adrenal gland increasing regularly, *pari passu* with age, to a plateau. This may provide a rough generalization for eutherians in general but, for example, changes occur in the human adrenal at birth (p. 112) and in the mouse at puberty and pregnancy (p. 109) which mar the smoothness of this type of curve (text-fig. 3). Also, in the guinea-pig at least, the curve is diphasic, there being a greater increase in adrenal weight relative to body weight above 500 g.—a weight roughly equivalent to 100 days of age (Bessesen and Carlson, 1923; Mixner, Bergman and Turner, 1943). The weight of the adrenal in addition to varying with age also alters in accord with such physiological episodes as oestrus and pregnancy, and, in proportion to the body weight, varies from strain to strain and from species to species. The mouse provides a good example of intra-strain variation (table 2). The laboratory rat has become a distinct subspecies and the wild rat has a much heavier adrenal, both relative to body weight and absolutely, than the albino and other domestic strains (Watson, 1907; Donaldson and King, 1929; Rogers and Richter, 1948).

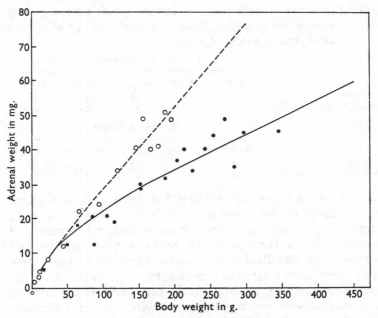

Text-fig. 2. Paired adrenal weight plotted against body weight for the rat (male figures below 50 g. obtained from ninety-two Jackson strain animals; above 50 g. from fifty-three Wistar strain; female figures, below 50 g. by eighty-four Jackson and above 50 g. by twenty-nine Wistar) (from Donaldson, 1924.) ● males; ○ females.

The guinea-pig is remarkable in possessing a very large adrenal for its size (Bessesen and Carlson, 1923; Kojima, 1928; Eaton, 1938; Mixner *et al.* 1943). Eaton, in inbred families of guinea-pigs, found: females (24), mean body weight 683 ± 14·2 g., mean paired adrenal weight 803 ± 19·9 mg., i.e. 117·5 mg. per 100 g. body weight; males (94), mean body weight 688 ± 10·4 gm., paired adrenal weight 868 ± 20·3 mg., i.e. 126 mg. per 100 g. body weight, and these figures may be compared with those for the rat and mouse (tables 1 and 2). Though Kojima's guinea-pigs had somewhat small adrenals for the species, examination of his data brings out the further point, that the left and right adrenals are not necessarily the same weight, namely, males (17), mean body weight 533.52 ± 8.8 g., mean left adrenal weight 105.29 ± 7.8 mg., mean right adrenal weight 89.41 ± 7.85 mg. (Kojima, 1928, recalculated). This is frequently true of mammals. Donaldson (1924) noted that the left adrenal was 10 % to 20 % greater than the right in the rat; on the other hand, in the dog, for example, in 56 %

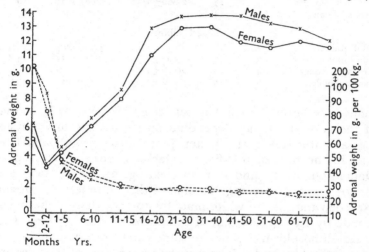

Text-fig. 3. Adrenal weight in man at different ages (from the data of Rössle and Roulet, 1932 in Bachmann, 1954, recalculated). ——×——×—— males, actual paired adrenal weight; O———O—— females, the same; × - - - - - ×- - - - males, adrenal weight in g. per 100 kg. body weight; O - - - - - -O - - - - females, the same.

of the males and in 54 % of the females examined by Baker (1937), the right adrenal was larger than the left. No functional significance need be attributed to such differences in the paired organs and one of the factors involved may well be their anatomical position, such as whether or not they are equally closely adpressed to the kidney.

In some eutherians the relative and absolute weight of the adrenals in the female is greater than in the male. In the male

9

The Adrenal Cortex

TABLE 2. *Body and adrenal weights of two strains of mice at different ages illustrating the sex difference in adrenal weight characteristic of many mammals, the increase in absolute adrenal weight with age (demonstrated here by the males), and the variations in adrenal weight as between un-mated and parous females irrespective of age (Chester Jones, 1955)*

	Age months	n	Body weight	Adrenal weight (mg.)	mg. %
	Strong A strain				
Adult males	2–4	45	23·00 ± 1·12	2·58 ± 0·03	11·4
Adult males	6–8	18	28·72 ± 0·59	2·83 ± 0·07	9·8
Unmated females	4	17	23·20 ± 1·38	3·88 ± 0·10	16·7
Parous females	6–7	24	24·62 ± 0·34	3·63 ± 0·18	14·7
	Bagg C albino				
Adult males	4	28	26·2 ± 1·23	3·50 ± 0·09	13·3
Adult males	8	27	25·8 ± 1·54	3·94 ± 0·11	15·2
Unmated females	4½	28	20·26 ± 1·58	5·19 ± 0·11	25·6
Parous females	8	15	25·22 ± 0·69	6·47 ± 0·20	25·6

mouse the adrenal weight is between 57 % and 77 % of that of the co-eval female, depending on strain, age and physiological status (table 2; Chester Jones, 1948; 1955), and likewise in the rat (cf. text-fig. 14; Jackson, 1913; Hatai, 1914; Donaldson, 1924), and in man (text-fig. 3), where Swinyard (1940) found the average volume of the white female adrenal exceeded that of the white male by 10·1 %, that of the negro female by 19·3 % and that of the negro male by 31·4 %. This sexual dimorphism depends to some extent on the gonads and gonadotrophins and will be discussed later. Sex differences are not so marked in other common animals. In the dog, Baker (1937) found little difference between adrenal weight of the mature male dog and that of females in dioestrus (see also Baker, 1938). The guinea-pig has no apparent sex difference in adrenal size. The golden hamster reverses the condition seen most strikingly in the mouse and rat, in that the weight of the adrenal of the non-hibernating female is 78·8 % of that of the male (cf. table 13).

Cortex-medulla relationship

A cut through the middle of the gland reveals that there are two types of tissue in its make-up. In the centre of the gland is the medulla consisting of cells derived from the sympathetic nervous system, and the peripheral cortex composed of cells derived from mesoderm (ch. VII). The proportion of cortex to medulla is different in different species and in a given species the amount of cortex depends very much on the functional state of the gland. Generally, throughout the Eutheria, the medulla of the adult animal does not change in size very much or very quickly, while, on the other hand, the adrenal cortex is prone to rapid and marked changes in volume. A brief glance at the general cortex-medulla relationship is of some interest. In the mouse the medulla is about the same size in adult male and female mice. However, the whole gland is about 25 % bigger in the female where the cortex accounts for $82 \cdot 09 \pm 0 \cdot 42$ % of the gland compared with $75 \cdot 64 \pm 0 \cdot 82$ % in the male (Chester Jones, 1948). After birth, the cortex increases in size more rapidly than the medulla. In the rat, the medulla is about 25 % of the whole gland at birth, about 7 % at ten weeks old and some 10 % at one year old (Jackson, 1919). Hett (1928) brings out a similar point for the mouse.

An idea of the actual weight of the cortex and medulla in rats is given by Donaldson (1928a, b). He found that in albinos the cortex and medulla weighed $30 \cdot 3$ mg. and $2 \cdot 8$ mg. respectively in males and $47 \cdot 1$ mg. and $3 \cdot 1$ mg. respectively in females, while in wild rats the respective figures were $75 \cdot 3$ mg. and $4 \cdot 7$ mg. in males, and $92 \cdot 7$ mg. and $4 \cdot 3$ mg. in females. In the human, Quinan and Berger (1933) found the medulla to be 10 % of the whole gland while Swinyard (1940) estimated that it occupied 8 % of the gland in white males and $4 \cdot 9$ % in white females. Where there is little sex difference in adrenal weight, the medulla occupies about the same percentage of the whole gland in males and females. For example, Baker (1937) found that in the dog the medulla was $17 \cdot 5$ % of the whole gland in the mature male and $18 \cdot 6$ % in the female in dioestrus.

(II) HISTOLOGY

Connective tissue surrounds the adrenal gland and also provides an internal supporting framework. Argyrophilic fibres predominate in the cortex as a whole and together with

The Adrenal Cortex

collagenous fibres (in so far as the distinction between silver-reducing and aniline-blue-staining fibres is nowadays maintained) form the bulk of the connective tissue, though elastic fibres can be found, especially in the outer capsule (da Costa, 1913; Sacarrão, 1943). The amount of connective tissue present in the gland varies from species to species. The framework is not a dense one in the rat and mouse, in man it is moderately marked, while in the sheep and cow it can be very pronounced. Histological examination of the eutherian adrenal cortex, after routine fixing and staining, allows three main concentric zones to be distinguished (Pl. I, fig. 1). Arnold (1866) gave to them the names 'zona glomerulosa' (dim. of Lat. *glomus*, a ball), 'zona fasciculata' (dim. of Lat. *fascis*, a bundle) and 'zona reticularis' (dim. of Lat. *rete*, a net). The names are descriptive of the patterns formed by the parenchyma or cortical tissue itself, in that the zona glomerulosa, lying against the outer connective tissue capsule, is formed into balls or loops of cells, the zona fasciculata comprises long columns of cells running centripetally and then breaking up into a less organized area, the zona reticularis, surrounding the medulla. The connective tissue framework contributes to the pattern, as it outlines the groups of cells in the zona glomerulosa and demarcates the lines of cells in the zona fasciculata. In the zona reticularis each cell is enclosed by heavier strands of connective tissue to form a basket-work of reticular fibres.

The blood system, too, falls into the same general pattern and it was on the arrangement of blood vessels and connective tissue that Arnold based his nomenclature. Flint (1900) described the distribution of the cortical blood vessels in the dog, Bennett and Kilham (1940) in the cat and Gersh and Grollman (1941) in the mouse and rat. Flint divides the arteries into three types: (i) arteriae capsulae, (ii) arteriae corticis and (iii) arteriae medullae (text-fig. 4). These latter turn abruptly down at right angles from the capsule, passing without branching through the cortex to supply the medulla. The cat is similar in principle though not in detail to the dog. Bennett and Kilham found that, in the cat, the arteries passing to the adrenal penetrate the capsule and assume a position directly under the connective tissue of the capsule and outside the parenchymal cells of the gland proper. Under the capsule, these arteries ramify and anastomose freely, forming an extensive subcapsular arterial plexus investing the gland and giving off vessels which supply the entire cortex. Apart from arterial loops of vessels which leave the subcapsular plexus, dip down and return without branching, and the arteriae medullae already noted, the only vessels in the cortex are capillaries. Therefore, the arteriae corticis of Flint in the dog (that is, actual vessels entering the cortex and there breaking up into capillaries)

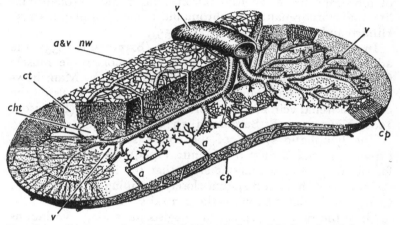

Text-fig. 4. Reconstruction of the left adrenal of an adult-dog, viewed from the ventral surface (from Flint, 1900). *v*, veins; *a*, arteries; *cp*, capillaries; *a* and *v nw*, arterial and venous network; *ct*, cortex; *ch.t*, chromaffin tissue (medulla).

are absent in the cat. Likewise Bennett and Kilham found that in the cat the capsule is served by capillaries rather than vessels sufficiently large to be named arteriae capsulae. The blood supply in the rat and mouse resembles that described for the dog. In addition, Gersh and Grollman note arteriolar loops which may or may not send off capillaries into the cortex as they run through it (cf. Lever, 1952).

In the cat, as in the Eutheria generally, the cortical capillaries embrace the knots of cells of the zona glomerulosa and pass centripetally in the connective tissue septa demarcating the groups. Once at the level of the zona fasciculata the capillaries become straighter and run parallel with the columns of cells, giving off branches and forming anastomoses. At the zona reticularis the capillaries widen and the capillary pattern becomes more plexiform and intricate. The main venous return from the cortex begins in the zona reticularis and at the cortico medullary border where there are large sinuses and venous branches which gather the blood from the capillaries and lead it into the extensive and highly branched venous tree of the medulla, whence it leaves by the adrenal vein. Thus, apart from minor subsidiary channels, nearly all the blood from the cortex traverses the medulla, a point of some importance in consideration of the production of adrenaline and noradrenaline by chromaffin tissue (Coupland, 1953; Hillarp and Hökfelt, 1954; West, 1955).

In the cat, as in many other species (Bargmann, 1933), the adrenal veins of the medulla have no demonstrable muscle fibres in their walls (Bennett and Kilham, 1940). Man, however, possesses longitudinal smooth muscle in the walls of the central adrenal vein and some of its larger branches (Ferguson, 1906; Kolmer, 1918; Bargmann, 1933).

The lymphatic system of the cortex seems to have been explored in detail only by Kumita (1909) whose illustrations for the dog adrenal make it clear that the lymphatics follow the course of the blood system closely, draining in the medulla to a plexus which surrounds the central vein.

Since the time of Arnold, histologists have found variations in the three main types of zones consistently enough to demand enlargement of the descriptive terminology. The terms themselves have synonyms, some of which I shall note later. The terms I find most useful (cf. Nicander, 1952) are:

outer connective tissue capsule
zona glomerulosa
zona intermedia
zona fasciculata $\begin{cases} (a) \text{ outer zona fasciculata} \\ (b) \text{ inner zona fasciculata} \end{cases}$
zona reticularis
medullary connective tissue capsule

In addition, other cortical layers are sometimes found, such as the X zone (Pl. II, fig. 6) and transient or foetal cortex (Pl. II, fig. 10) and these will be discussed later (p. 109 et seq.).

Histology

The zona intermedia is a band of narrow, cuboidal cells with small, darkly staining nuclei lying between the zona glomerulosa and the zona fasciculata (Pl. I, fig. 1). The zona intermedia is also referred to as the 'transitional zone' (Greep and Deane, 1947), 'zone of compression' (Mitchell, 1948), 'intermediary zone' (Nicander, 1952), 'sudanophobic zone' (Reiss, Balint, Oestreicher and Aronson, 1936). The presence and extent of the zone depends on the functional state of the gland, as indeed does the appearance of all the cortical zones (ch. VIII).

Two other terms should be noted: firstly 'zona spongiosa', spongy zone and spongiocytes (Guieysse, 1901) are often employed for the zona fasciculata and its cells, particularly in the French literature, and secondly 'zona arcuata' for the zona glomerulosa in those species, such as the horse, where the looping of the groups of cells is particularly pronounced. It is true that other descriptive terms exist for divisions of the cortex, such as those of Bennett (1940), but these prejudge the issue as to their function, discussion of which is reserved until ch. VIII. This section is a general account solely of histological appearance.

Variations in the histological appearance of the cortices in different eutherian species are only alterations of a common ground plan (Kolmer, 1918; Nicander, 1952). There is varying predominance of the three major zones but these are always present, except in disease. For example, both the horse and the dog have a prominent zona glomerulosa, while in the monkey, hamster and mouse it is less marked. Both the hamster (Holmes, 1955) and the adult male mouse (Chester Jones, 1948) have an ill-defined zona reticularis while in the pig it is an obvious layer.

The zona fasciculata is the widest zone of the three main ones, the zona glomerulosa takes up a narrow concentric band underneath the capsule, while the zona reticularis, subject to much intraspecific variation, can be quite wide though never normally as wide as the zona fasciculata. As a rough guide we can take Swinyard's (1940) figures for man which gave the zona glomerulosa as occupying 15% (using six glands of his material), the zona fasciculata 78·4% and the zona reticularis 6·4% of the total cortical volume.

The more detailed histological structure of the zones has frequently been described. Readers are referred to Hoerr (1931; 1936a, b), Bennett (1940), Deane and Greep (1946), Greep and Deane (1949a, b), Bourne (1949), Weber, McNutt and Morgan (1950) and Bachmann (1954), for a guide to the literature, while the older literature on this and all other endocrinological matters is to be found in Biedl (1913).

Zona glomerulosa. The cells of the zona glomerulosa, arranged in groups by the connective tissue trabeculae, possess basophilic nuclei which show a few chromatin masses and one or two nucleoli (Pl. I, fig. 3). The cytoplasm after routine methods involving the use of fat solvents is often vacuolated. Between the vacuoles it is lightly acidophilic and after Masson's stain, for example, finely granular and has a reddish or purplish appearance. The overall colour after routine staining rests on the degree of cytoplasmic vacuolation which depends in turn on the number and size of the fat droplets. The nuclei vary in shape in this zone from round, oval to sausage-shaped with intermediate forms. The cells of the zona glomerulosa frequently contain a large number of mitochondria: in the rat they are fine granules (Deane and Greep, 1946), though in the mouse seemingly granular forms are in fact rods and filaments seen end on (Miller, 1950); in the guinea-pig they take the shape of straight thin filaments, short rods and granules (Hoerr, 1931); and in the cat they are densely packed rods arranged throughout the cytoplasm. The mitochondria in the zona glomerulosa of all species are in the interstices between vacuoles so that, again, variation in the latter may be a deciding factor in their appearance, and indeed the number of mitochondria and the concentration of vacuoles (i.e. lipid droplets) may be inversely correlated (Miller and Riddle, 1942; Miller, 1950). The Golgi apparatus, revealed by da Fano's method, consists of a network of black connecting channels in the form of a conical or semilunar mass with a concave base adjacent to the nucleus. Generally, the apparatus has no particular orientation towards the nearest capillary (Bourne, 1934a; Reese and Moon, 1938), though it apparently has in the cat (Bennett, 1940).

Zona fasciculata. The columnar cells of the zona fasciculata, arranged in radial rows, have nuclei which are round in section with one or two nucleoli (Pl. I, fig. 4). The cells of the outer region of the zona fasciculata are larger than those of other zones, often attaining two to three times the diameter, as seen in section, of a glomerulosal cell. The abundant cytoplasm contains many lipid droplets. In some species, when the vacuoles are ordinarily plentiful, the zone has a foamy appearance from which arose the term 'spongiocyte' noted above. There seems to be general agreement that, in all species studied, the mitochondria lying in the cytoplasm are spherical or rod-like in shape. The Golgi apparatus (after da Fano's method) in the outer zona fasciculata cells of the cat shows as a cap of tangled black threads attached like a cone to the nucleus and sending branches and projections between the neighbouring lipid vacuoles (Bennett, 1940), a form which is commonly found (Greep and Deane, 1949 *b*). The outer zona fasciculata merges into the inner region where the cells are smaller though the radial arrangement is still apparent. In many species the cytoplasm here takes up acidophilic stains more intensely and this is correlated with a lessening in the number of vacuoles. Connective tissue is more apparent in this region than in the outer zona fasciculata. The mitochondria of the cells are of the same form, but the Golgi apparatus tends to be more compact with fewer projections.

Zona reticularis. The zona reticularis appears where the radial rows of the inner zona fasciculata cease and, becoming disrupted and broken, surround large blood sinuses and are demarcated by heavy connective tissue strands, fibres of which tend to encircle individual cells. The cells themselves are smaller than those of the rest of the cortex, with the exception of those of the zona intermedia, and are of several different types (Pl. I, fig. 5). Species vary in the appearance of the zona reticularis but in general it is characterized by the small size of its cells together with more acidophilic cytoplasm and the absence of cell vacuoles or, if present, their large size.

In the guinea-pig, Hoerr (1931), confirming Dostoiewsky (1886) and others, found two types of cell in the zona reticularis, called 'light' and 'dark' cells. Dark cells are, for the most part, smaller than the light cells, have excavated contours, numerous mitochondria, which may be enlarged, angular, of unequal sizes and clumped. The cytoplasm of the dark cells stains intensely with all the usual cytoplasmic dyes and gives a deep black colour with iron haematoxylin, for which reason they are termed 'siderophil' cells. The dark cells may contain abundant pigment granules and fat vacuoles. The light cells are round, the cytoplasm is finely granular, the nucleus is large, usually vesicular and pale-staining, there are few mitochondria and fat vacuoles; there is little pigment. Between these two cell types there are many intermediate forms. The Golgi apparatus in all the cells is usually deeply staining, compact and often fragmented. In the zona reticularis, too, are cells which, with pycnotic nuclei (blackened with such stains as haematoxylin), shrunken cytoplasm and avidity for trypan blue (Darlington, 1937), can be regarded as dying or dead cells (Hoerr, 1931; Bennett, 1940). Hoerr feels too that the light and dark cells die 'as they have lived, retaining their identity to the end' (p. 155).

Although no worker denies that necrotic cells can be found in the zona reticularis, their abundance and possible significance is disputed (ch. VIII). Dead cells, when they do occur, are most likely to be found in the zona reticularis but they have been occasionally found in the zona fasciculata (Bennett, 1940). It has been claimed that necrotic cells also occur in the zona glomerulosa (Vaccarezza, 1945, 1946; Bachmann, 1941; 1954) but this is not the usual finding.

Histochemistry

The histological appearance of the adrenal cortex does not give a great deal of information about the functional state of the gland. For this reason, variation in the amount of fat present and differences in the size of fat droplets are used to estimate the activity of a gland. In the adrenal cortex the fat is made up of a mixture of cholesterol and its esters, the palmitic, stearic and oleic esters of glycerol and the actual steroid hormones themselves (p. 29; and Pearse, 1953). The most usual colorants in routine use for lipid are the sudans and osmic acid. The sudanophilia and the osmophilia of the cortical lipid, the extent of the colouring of the various zones and the size of the stained cytoplasmic droplets can be interpreted to give an estimate of the functional activity of the gland if due caution is used. In the rat, for example, end-point criteria can be established so that large globules of fat tending to run together are indicative of low activity, while fine and numerous droplets are typical of active glands. Arguments from the particular to the general must be avoided and the criteria should be used in an *ad hoc* way in the light of the experimental conditions (Deane, Shaw and Greep, 1948).

In an effort to obtain more precise information on the activity of a gland, recourse has been made to histochemical techniques. Bennett (1940) was a pioneer in the application of histochemical methods to the adrenal cortex. Some methods depend on the adrenocortical hormones possessing a reactive ketone group, but in any case the amount of hormone present at any one instant of time in an adrenal is probably so small that it could not be revealed by precise histochemical tests, did they exist. Among the methods used on the adrenal cortex are the phenylhydrazine reaction and the plasmal reaction. The former is not specific for adrenocortical steroids, as Bennett (1940) thought, and those substances which give the plasmal reaction probably also stain after phenyl-hydrazine (Gomori, 1942; Albert and Leblond, 1946, 1949). The plasmal reaction (Feulgen and Voit, 1924) indicates the acetal phosphatides (see Cain, 1949*a*, *b*; Pearse, 1953; Fetzer, 1953); it is doubtful if the method reveals ketone groups of steroid hormones even when used with rigorous care. Despite the lack of specificity, those areas in the cortex which stain with the sudans are not necessarily co-extensive with those giving the phenylhydrazine and plasmal reactions and the differing distributions often help to throw light on the functional state of the gland, although imprecisely (Yoffey and Baxter, 1949; Cain, 1950).

Another reaction often used is the Schultz reaction, and the blue-green colour obtained with this histochemical modification of the Liebermann-Burchardt reaction seems to distinguish cholesterol and its esters. Cholesterol and its derivatives are the possible precursors of the adrenocortical hormones. Furthermore, variations in total cholesterol content of the adrenal estimated chemically has been shown in the rat to correspond to functional changes (Sayers and Sayers, 1948).

Another method useful for showing cholesterol, but useless for distinguishing adrenocortical steroids, is the Windhaus (1910) digitonin reaction. Digitonin precipitates those steroids including cholesterol and some corticosteroids with 3β-hydroxyl groups *trans* to the carbon-10 methyl group and the birefringent crystals so formed can be examined with a polarizing microscope. This latter is sometimes used directly on frozen sections, either fresh or after formalin fixation, when the cytoplasmic lipid droplets form birefringent crystals, and their type (whether fine or coarse) and their number may be shown to vary with the activity of the gland (Weaver and Nelson, 1943; Yoffey and Baxter, 1947). Sayers (1950) considers that 'the battery of specialized stains has contributed no more to our knowledge of the chemistry and physiology of the adrenal gland than the Sudan stains'.

In most eutherian species the cells of the zona fasciculata contain numerous small droplets of lipid which are plentiful in the outer and sparse in the inner region. The lipid content of the zona glomerulosa is subject to considerable inter-specific variation (Pl. I, fig. 2). This zone is particularly rich in lipid droplets in the dog and the rat and these are generally larger and more sudanophilic than those in the zona fasciculata. On the other hand, in the horse, mouse, pig, rabbit and cat, the zona glomerulosa has little lipid (Nicander, 1952; Bennett, 1940; Chester Jones, 1949*a*, *b*; Schweizer and Long, 1950). The zona reticularis is very variable in its lipid content. On some occasions no lipid can be seen. At other times, it is present in the cells as large globules which tend to run together. The lipid can become pigmented to form lipochromes and a type of fatty degeneration, 'brown degeneration', can occur, especially in the mouse (Chester Jones, 1948; 1950).

In those glands which have normally an obvious amount of lipid there is often an area between the zona glomerulosa and the outer zona fasciculata, where the cells do not take up the Sudan stains, and for this reason it is sometimes referred to as the 'sudanophobic zone'. It is roughly co-extensive with the zona intermedia but the variation in lipid content is such that the latter term is preferable and is capable of more precise definition. The sudanophobic zone is more particularly applicable to the adrenal of the rat, an animal upon which we base, to an unfortunately large extent, much of our endocrinological knowledge and assumptions. Even in the rat, while the zone is present in the adrenal of the immature of both sexes, it is normally absent in the adult female and present in the adult male gland. Further observations on the zona intermedia will be made in ch. viii.

The histologist, when examining cortical lipid, is able to see some of its components but not the ones he would really like to estimate, the steroid hormones. Nevertheless, he is able to make some deductions about gland activity from the reactions employed. In the less investigated lower vertebrates it is helpful to know whether or not the cytoplasmic lipid droplets contain similar sudanophilic, osmophilic, Schultz- and Schiff-positive and birefringent material. It is also important to remember that the golden hamster (Alpert, 1950; Knigge, 1954*a*), some other hibernants (Deane and Lyman, 1954) and some ruminants (Nicander, 1952) apparently secrete the normal adrenocortical steroids without the occurrence of lipid droplets.

The adrenal cortex is rich in another substance, namely ascorbic acid. It can be determined chemically by Roe and Kuether's (1943) method and the variation in ascorbic acid content is used as an index of gland activity (Giroud and Ratsimamanga, 1942; Sayers, Sayers, Liang and Long, 1945, 1946). No role has been assigned to vitamin C in the adrenal, though at one time it was suggested as a link in the formation of adrenocortical steroids (Lowenstein and Zwemer, 1946). In the scorbutic guinea-pig (requiring dietary vitamin C as opposed to the rat which synthesizes its own), the adrenocortical hormones are produced until death (Prunty, Clayton, McSwiney and Mills, 1955). The specificity of the histochemical method for ascorbic acid is in some doubt (see Pearse, 1953; Bourne, 1955), but silver-nitrate-reducing substances occur throughout the cells of the cortex; in the zona glomerulosa they occur as fine granules, and as coarse ones in the zona fasciculata and the zona reticularis (Greep and Deane, 1949*b*).

The Adrenal Cortex

Accessory cortical bodies

Small pieces of cortical tissue can be found against the adrenals themselves where they form nodules (Waring and Scott, 1937). In addition, they may occur in the peri-renal fat, and throughout the abdomen in the retro-peritoneal tissue. Occasionally accessory cortical tissue may be found in the gonads, particularly the testes, and are referred to as 'adrenal rests'. The accessory bodies represent the cortical tissue which did not join in the general coalescence of the original primordia (ch. VII). They are generally microscopic in size and become of significance in experiments when the adrenal glands are removed. In these cases the accessory bodies may be stimulated to function and so alleviate the expected signs of adrenal insufficiency (p. 58).

Nervous system

There is general agreement that the adrenal cortex lacks a nerve supply (Greep and Deane, 1949*b*). The only nerves to be seen in the cortex are radial bundles of nerve fibres traversing the cortex to pass to the medulla which is innervated by preganglionic fibres (Hollinshead, 1936; Swinyard, 1937; Bennett, 1940). Descriptions of nerve endings in the adrenal cortex, however, do exist (Stöhr, 1935; Alpert, 1931; Kiss, 1951) and probably are a reflexion of the difficulty of differentiating between connective tissue and nerve fibres by the usual histological techniques. The problem of the innervation of the anterior lobe of the pituitary is beset with the same type of difficulty (Vazquez-Lopez and Williams, 1952). Recently, Lever (1952, 1955) has described fine nerves, with occasional boutons, in relation to vessel walls in the outer layers of the rat adrenal cortex. In addition he observed button-shaped nerve endings in relation to parenchymal cells especially in the zona glomerulosa. Lever considers that sudden alterations in the cortical capillary flow are effected by nervous control of the calibre of the cortical arteries.

Effects of hypophysectomy

The anterior lobe of the pituitary secretes several hormones of which adrenotrophin (ACTH), governing the adrenal cortex, is one (p. 47). The pituitary can be removed surgically from an animal (Smith, 1927). This operation is referred to as hypophysectomy, and means that the anterior, intermediate and posterior lobes of the pituitary have been removed. When the anterior lobe alone is removed, the term 'adenohypophysectomy' may be used, while 'neurohypophysectomy' denotes the removal of the posterior lobe alone. The pars tuberalis is generally not removed in the operation as practised on rodents although it may be in other animals, such as the dog. It is conceivable that the pars tuberalis has some function though none has yet been demonstrated. Hypophysectomy will be used here as the general term unless a special point has to be made about responsibility of action.

The results of hypophysectomy of the common laboratory animals are now well documented: the rat (Smith, 1930; Crooke and Gilmour, 1938; Simpson, Evans and Li, 1943; Sarason, 1943; Deane and Greep, 1946, among others), the rabbit (Ikeda, 1932), the cat (McPhail, 1935), the dog (Houssay, Biasotti and Magdalena, 1932; Houssay and Sammartino, 1933; Lane and de Bodo, 1952), the guinea-pig (Schweizer and Long, 1950), the hamster (Knigge, 1954*b*), the mouse (Leblond and Nelson, 1937; Chester Jones, 1948; 1949*a*, *b*, 1950; Miller, 1950). The operation is now used in man for some severe conditions (e.g. advanced carcinoma of the breast; Luft *et al.* 1954, 1955).

The changes that occur in the adrenal cortex after hypophysectomy are similar in both sexes. The medulla is not believed to change in size after hypophysectomy (Collip, Selye, and Thompson, 1933; Houssay and Sammartino, 1933; Deane and Greep, 1946). Christensen (1954) recently concluded, however, that there is some medullary atrophy following this operation, confirming Smith's (1930) original observation and in line with the trend of Cutuly's (1936) results. It is quite clear, in any case, that the main decrease in adrenal weight after removal of the pituitary is due to changes in the cortex. Deane and Greep (1946) reported that in young male rats the normal paired adrenal weight was 21·6 mg. per 100 g. body weight and that after hypophysectomy there was a rapid drop in the first 10 days after operation, after which the decrease was more gradual (text-fig. 5). At 56 days post-operatively, the adrenal weight was 9 mg. per 100 g. body weight, a drop of 58·3 % and at 136 days it was 5·7 mg. per 100 g. body weight (see also Smith, 1930; Crooke and Gilmour, 1938).

The changes in the histology of the adrenal cortex following hypophysectomy are very clear indications that this part of the gland is a 'target organ', that it is a member of the balanced endocrine system of the body depending on the

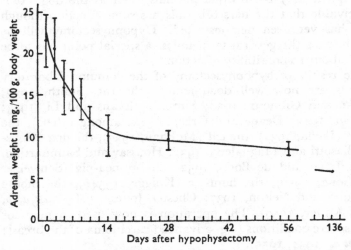

Text-fig. 5. Decline in paired adrenal weight in the rat after hypophysectomy; standard deviation shown at the mean of each age group (from Deane and Greep, 1946).

adenohypophysis for normal function. After removal of the anterior lobe of the pituitary the zona reticularis, together with the inner and outer zona fasciculata, atrophy rapidly. It is a common finding, however, that the zona glomerulosa and the zona intermedia do not undergo degeneration (Pl. II, figs. 7, 8 and 9) (dog, Houssay and Sammartino, 1933; mouse, Chester Jones, 1949a, b, 1950; guinea-pig, Schweizer and Long, 1950; rat, Deane and Greep, 1946; Lever, 1955). Indeed they both enlarge, and for two reasons, firstly, because the atrophy of the inner zones results in general shrinkage of the gland and a consequent rearrangement of the cells of the

outer layers and secondly, because some cell division continues (Chester Jones, 1950). Lane and De Bodo (1952)
reported that the zona glomerulosa in the dog also undergoes
atrophy, although not so quickly or extensively as the inner
zones, since a post-operative interval of 75 days or more was
necessary. This finding does not agree with the general view
and it may be that the change in the basic connective-tissue
architecture of the zona glomerulosa due to general shrinkage
of the gland gives a false appearance of generalized atrophy. In
any event there are indications of continued cell function and
the position is reminiscent of that obtaining in reptiles (ch. IV).

The atrophy of the inner zones is, for the most part, accompanied by pycnosis of the nuclei and shrinkage of cytoplasm;
the number of mitochondria diminish and they may fuse; the
Golgi apparatus becomes compact, losing its digital processes
(Reese and Moon, 1938). The mitochondria in the zona
glomerulosa remain essentially normal or may increase in
number (e.g. the mouse, Miller, 1950). The precise make-up
of the cortex after hypophysectomy seems to be in some doubt.
Knigge (1954 b) finds in the hamster that the cells of the zona
fasciculata undergo extensive atrophy, the cords of cells becoming somewhat distorted and compressed while the nuclei
remain spherical and vesicular. My own observations on the
rat and the mouse show that after hypophysectomy, cells of
the zona glomerulosa type, healthy by all the usual criteria,
continue to form a layer against the outer connective-tissue
capsule, underneath which are narrower cells forming a zone,
whose appearance varies according to the time after hypophysectomy, similar to the zona intermedia in appearance.
Cells on the inner side of the zona intermedia may show more
cytoplasm than is typical of the true intermediate cell and the
nuclei may be more nearly spherical. These quickly merge
into degenerating and dead cells which form a layer round the
medulla. The necrotic cells are removed by phagocytosis; in
the mouse they tend to form areas of 'brown degeneration'
(Chester Jones, 1950). The overall appearance of the cortex
of the hypophysectomized animal changes with time, so that
in about six weeks in the rat (Lever, 1954) and 75 days in the
dog (Lane and de Bodo, 1952), the zona glomerulosa may

appear to shrink in size, or indeed may actually do so. Its appearance may well depend on shifting mechanical forces brought about by the disappearing inner zones (ch. VIII).

The results after staining with the sudans vary from species to species. In the rat, sudanophilia of the zona glomerulosa is unchanged after hypophysectomy (Pl. II, fig. 9) (Simpson, Evans and Li, 1943; Greep and Deane, 1949 b). Some other reactions of lipids in this zone in the normal animal are still present, for example the Schultz and plasmal reactions and the formation of birefringent crystals. The lipid droplets in the declining zona fasciculata of the hypophysectomized rat diminish in number and the sudanophilia consequently decreases. In the mouse on the other hand, the lipid droplets rapidly disappear from the zona glomerulosa, though some may occur deeper in the persistent healthy cortex (Pl. II, fig. 8) and this division into sudanophobe and sudanophil regions has been seen by both Chester Jones (1950) and Miller (1950). The degenerating inner zones, especially the outer zona fasciculata, become intensely sudanophilic as they disintegrate, the lipid droplets enlarging and running together. The lipid begins to disappear some 40 days after the operation and by 98 days, with the total atrophy of the inner zones, there are only traces of sudanophilia (Chester Jones, 1950). In the dog, the zona glomerulosa remains sudanophilic after hypophysectomy although the intensity of staining may lessen, and Lane and de Bodo (1952) observed a comparable sudanophilia in the outer zona fasciculata. The hamster does not show lipid in the cortex of the normal animal nor does it appear after the removal of the pituitary either in the persistent zona glomerulosa or the degenerating zones. Phospholipids can be demonstrated by acid haematein (Baker, 1946) in the zona glomerulosa of the hypophysectomized animal as they can be in the normal. In the zona fasciculata of the hypophysectomized hamster the plaques and spheroid bodies, of unknown significance, also persist (Knigge, 1954 b).

It is worth noting, before leaving the subject of cortical histology, that there is a sex difference in alkaline phosphatase distribution in the mouse (Elftman, 1947) and in the hamster (Knigge, 1954 a). In the latter, the zona glomerulosa does not show any activity in either sex but the enzyme can be demonstrated in the zonae fasciculata and reticularis in the male, while it is restricted to the zona reticularis in the female. Hypophysectomy brings out a marked decrease in the reaction in both sexes.

Effects of stimulation

A hormone secreted by the anterior lobe of the pituitary (ACTH) has a direct effect on the adrenal cortex (see p. 47 for further discussion). This effect can be seen in the simple experiment of removing one adrenal gland, when the other increases in size, showing 'compensatory hypertrophy'. The cortex of the remaining adrenal is activated by increased ACTH secretion until it attains an output equivalent to that of the original two glands. This 'compensatory hypertrophy' can be prevented by the injection of adrenocortical hormones which, indeed, when given to the normal animal, lead to cortical atrophy (Ingle, Higgins and Kendall, 1938; Ingle, 1951). An adequate exogenous supply of cortical steroids therefore suppresses endogenous formation in the animal's adrenal cortices. Conversely, as embodied in Halsted's law, an autotransplanted adrenal cortex will not regenerate in the presence of an intact adrenal gland, though this dictum has recently been challenged by Dempster (1955). Clearly there is an interrelationship between the amount of adrenotrophin secreted at any one time and the amount of circulating corticosteroids, but the actual mechanisms involved in the regulation of secretion have not been elucidated.

It is sufficient in this section to note that stimulation of the adrenal by ACTH, whether endogenous or injected, leads to cortical hypertrophy, hyperplasia and to increased secretion of adrenocortical hormones. The histological changes which accompany this are, with mild stimulation, increase in cell size in the zona fasciculata, i.e. hypertrophy, and an increase in cell division with consequent increase in the number of cortical cells, i.e. hyperplasia. The mitotic figures occur principally in the outer portion of the zona fasciculata, but they may also be found in the other parts of the cortex, including the zona reticularis (Baxter, 1946; Baker, 1952; see text-fig. 33).

In the rat and in those animals with similar cortical fat distribution, the lipid droplets in the zona fasciculata become less concentrated and finer. The zona glomerulosa narrows (Baker, 1952; Krohn, 1955), though the appearance of the lipid droplets in this zone may not change quickly in the rat (Baker, 1952). In the mouse, on the other hand, where glomerulosal lipid is normally often scanty, ACTH administration may cause an increase in the amount (Chester Jones, 1949*b*). ACTH, then, produces (i) overall lipid depletion in the cortex of the rat (Bergner and Deane, 1948; Sayers and Sayers, 1948; Ducommun and Mach, 1949); (ii) hypertrophy of the Golgi apparatus, particularly in the cells of the outer zona fasciculata (Reese and Moon, 1938); (iii) an increase in the number of mitochondria especially in the latter area (Miller, 1950); (iv) fine crystalline birefringence (Weaver and Nelson, 1943).

In more actively stimulated glands the lipid can almost entirely disappear from the greater part of the cortex. At the same time, the zona glomerulosa may disappear (Baker, 1952; Chester Jones and Wright, 1954*a, b*; Krohn, 1955; Chester Jones, unpublished), together with a further widening of the zona fasciculata. Baker (1952), at the stage when only small clusters of cells underneath the capsule could be recognized as zona glomerulosa, found a narrow layer underneath the capsule packed with lipid droplets, then a region of large cells with fine sparse lipid droplets followed by a broad inner zone with cells filled with large fat droplets negative to the Schultz test for cholesterol. On one interpretation (ch. VIII) these are cells in the post-secretory phase. Under intense stimulation, together with the complete disappearance of the zona glomerulosa, both the inner zona fasciculata and the zona reticularis show increased non-cholesterol lipid. Accompanying the changes produced by ACTH stimulation is a vast increase in the vascularity of the gland, the adrenal venous effluent being greatly increased. Indeed, massive administration of ACTH can result in a haemorrhagic breakdown of the cortex (Ingle, 1951) reminiscent of the glands in the Waterhouse-Friderichsen syndrome (p. 115).

(III) ADRENAL STEROID HORMONES

Biochemists, working with whole glands and with blood, particularly from the adrenal vein, have established that the adrenal cortex manufactures substances which come under the general heading of 'steroids'. These compounds have a wide distribution and include the plant and animal sterols such as ergosterol, cholesterol, cholic acid, strophanthidin (cardiac aglycone), toad poisons and the sex and adrenocortical hormones.

The first steroid hormone to be isolated was an ovarian one, oestrone, obtained in 1929 by Doisy, Veler and Thayer and by Butenandt, and its structure established in 1932 (Bernal; Butenandt; Marrian and Haslewood). The corpus luteum hormone, progesterone, was isolated in 1934 and the androgenic hormone, testosterone, in 1935. Three groups of workers, Reichstein at Basel, Kendall at the Mayo Clinic and Wintersteiner and Pfiffner at Columbia, commenced the extraction and identification of the steroids from the adrenal cortex and by 1938 twenty-one types had been isolated; now twenty-nine are known (Reichstein and Shoppee, 1943; Simpson, Tait, Wettstein, Neher, von Euw, Schindler and Reichstein, 1954*a*).

Chemistry of steroids

The terms 'steroid', 'sterol', 'sterane' and so on have their common origin in the 'cholesterine' (χολή, 'bile' and στερεός, 'solid') coined by Chevreul in 1815 for the first member of the group found by Conradi in 1775 in bile calculi (Devis, 1951). In 1936 Callow and Young proposed the name 'steroid' for all those substances based on the reduced *cyclo*pentenophenanthrene system. The parent system consists of four fused ring structures assembled as illustrated in text-fig. 6, in which the group R is a saturated or unsaturated hydrocarbon radical and is referred to as the side chain, the remainder of the molecule being termed the nucleus. The four ring systems consist of three six-sided carbon rings—phenanthrene—lettered A, B and C, to which is attached a fourth five-sided ring—*cyclo*pentane—lettered D. For easy reference, each carbon atom is numbered (text-fig. 6), but by common usage, when formulae are written, the nucleus is depicted in outline only. The

29

systems exist in three dimensions and stereoisomeric forms are possible. The molecule contains eight asymmetric carbon atoms at positions 3, 5, 8, 9, 10, 13, 14 and 17, so that 2^8 or 256 stereoisomeric forms are possible (Hormones, 1951); however, we are fortunately only concerned with a few of these.

Conventions have been adopted to describe the spatial configuration of any one steroid. The nucleus as it is drawn is considered to lie in the plane of the paper (though it does not in fact do so, Shoppee, 1946); and the two methyl groups at carbon-18 and carbon-19 which are attached to carbon-10 and carbon-13 (both being usually but not always present, e.g. oestrone, no methyl group at carbon-10; aldosterone, no methyl group at carbon-13) are the reference groups; they are angular and rise from the plane of the paper; they are then represented as being attached by a full line and are referred to as being β-orientated (Shoppee, 1946). Thus, all those substituent groups which lie on the same side of the nucleus as the carbon-10 and carbon-13 methyl groups are joined to the nucleus by a full line, and they are β-orientated. Substituent groups which lie on the side of the nucleus away from the angular carbon-10 and carbon-13 methyl groups, that is those that are considered to lie below the plane of the paper, are represented as joining the nucleus by a dotted line and are referred to as being α-orientated. When two groups lie on the same side of the steroid molecule they are said to be *cis*, and when they lie on opposite sides they are referred to as being *trans* to each other.

A main centre of asymmetry is at carbon-5 which is attached to three carbon atoms and one hydrogen atom. This latter can lie on the same side of the molecule as the carbon-10 and carbon-13 methyl groups, that is *cis* to them and be represented as attached by a full line and this spatial configuration is called the 'normal' series. On the other hand, the hydrogen atom at carbon-5 can be considered as lying below the plane of the paper, *trans* to the methyl groups, depicted by a dotted line and this gives the 'allo' series. Among other variations to be noted are α- or β-orientated hydroxyl (—OH) substituent groups at carbon-3, and α- or β-orientated side chains from carbon-17, only the latter occurring in corticosteroids.

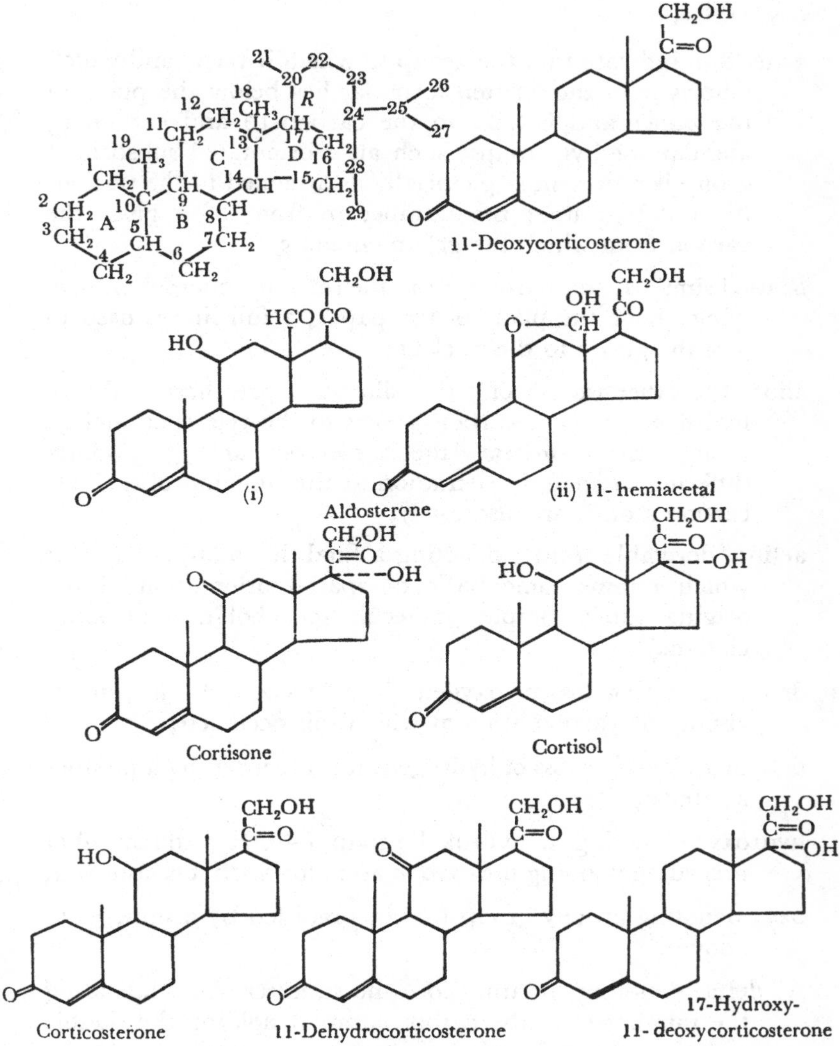

Text-fig. 6.

31

Some of the conventional terms employed can be briefly summarized:

α: used to indicate that the group to which it refers and which follows it in the written formula, lies below the plane of the paper and is *trans* to the carbon-10 and carbon-13 angular methyl groups; such an α-orientated substituent group is represented pictorially as attached to the nucleus by a dotted line; the number in front of α gives the carbon from which the group emanates;

β: used similarly for a group, *cis* to the reference methyl groups, lying above the plane of the paper; a full line is used to join the group to the nucleus;

allo: 'the other isomer of'; the allo series are those with the hydrogen atom at carbon-5 *trans* to the reference methyl groups (the terms *iso*, 'the isomer of', and *epi*, 'isomer differing in spatial orientation at the junction of a single carbon atom', are also used);

aetio: 'derivable from'; denoting a final degradation product which retains some basic or spatial orientation of the original more complex molecule (cf. cholane and aetio-cholane);

deoxy: denoting loss of oxygen; it is preceded by a number giving the carbon atom at which this occurred;

dehydro: denoting loss of hydrogen; it is preceded by a number as above;

hydroxy: denoting a hydroxyl group (—OH), the number preceding it giving the carbon atom to which it is attached;

oxy: denoting an oxy group (=O), preceded by a number as above;

Δ: delta, denoting unsaturation, the superscript(s) indicating the number(s) of the carbon atom(s) holding the double bond(s); the delta sign is sometimes dropped and the position of the double bond(s) is given by a number or numbers followed by the suffix -ene (see below);

suffix -ane: denoting the nucleus is fully saturated;

suffix -ene: denoting the nucleus is unsaturated, and -ene, -diene, -triene, etc. give the number of double bonds; in addition, a further number (or numbers) will give the position of the double bond(s) (see above);

suffix -ol: denoting the substitution of a hydroxyl group for one of the hydrogen atoms attached to a carbon atom; and -ol, -diol, -triol, etc. give the number of occurrences of such substitution; a further number (or numbers) gives the carbon atom(s) at which the substitution took place;

suffix -one: denoting the substitution of a bivalent oxygen for two of the hydrogen atoms attached to a carbon atom, giving a ketone group; -one, -dione, -trione, etc. follow the same scheme as above.

Main types of adrenal steroids

Seven of the most important, or so far regarded as the most important, steroids which resemble, if they are not actually identical with, adrenocortical hormones are depicted in text-figs. 6 and 7 and listed in table 3. The systematic name defines the substance chemically according to the rules summarized above. The trivial names are those used in the everyday speech and writing of the endocrinologist and can be of two kinds: those that give some idea of the chemical composition, such as 11-dehydrocorticosterone, which indicates that the compound is the same as corticosterone except for the loss of two hydrogen atoms at carbon-11; and the second type, shorter more compact names, as the compound enters into general, more particularly clinical, use, such as cortisone or cortone for 17-hydroxy-11-dehydrocorticosterone. The alphabetical designations were given by teams of workers (Reichstein, Kendall, Wintersteiner and colleagues) for their convenience when they found a compound but had not identified it rigidly. Despite the feeling that their usefulness has passed (Reichstein, 1953), some have become entrenched in common usage in which case the name of the author should be included (Klyne, 1953).

All these seven compounds have two features in common (text-fig. 6).

TABLE 3

Empirical formula	Trivial name	Systematic name	Alphabetical designation*		
			I	II	III
$C_{21}H_{30}O_5$	Hydrocortisone; cortisol	Δ^4-Pregnene-11β,17α,21-triol-3,20-dione	M	F	F
$C_{21}H_{28}O_5$	17-Hydroxy-11-dehydrocorticosterone; cortisone; cortone	Δ^4-Pregnene-17α,21-diol-3,11,20-trione	Fa	E	F
$C_{21}H_{30}O_4$	17-Hydroxy-11-deoxycorticosterone	Δ^4-Pregnene-17α,21-diol-3,20-dione	S	—	—
$C_{21}H_{30}O_4$	Corticosterone	Δ^4-Pregnene-11β,21-diol-3,20-dione	H	B	—
$C_{21}H_{28}O_4$	11-Dehydrocorticosterone	Δ^4-Pregnene-21-ol-3,11,20-trione	—	A	—
$C_{21}H_{30}O_3$	11-Deoxycorticosterone (DOC); cortexone; deoxycortone	Δ^4-Pregnene-21-ol-3,20-dione	Q	—	—
$C_{21}H_{28}O_5$	Aldosterone; electrocortin	Δ^4-Pregnene-11β,21-diol-3,20-dione-18-al	—	—	—

* I, Reichstein et al.; II, Kendall et al.; III, Wintersteiner et al.

(i) The α, β-unsaturated carbonyl group in ring A; that is, they possess a Δ^4-3-ketone grouping thus:

This configuration is regarded as essential for any steroid with properties characteristic of the adrenocortical hormones.

(ii) A two-carbon side chain at carbon-17, that is carbon atoms 20 and 21. At carbon-20 is a ketonic substitution ($C{=}O$) and at carbon-21 a primary hydroxyl group and this type of carbon-17 side chain is referred to as an α-ketol side chain; it has strong reducing properties which are used as the basis of tests. Thus, when treated with periodic acid the α-ketols are oxidized with the formation of formaldehyde which can be determined colorimetrically, and steroids with side chains which will do this are called formaldehydogenic steroids (FS). All adrenocortical hormones are formaldehydogenic steroids but only some FS substances are adrenocortical hormones. This has been overlooked by some investigators of urinary metabolites (see Marrian, 1951). In addition to the α-ketol side chain, three of the seven compounds have a tertiary α-orientated hydroxyl group at carbon-17. Other differences lie in the presence or absence of a ketonic or β-orientated hydroxyl group at carbon-11. Aldosterone so far stands alone in subtending an aldehyde CHO not a methyl group at carbon-13.

It is convenient at this point to give a few notes on the seven steroids listed above so that discussion on their formation and on the functions of the adrenocortical hormones will be facilitated (text-fig. 6; tables 3 and 4).

(i) *11-Deoxycorticosterone* (DOC). This is generally available as the carbon-21 acetate (DCA) or as the sparsely water-soluble glycoside. This substance was obtained by partial synthesis in 1937 by Steiger and Reichstein (Reichstein and Shoppee, 1943; Dorfman and Ungar, 1953). Despite the fact that

3-2

TABLE 4. *Comparative physiological activities of adrenal preparations*

	Survival test in rats	Test in adrenal-ectomized dogs	Na and Cl retention in normal dogs	Everse-de Fremery test	Ingle's work test	Influence on carbo-hydrate metabolism	Mouse eosinophil test
Deoxycorticosterone	85	100 e	100	100	1 l	40	?
Cortisone	55	5 e	?	40	52 l	100	100 b
Aldosterone	100 or >100 k	2500 e and h	2500 d and g (rats)	?	?	?	50 a
Corticosterone	70	?	70	70	34 l	70	37 b
Hydrocortisone	55	5 e	?	40	100 l	100	78 b
Kendall's compound A	55	?	70	?	25 l	55	24 b
Reichstein's substance S	10	?	?	70	1 l	25	Inactive n
Amorphous fraction	100	100	85	85	21	?	?

	Cold stress test	ACTH suppression test	Liver glycogen test (Pabst)	24Na/42K	K excretion	Liver glycogen deposition (Venning)
Deoxycorticosterone	10 i	4 k	1 c	100 f	100 g	1 j
Cortisone	100 a	100 k	67 c	?	?	100 j
Aldosterone	100 or >100 a	33 k	?	12000 f	500 g	33 j
Corticosterone	9 m	?	35 c	?	?	?
Hydrocortisone	?	?	100 c	?	?	?
Kendall's compound A	33 m	?	32 c	?	?	33 o
Reichstein's substance S	?	?	?	?	?	?
Amorphous fraction	?	?	?	?	?	?

Relative potencies of adrenal steroids in different types of tests: 0=no effect; 100=very effective. The potency of aldosterone is so great in some tests that it is given in multiples of 100.

a Gaunt et al. (1954a, b; 1955).
b Bibile (1953).
c Pabst et al. (1947).
d Mattox et al. (1953).
e Swingle et al. (1954).
f Speirs et al. (1954).
g Desaulles et al. (1953).
h Gross and Gysel (1954).
i Zarrow (1942).
j Schuler et al. (1954).
k Renzi et al. (1954).
l Ingle (1944).
m Deming and Luetscher (1950) and Dorfman et al. (1946).
n Grad et al. (1953).
o Venning et al. (1946).

Adrenal Steroid Hormones

Reichstein and von Euw (1939) obtained it in small yield from beef adrenals, it is not a normal secretion of the adrenal cortex or, at best, only in very small amounts. The substance has, however, been used in countless experiments because it has properties in common with the known adrenocortical hormones and cortical extracts, though many descriptions of its action are in the realm of pharmacology rather than physiology. Deoxycorticosterone exerts its chief effect on salt-electrolyte metabolism (p. 72), but it will influence carbohydrate metabolism, though relatively less effectively (Verzár, 1952, and table 4). Adrenocortical steroids which exert their main effect on water and salt-electrolyte metabolism are sometimes (in these days more frequently) referred to as 'mineralocorticoids' (Selye, 1947). The term has its uses; however, (a) water and salt-electrolyte metabolism should be treated separately, because the steroid which has a primary effect on sodium, for example, does not necessarily alter water metabolism directly; 'mineralocorticoid' may be conveniently used as a brevity for those steroids which alter the metabolism of sodium or potassium (or both); and (b) no steroid is exclusively mineralocorticoid; to a greater or lesser extent the known adrenocortical steroids have an effect on carbohydrate metabolism, either primarily or secondarily; mineralocorticoid cannot therefore be rigorously defined.

(ii) *Aldosterone.* This is the most recently characterized adrenocortical steroid and is the 18-aldehyde of corticosterone (Simpson, Tait, Wettstein, Neher, von Euw and Reichstein, 1953; Simpson, Tait, Wettstein, Neher, von Euw, Schindler and Reichstein, 1954a, b). It exists in solution as the 11-hemiacetal. It has been demonstrated in beef adrenal extracts, in dog and monkey adrenal venous effluent (Grundy, Simpson and Tait, 1952; Simpson, Tait and Bush, 1952; Simpson and Tait, 1953). Material similar to aldosterone has been obtained from beef and pig adrenal extracts by Mattox, Mason, Albert and Code (1953), and Knauff, Nielson and Haines (1953), and in dog adrenal venous blood by Farrell and Richards (1953), Farrell and Lamus (1953), Farrell and Werle (1954), and in human urine by Luetscher and Johnson (1953, 1954). It has an effect many times more powerful, on a molar basis, than

37

The Adrenal Cortex

DOC on salt-electrolyte metabolism and is thus regarded as the naturally occurring mineralocorticoid. Nevertheless, it has an effect on carbohydrate metabolism (Schuler, Desaulles and Meier, 1954), though more recently Prunty, McSwiney and Mills (1955) think the effect is not a significant one.

(iii) *17-Hydroxy-11-dehydrocorticosterone*. This is known by the trivial names 'cortisone' or 'cortone' and as Kendall's compound E. It was, however, described in the first place by Wintersteiner and Pfiffner (1935) as their compound F and subsequently by Reichstein (1936), Mason, Myers and Kendall (1936) and Kuizenga and Cartland (1939). One of the substance's main effects is on carbohydrate metabolism (p. 60) and it is for this reason placed with the glucocorticoids (Selye, 1947) in contrast to the mineralocorticoids. It has, however, an effect on water and on salt-electrolyte metabolism as well as being of value in rheumatoid arthritis, status asthmaticus and other diseases (p. 97).

(iv) *17-Hydroxycorticosterone*. Hydrocortisone, cortisol or Kendall's compound F was isolated and described by Reichstein (1937), by Mason, Hoehn and Kendall (1938) and by Kuizenga and Cartland (1939). It has very similar properties to cortisone, but has the additional distinction of being probably one of the major components of cortical secretion in many species of Eutheria, whereas Kendall's compound E is not.

Both cortisone and hydrocortisone can be substituted at carbon-9 by halogens (Fried and Sabo, 1953, 1954). This gives increased potency to the compounds. It seems that by substitution with halogen atoms of decreasing volume increasing activity is obtained. Therefore 9-α-fluoro-hydrocortisone acetate has been found to be twenty times, and 9-α-chloro-hydrocortisone acetate eight times, as biologically active, on an equimolar basis, as the acetate of compound F (Liddle, Pechet and Bartter, 1954). It may be that substitution at carbon-9 stabilizes the 11-hydroxyl group, or that the halogen atom facilitates some condensation reaction, e.g. with sulphydryl compounds (Callow, Lloyd and Long, 1954).

(v) *Corticosterone*. This substance was isolated by Reichstein (1937) and by Mason, Hoehn, McKenzie and Kendall (1937) and by Kuizenga and Cartland (1940) from beef adrenals. Cortico-

38

sterone (Kendall's compound B) is moderately effective both in carbohydrate and salt-electrolyte metabolism. It was pushed rather into the background in discussions on the effective corticosteroids but has recently come into its own again on its identification as one of the major compounds actually secreted.

(vi) *11-Dehydrocorticosterone.* This substance was isolated by Mason, Myers and Kendall (1936), and Mason, Hoehn, McKenzie and Kendall (1937), who called it compound A, and by Reichstein and von Euw (1938) who called it 11-dehydrocorticosterone, and by Kuizenga and Cartland (1939). It is very similar in its effects to corticosterone; it differs chemically from Kendall's compound E only in the absence of an angular hydroxyl group at carbon-17. It can be relatively easily produced by partial synthesis, and Sarett (1948) originated a method whereby it could be 17-hydroxylated to form cortisone, a much more physiologically effective substance.

(vii) *17-Hydroxy-11-deoxycorticosterone.* This substance was isolated by Reichstein and von Euw (1938) and called substance S. It is not a particularly active compound and not a major component of normal cortical secretion although Haines (1952) isolated considerable amounts of it from hog adrenal tissue.

(viii) *'Amorphous fraction.'* When the chemist has taken adrenal glands and applied extraction procedures to them, there is still left a large amount of liquor which contains physiologically active material, chemically unknown. The recent extraction of aldosterone may account for the activity of the 'amorphous fraction', but whether all or the major portion of it has yet to be determined.

Nature of circulating adrenal hormones

Among current problems are: (i) what is the chemical nature of the adrenocortical hormones actually secreted? and (ii) how are these hormones made in the cortical cells? With the advent of new micro-techniques, such as methods for the isolation of trace amounts of steroids from blood, paper chromatography for their separation and infra-red spectroscopy for their determination, attempts are being made to elucidate the exact nature of the adrenal secretory products, and to determine which of the twenty-nine steroids which have been isolated from adrenal tissue are elaborated by the gland and which are products of autolysis or artifacts of the method of isolation (Lieberman and Teich, 1953).

It seems clear that one answer to the first question is that in many species the adrenal gland secretes Kendall's compounds B and F and aldosterone. Bush (1953) found singly or together 17-hydroxycorticosterone and corticosterone made up 85–100 % of the total quantity of $\alpha\beta$-unsaturated 3-ketosteroids in the adrenal venous blood in all species he examined except cats in which they made up only 74–90 % of the total. Hechter and Pincus (1954) note that the monkey, sheep, guinea-pig and man secrete primarily 17-hydroxycorticosterone, the rabbit and the rat primarily corticosterone, the dog and cat mainly compound F but some B, and the ferret, ox and cow a mixture of both in roughly equal parts; but they point out that there is considerable individual variation within a single species. They illustrate this with data from eleven male dogs' adrenal venous blood which gave hydrocortisone/corticosterone ratios varying from > 20 to 1·2. Aldosterone is another major component of adrenal secretion. It is secreted in very low concentrations, at which it is, however, physiologically potent.

There are other steroids in adrenal venous blood but these are not of major importance nor do they occur in high concentrations. Bush (1953) found that the quantity of $\alpha\beta$-unsaturated 3-ketone material left after estimation of compounds B and F in adrenal venous blood was frequently only about 2·5 % of the total. In the cat, on the other hand, he found that 11-dehydro-17-hydroxycorticosterone and 11-dehydrocorticosterone made up 13 % of the total secretion. Nelson, Samuels and Reich (1951) found that after intravenous administration of ACTH to a dog there were two other compounds in the adrenal venous blood in addition to compounds F and B. In similar preparations Zaffaroni and Burton (1953) noted seven unidentified α-ketols as minor components. The adrenal cortex certainly has the capacity to produce a whole host of steroids. This has been demonstrated by *in vitro* methods, particularly by the use of the isolated cow adrenal. This is arterially perfused or the adrenal vein is cannulated with the outer tissue lacerated and citrated homologous blood circulated through it (Hechter, 1949, 1950; 1953; Hechter, Zaffaroni, Jacobsen, Levy, Jeanloz, Schenker and Pincus, 1951; Macchi and Hechter, 1954a, b, c). The numerous steroids found in addition to compounds F and B are shown in table 5. Bush (1953) points out that the additional steroids, making a high proportion of the total α-ketol output, may be the ones released by the abnormal *in vitro* conditions. He found, using large samples of dog and sheep adrenal venous blood, that in terms of the content of 17-hydroxycorticosterone less than 2 % of each of the unknown ketols I–V and VII–IX and less than 0·5 % deoxycorticosterone were present. This contrasts with a total of 32 % of unknowns I–V, and 1·2 % of VII–IX, shown in table 5. Results obtained from work *in vitro* are of great value in providing clues to biosynthetic pathways; their physiological significance must, of

TABLE 5. *The α-ketols present in adrenal perfusates*
(after Hechter, 1953)

	Micrograms per 2-litre samples of blood			Micrograms per 20 g. beef gland extracts
	Not perfused through adrenal	Adrenal perfusates		
		No ACTH	ACTH	
Unknowns I–V	220	80	700	70
17-Hydroxycorticosterone (compound F)	360	145	1100	40
Cortisone	—†	25⎫	200	20
Unknown VI*	—†	40⎭		
Unknowns VII–IX	110	45	250	25
Corticosterone (compound B)	400	230	1100	70
Unknown x	—†⎫	60	300⎫	35
Dehydrocorticosterone	—†⎭		250⎭	
Deoxycorticosterone	120	35	140	—‡

* Cortisone and unknown VI move to approximately the same positions on paper; a similar situation obtains for unknown x and dehydrocorticosterone. When acetate or propionate esters are formed, the unknown free steroids are widely separated from their respective unknowns. In some cases, the mixtures were not resolved, and data were obtained for the complex of cortisone plus unknown VI, or dehydrocorticosterone plus unknown x.

† Not detectable.   ‡ Not determined.

The Adrenal Cortex

course, be assessed in the light of appropriate *in vivo* studies on adrenal venous samples (Hechter and Pincus, 1954).

Levels of circulating hormones. The steroids produced by the adrenal cortex circulate in very small amounts under normal conditions. Whether they circulate free or as 'protein complexes' is unknown. In human peripheral blood, Morris and Williams (1953) found compound A at 4·5 μg./100 ml. plasma, B at 8·5, E at 4·0 and F at 7·0 μg./100 ml. plasma, to take their values for one male as generally representative. Nelson and Samuels (1952) found 3–10 μg./100 ml. of '17-hydroxycorticosteroids' in human peripheral blood. Sweat, Abbott, Jeffries and Bliss (1953) found normal values of 4·9 μg. of compound B per 100 ml. and 11·0 μg. of compound F per 100 ml. plasma. Other adrenal steroids are generally present in amounts too small to be measured in peripheral blood. Simpson and Tait (1955) found that bulked blood from twelve normal male patients had an activity of 0·085 μg. of aldosterone per 100 ml. of whole blood.

The normal rate of secretion of adrenal steroids seems to be subject to a diurnal variation (Tyler, Migeon, Florentin and Samuels, 1954). Levels of 17-hydroxycorticoids in plasma reach their maximum in the early morning in man, usually about 6 a.m. when average values of 19 μg./100 ml. of plasma were observed. The levels fell rapidly until noon and slowly thereafter to reach a minimum about 10 p.m. with values of 6 μg./100 ml. plasma. During the night little change was observed until after 2 a.m. when a rapid rise to the early morning levels followed. The physiological implications of this secretory rhythm are not known.

The adrenocortical hormones together with the sex steroids are changed rapidly by the liver and to a less extent by other organs such as the kidney. Some hundred steroid metabolites have been found in the urine and the gland of origin has been assigned with varying success (see Lieberman and Teich, 1953; Dorfman, 1955).

42

Biosynthesis of Cortical Hormones

The method of the formation of corticosteroids in the cells of the adrenal cortex is not fully known (text-fig. 7; Hechter and Pincus, 1954; Dorfman, 1955). Cholesterol, which is widespread in the body, seems to be an obvious potential precursor of the adrenocortical hormones. Zaffaroni, Hechter and Pincus (1951 a, b) were the first to show that cholesterol could, in fact, be converted into corticosteroids by the adrenal cortex. They found that perfusion of the isolated cow adrenal with radio-active cholesterol led to the formation of radio-active compounds F and B. Furthermore, using the same system, they demonstrated that ^{14}C-acetate underwent transformation to corticosteroids. The possibility of this type of synthesis was confirmed by the production of corticosteroids from the incubation of adrenal slices with ^{14}C-acetate (Hechter and Pincus, 1954). For example, Haines (1952) showed that incubation of slices of hog adrenal with ^{14}C-acetate together with ACTH gave predominantly radio-active Kendall's compound F with some E and perhaps B and Reichstein's substance S. This might be the case of 2-carbon fragments arising from acetate condensing to form cholesterol (Bloch, 1951).

The situation is not, however, as simple as $C_2 \rightarrow$ cholesterol $\rightarrow$ corticosteroids (Hechter, 1953). It seems that utilization of 2-carbon fragments in corticosteroidogenesis does not necessarily involve cholesterol as an intermediate (Hechter, 1951, 1953; Hechter and Pincus, 1954). Hence we can suppose either (i) that acetate or 2-carbon fragments can be transformed into an unknown intermediary X which in turn produces corticosteroids and cholesterol (Hechter, 1953), or (ii) that 2-carbon fragments can condense to form either X or cholesterol and the latter can also be converted into X (Lieberman and Teich, 1953). But 2-carbon fragments may not be essential, since Hofmann and Davison (1954), using rat adrenal slices, found that acetoacetate could cause a significant increase in adrenal steroid formation. And Brady and Gurin (1951) showed rat liver slices readily incorporated acetoacetate into cholesterol without the prior formation of 2-carbon fragments. The low content of cholesterol in the hyperfunctioning gland of some species and in the normal gland of others suggests, too, that cholesterol need not be formed for adrenocortical secretion to take place. There is no doubt, however, that cholesterol is an efficient precursor of corticosteroids and may be the main one in many animals. It is significant in this respect that in perfusion experiments, cholesterol is a more potent precursor of compounds B and F than is acetate. In the rat, where ACTH causes a decrease in adrenal cholesterol content together with increased corticosteroid secretion (Long, 1947), the initial depression in cholesterol content can be interpreted as a conversion of this substance to corticosteroids.

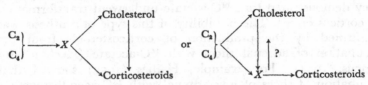

The conversion of cholesterol to the corticosteroids involves the following reactions: (i) side-chain scission from a carbon-27 to a carbon-21 steroid; (ii) formation of an α, β-unsaturated ketone in ring A; (iii) sometimes 11-hydroxylation in ring C; (iv) carbon-21 hydroxylation to form the α-ketol side chain at carbon-17, with and without 17-hydroxylation (Hechter and Pincus, 1954); to which may be added (v) carbon-18 oxidation to form the carbon-18 aldehyde group in aldosterone. Evidence from the perfusion of steroids through the isolated gland and by work *in vitro* with adrenal homogenates and extracts has suggested key substances in the lines of production of compounds B and F (text-fig. 7). Essentially, what happens in the adrenal cortex depends on the available enzyme systems. Using adrenal slices, breis and homogenates, Sweat (1951) and Hayano and Dorfman (1952) have described the characteristics of the enzyme which catalyses oxidation at position 11; Hayano and Dorfman have described that at position 21, and Plager and Samuels (1952) that for oxidation at carbon-17. The enzyme systems so far identified seem to have diphosphopyridine nucleotide (DPN) as the hydrogen acceptor for the system and the presence of fumarate of the Krebs citric-acid

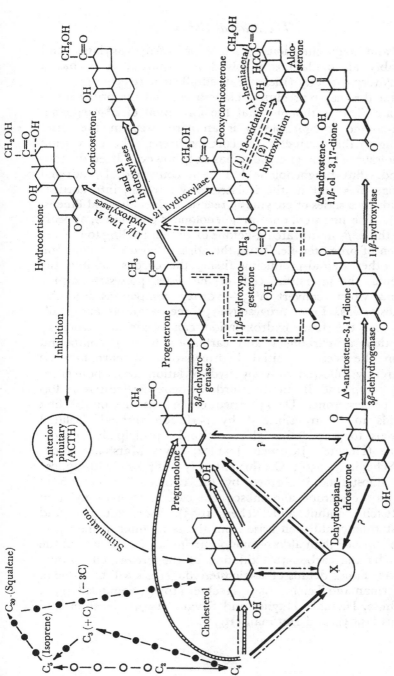

Text-fig. 7. Possible pathways in the formation of adrenocortical steroids (based on Hechter and Pincus, 1954; Dorfman, 1955). ⟶ Dorfman, 1955; ●—●—● Bloch, 1951; Hechter, 1953; ------⟶ Lieberman and Teich, 1953; ▭▭▭▭⟶ Hechter, Solomon, Zaffaroni and Pincus, 1953; ▬▬▬⟶ Hechter and Pincus, 1954.

The pathway for aldosterone is put in for convenience as arising from deoxycorticosterone (Wettstein, Kahnt and Neher, 1955) but it is probable that this is not so.

cycle and magnesium ions certainly for 11β-hydroxylation and probably all oxidations involved is required; ATP has a stimulatory capacity (Lieberman and Teich, 1953).

After the major precursor, cholesterol, and the prime intermediate X, it is thought that the formation of Δ^5-pregnene-3β-ol-20-one (pregnenolone) is an early step in the series leading to the production of corticosteroids. The conversion of cholesterol to pregnenolone has not, however, been demonstrated. But pregnenolone has been converted, by adrenal homogenates and in the isolated cow adrenal, into corticosteroids by a series of enzymatic reactions (Hechter and Pincus, 1954). The first step, once pregnenolone is postulated, appears to be the α, β-unsaturated 3-ketone oxidation leading to the formation of progesterone. Thus, the adrenal cortex, in common with other steroid-producing tissue, the corpus luteum, the placenta, the interstitial cells of the testis, possesses enzyme systems which convert Δ^5-3β-hydroxy compounds into Δ^4-3-ketosteroids. From progesterone, three pathways are available. In the first, hydroxylation can occur at carbon-17 and then at carbon-11 and carbon-21 to give compound F; in the second, initial hydroxylation at carbon-21 is apparently followed only by hydroxylation at carbon-11 to yield compound B; in the third, 11-deoxycorticosterone follows progesterone. Deoxycorticosterone was one of the first steroids to be transformed by perfusion through isolated adrenals to give active glucocorticoids, principally corticosterone (Hechter, Jacobsen, Jeanloz, Levy, Marshall, Pincus and Schenker, 1950). On this pathway it is possible to suppose that aldosterone is formed by the carbon-18 oxidation of deoxycorticosterone and subsequent carbon-11 hydroxylation (Wettstein, Kahnt and Neher, 1955). Corticosterone and aldosterone could then arise from the same intermediate precursor, rather than aldosterone occurring as the last step in the series by the carbon-18 oxidation of the preformed corticosterone. Fuller details of corticosteroidogenesis will be found in Lieberman and Teich (1953), Hechter (1953), Samuels (1953), Dorfman, Hayano, Haynes and Savard (1953), Hechter and Pincus (1954), and Dorfman (1955).

(V) CONTROL OF ADRENOCORTICAL SECRETION

Corticotrophin

The atrophy of the adrenal cortex following hypophys-ectomy can be prevented by administration of a fraction of the anterior pituitary called adrenocorticotrophin, or cortico-trophin, or adrenotrophin, and most frequently designated by initials, ACTH. This hormone when given to normal animals causes stimulation and enlargement of the adrenal cortex with consequent increase in adrenal weight. Two groups of workers in 1943 (Li, Evans and Simpson; Sayers, White and Long) reported the isolation of an apparently homogeneous protein with a molecular weight of about 20,000 from the pituitary glands of the sheep and hog respectively, which was considered to represent the natural form of ACTH. Since this time, the use of various procedures involving, for example, ultrafiltra-tion, oxycellulose adsorption, ion exchange resins, and counter-current distribution, has resulted in corticotrophin prepara-tions, many times the potency of the original 'protein-hormone', consisting of proteins of low molecular weight or of polypep-tides (reviews in Stack-Dunne and Young, 1954; Hays and White, 1954).

As there are several groups working on the chemistry of ACTH, the nomenclature of the sundry preparations has become involved, their equivalences equivocal and their bio-logical relevance uncertain. Following Stack-Dunne and Young (1954), the original ACTH can be referred to as 'crude ACTH' and this preparation can be divided into two types depending on the subsequent procedures: the term 'cortico-trophin A' denotes the active material concentrated from *unhydrolysed* preparations of crude ACTH by a process involving adsorption on and elution from oxycellulose or ion-exchange resins, while 'corticotrophin B' is obtained from crude ACTH by *hydrolysis* principally by pepsin and then by subjection to methods for purification. Both corticotrophin A and B are general categories of active material from which further frac-tions can be obtained by various purification methods. Hays and White (1954) propose a classification of four 'preparative' types of corticotrophin based on the treatment used to isolate

47

them, namely type I, subjected to little or no hydrolysis; type II, subjected to pepsin hydrolysis; type III, subjected to pepsin and acid hydrolysis; and type IV, subjected to acid hydrolysis. These types can then be further purified in various ways.

Much work has been directed towards characterizing these corticotrophins. The molecular weight of corticotrophin B has been estimated as 5200 by direct measurement and as 6000–7000 when calculated from molar amino-acid ratios by Brink, Boxer, Jelinek, Kuehl, Richter and Folkers (1953), while Hays and White (1954) give 2300 as the figure obtained by means of the ultracentrifuge at low pH. Corticotrophin A has been assigned a calculated minimum molecular weight of 3500 (Hays and White, 1954). The constituent amino acids of the corticotrophins have been determined and their relative positions in the molecule in part assigned. Various authors have suggested the sequence of the amino acids at the N terminus (the free α-amino group at the end of the peptide chain) and at the C terminus (the free carboxyl group end of the peptide chain) (Stack-Dunne and Young, 1954; Hays and White, 1954). Bell (1954) has produced an extended analysis of the amino-acid structure of β-corticotrophin with a free base molecular weight of 4500. In this case, β-corticotrophin is obtained from 'crude ACTH' as corticotrophin A (type-I corticotropin) and then resolved into eight components by an extended counter-current distribution, β being the fraction obtained in the largest amount.

The chemical fractionation of the original crude ACTH brings in many difficulties of biological interpretation. The nature of the ACTH actually secreted by the pituitary is still far from clear. The position is the more complicated because, before even the refined purification methods can be used, the original extraction method and its concentration produces crude ACTH altered in unknown ways, by these very procedures, from the supposed 'natural' ACTH. Beyond this again are the different biological activities of the various ACTH preparations. Corticotrophin is a substance extracted from the anterior lobe of the pituitary and can only be further defined in terms of the adrenal cortex. The adrenal cortex can itself be altered by ACTH and these variations estimated, or recourse can be made to changes in target organs or substances known to be dependent on or emanate from the adrenal cortex.

Stack-Dunne and Young (1954) give the following list of the methods of detection and assay of ACTH, which can be compared with table 4: (*a*) various forms of the 'maintenance' and 'repair' test, that is the prevention of atrophy of the adrenal cortex after hypophysectomy by injection of ACTH; (*b*) the Sayers test, by which is meant a test or assay based on the fall of the adrenal ascorbic acid in hypophysectomized rats which follows intravenous administration of ACTH; (*c*) the reduction of thymus weight in the nestling intact rat when ACTH is administered subcutaneously in a medium designed to delay absorption, for example mixed with beeswax and arachis oil or precipitated with alum; (*d*) the fall of blood eosinophils in intact or hypophysectomized mice; (*e*) the increase in urinary excretion of 17-hydroxycorticosteroids in man or guinea-pig; and (*f*) the *in vitro* stimulation of the metabolism of isolated adrenal preparations or slices with rise in oxygen consumption or an increase in adrenal steroid production.

The various preparations of ACTH now produced do not influence all these things equally, so that the interpretation of the chemically obtained fractions in terms of their natural biological function is obscure. Not only this, but supposed equivalent fractions prepared in different laboratories often have markedly different potencies (Hays and White, 1954). Furthermore, one and the same preparation evinces different potencies depending on the site of administration. Thus crude ACTH and corticotrophin B is liable to inactivation at the site of injection when given subcutaneously or intramuscularly (Stack-Dunne and Young, 1954).

Young and his co-workers (Stack-Dunne and Young, 1951; Dixon, Stack-Dunne, Young and Cater, 1951; Dixon, Stack-Dunne and Young, 1953) have, in particular, been in the forefront in the demonstration of different fractions of ACTH with different biological properties. These authors consider that there are at least two types of corticotrophin A. One fraction is active in the Sayers test and is termed the Ascorbic Acid Factor (AAF), while the other, poor in ascorbic-acid-reducing property, nevertheless causes the adrenal of

49

hypophysectomized rats to gain in weight (Adrenal Weight Factor, AWF). It is possible, though not proved, that corticotrophin B is rich in AWF, with little AAF (Stack-Dunne and Young, 1954). All the corticotrophin preparations, however, show some activity in all the tests (Li, 1953a) and, indeed, it is sometimes difficult to maintain such a sharp division between preparation and biological activity.

The picture is further obscured by the demonstration that purified growth hormone is, or contains, the strong Adrenal Weight Factor (AWF) while possessing little Ascorbic Acid Factor (AAF) (Cater and Stack-Dunne, 1953). Growth hormone does not seem to have any significant effect on the quantity of adrenal secretion (Singer and Stack-Dunne, 1955), but only, or mainly, on the cortical cell number by increasing mitotic activity. Cater and Stack-Dunne suggest that corticotrophin, on the other hand, has its effect not so much on mitotic activity but on cell size: that is, growth hormone produces cortical hyperplasia, corticotrophin cortical hypertrophy. Those corticotrophin preparations which produce both cortical hypertrophy and hyperplasia might, on these grounds, be thought to be contaminated with growth hormone. Equally well it might be maintained that the corticotrophic activity of growth hormone is due to contamination. Cater and Stack-Dunne (1953) point out, however, that when purified growth hormone was submitted to ion-exchange chromatography (under the conditions used for crude corticotrophin by Dixon and his co-workers, 1951) the fraction showing the mitosis-stimulating activity was unretarded: that is to say it could not be separated from growth-hormone protein or from the somatotrophic activity which was also unretarded under the conditions used. By definition, growth hormone is a substance which promotes growth by increase in cell number in all parts of the body at certain times during the maturation of the individual. It may be, then, that the action of growth hormone on the adrenal cortex is only a manifestation of this general property and therefore somatotrophin need not be regarded as specifically corticotrophic.

These considerations, however, still leave us a long way from knowing what adrenocorticotrophin is and in what form it is secreted. It should be remembered that the original ACTH, isolated as a homogeneous protein hormone of large molecular weight, occurred in this form due to the chemical manipulations involved. Now it is true that 'evidence is lacking that the protein-hormone has either chemical or biological significance' (Stack-Dunne and Young, 1954) but it has yet to be shown that the multiplicity of corticotrophic and corticotrophic-like substances are not also merely chemical artifacts of little importance in the biological story. Clearly the identi-

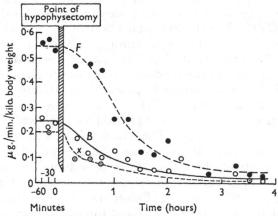

Text-fig. 8. Decline in adrenal steroid production in the dog after hypophysectomy (from Sweat and Farrell, 1954). F, 17-hydroxycorticosterone; B, corticosterone; X, unknown 17-hydroxycorticosteroid.

fication and characterization of 'natural' ACTH rests upon advances in protein chemistry as does indeed the whole problem of the form of secretion of all the anterior pituitary hormones.

Corticotrophin may lack precise biological and biochemical definition but its overall action on the adrenal cortex is not in doubt. In the absence of the pituitary, the secretion of compounds F and B by the dog adrenal falls rapidly to low levels but nevertheless persists (de Gurpide, 1953; Sweat and Farrell, 1954; text-fig. 8). Corticotrophin acts very quickly on the adrenal cortex. Bush (1953) found, using dog *in vivo* preparations, that infusion of ACTH gave an increase in corticosteroid secretion rate within two minutes. The peak

The Adrenal Cortex

rate was reached within five to ten minutes of the beginning of the infusion. *In vitro* preparations demonstrate the rapidity with which ACTH can affect the production of formaldehydogenic material. Using the isolated cow adrenal it was shown that the concentration of corticosteroids before ACTH was 180 μg./100 ml.; addition of ACTH increased the titre to 406 μg./100 ml. within 30 seconds and to 2606 μg./100 ml. 90 minutes later (Hechter, 1949; Pincus, Hechter and Zaffaroni, 1951; Hechter *et al.*, 1951). Likewise Macchi and Hechter (1954*a*), using a similar preparation, obtained a 19-fold increase in the rate of corticosteroidogenesis by the addition of ACTH. For the maximal effect to be achieved, there is a latent period of at least 18 minutes (Macchi and Hechter, (1954*b*). On withdrawal of ACTH from the perfusing medium of their preparation, the residual activity of the gland declined exponentially; corticosteroid production, relative to that obtained with maximal ACTH stimulation, was 45 % at 18–36 minutes and 18 % at 126–144 minutes after withdrawal of ACTH from the system. The effect of ACTH on corticosteroidogenesis can be seen from the use of other *in vitro* methods, particularly using slices of adrenal cortex from different animals, for example the rat (Saffron, Grad and Bayliss, 1952; Birmingham, Elliott and Valère, 1953; Hofmann and Davison, 1953, 1954) and the cow (Haynes, Savard and Dorfman, 1952).

Aldosterone secretion may be dependent on ACTH but not in such a direct and clear-cut way as compounds B and F. In the rat after hypophysectomy, Singer and Stack-Dunne (1955) showed that aldosterone secretion did not decline to anything like the same extent as corticosterone did. Before hypophysectomy the ratio of aldosterone to corticosterone was 0·30/1 and after operation it was 3·65/1. Other factors, in addition to ACTH, apparently control the rate of secretion of aldosterone. It is possible that one of these factors is the level of plasma sodium and potassium or the proportion of these two ions. Singer and Stack-Dunne found that treatment with DCA for seven days or feeding a high potassium diet for four weeks reduced aldosterone secretion by more than 90 % without altering that of corticosterone. It is probable, however, that the marked sodium retention occurring after ACTH administration, often seen clinically, is due to the increased secretion of aldosterone (Prunty, McSwiney and Mills, 1955).

It is possible that adrenotrophin increases the rate of formation of corticosteroids either (i) by a general stimulation of all the reactions involved or (ii) by acting at an early stage, after which the subsequent steps would occur due to the intrinsic properties of the cell enzyme system. The latter is thought to be the case (Hechter and Pincus, 1954). Even if we suppose, however, that ACTH acts only by making an early intermediate available for the adrenal enzymatic apparatus, it should be remembered that this hormone provides the cells in which the later biosynthetic steps take place. In the hypophysectomized animal, cells remain which produce corticosteroids in the absence of ACTH. It is thought that in this case the alternative pathway of acetate (or 2-carbon fragments) conversion directly to corticosteroids is used exclusively. Moreover, it is considered that the rate of this conversion is not influenced by ACTH in the normal animal. The role of ACTH cannot, however, be solely ascribed to effecting a side-chain fission of cholesterol to produce intermediate X and pregnenolone. For this reaction seems to be the primary one in the corpus luteum and the gonads in general under the influence of gonadotrophins. There is, of course, some degree of overlap in the relationship of the adrenal cortex and the gonads to the pituitary trophic hormones (p. 99 below). Nevertheless, ACTH must have some additional specific effect peculiar to its relationship with its end organ, the cortex (Hechter, 1953).

If ACTH does act at the level of condensation of 2-carbon fragments, or of cholesterol, or of some unknown intermediate, it would follow, all things being equal, that the three major components of cortical secretion, compounds B and F and aldosterone, would normally appear in similar proportions even under different conditions. Variations in the proportion of compounds B and F as between species, and between individuals of one species can be explained by postulating different relative activities of the carbon-17-hydroxylating and carbon-

The Adrenal Cortex

21-hydroxylating systems. Predominance of the former leads the major part of the synthesis along the pathway to compound F; predominance of the latter leads to relatively higher concentrations of compound B. Bush (1953) feels that the properties of these two enzyme systems are fixed genetically, thereby determining the ratio of the two major secretory products. We noted earlier, however, that, although a characteristic ratio of compounds B and F could be given for each species, nevertheless variation of this ratio could appear within one species. The factors influencing the activation and inhibition of these two enzyme systems are unknown.

Aldosterone presents a further problem. As noted earlier, after hypophysectomy in the rat, the decline in aldosterone secretion is not so great as that of corticosterone (Singer and Stack-Dunne, 1955). It would be necessary to suppose, on present concepts, that, in the absence of ACTH, formation of pregnenolone to progesterone and to deoxycorticosterone went on and that the carbon-18 oxidation system was little affected. Alternatively, the biosynthetic pathway of aldosterone is completely different and not yet known. The problem is complicated by the decline in the number of functional cells after hypophysectomy and involves a consideration of the cortical zones responsible for adrenocortical secretion (see ch. VIII).

Control of corticotrophin secretion

We have seen that the normal secretion of the corticosteroids depends on ACTH. On what, in turn, does the secretion of ACTH depend? Three hypotheses are available but they are by no means mutually exclusive. The first hypothesis supposes that on demand, such as in stress (p. 120), an increased secretion of adrenaline and noradrenaline acts upon the pituitary to effect release of ACTH which stimulates the adrenal cortex to a higher rate of secretion. Thus it has been shown that the injection or infusion of adrenaline may increase the amount of cortical secretion and that this is due to a release of ACTH (Long, 1950, 1952; Vogt, 1955). It may be that the site of action of adrenaline is not the cells of the anterior lobe of the pituitary themselves but those of the hypothalamus (Vogt, 1955). The hypothalamic control of pituitary secretion is a further complication which is considered below in the third hypothesis. In any case, even if the mechanism of adrenaline-ACTH be true, it accounts for only the short-term response. For the body quickly accommodates itself to adrenaline administration, the drug losing its effectiveness on repeated administration. Thus repeated injections of adrenaline do not lead to hypertrophy of the adrenal cortex as do those of ACTH (Vogt, 1955). Although the suggestion has been made that adrenaline has a direct effect on the adrenal gland in promoting corticosteroid production without ACTH mediation, this is not at present regarded as a normally occurring mechanism (Speirs, 1953; Vogt, 1955).

The second hypothesis postulates a reciprocal relationship between pituitary ACTH on the one hand and the adrenal corticosteroids on the other. Sayers and Sayers (1948) and Sayers (1950) consider that there is a basic adrenotrophin-corticosteroid relationship when the pituitary secretion and the cortical output are in an optimal balance, in the same way as the gonadotrophins have a reciprocal relationship with the gonadal hormones (see Greep and Chester Jones, 1950a, b). In times of demand the rate of 'utilization' of cortical hormone by the tissue cells is increased and hence the quantity in venous blood is temporarily reduced. The adenohypophysis responds by discharging ACTH at an accelerated rate. Increased secretory activity of the adrenal cortex occurs and thus the increased needs of the peripheral tissue cells for cortical hormones are met (Sayers, 1950). Essentially this theory depends on the concept of tissues making use of cortical hormones at different rates, the consequent varying titres of circulating corticosteroids being the determining mechanism in ACTH release from the pituitary.

The third hypothesis assigns the mechanism controlling the rate of secretion of ACTH to the hypothalamus which may indeed regulate all adenohypophysial secretion. A guide to the literature and reviews on this problem can be found in Harris (1948; 1952a, b), Hume (1952), Porter (1954) and Zuckerman (1952, 1955). Exactly how the hypothalamus exerts this control is not known; it could be by innervation of the anterior lobe (Vazquez-Lopez and Williams, 1952), and

The Adrenal Cortex

though generally denied the possibility still exists (Zuckerman, 1955). Harris is the proponent of the theory that the posterior region of the hypothalamus exercises its control over the anterior pituitary through the route of the hypophysial portal vessels. These latter comprise a primary capillary network in the median eminence, fed by branches of the superior hypophysial arteries, draining into large portal vessels which run along the infundibular stalk to open into the sinusoids of the pars distalis (Zuckerman, 1955). Harris favours the idea of a neurohumour elaborated by the hypothalamic nerve fibres and liberated into the primary plexus whence it is carried down the portal vessels to excite the cells of the pars distalis. Hume (1954) and Ganong and Hume (1954) carry the conception of neurohumours further by supposing that there are, in fact, two such substances: one is carried to the pituitary through the hypophysial portal system and leads to the release of 'maintenance' ACTH in the resting animal; the second, released only after stress, results in increased ACTH secretion.

The theory has not so far been substantiated by identification and isolation of a hypothalamic neurohumour specific for the adenohypophysis. Various conjectures as to the nature of such a substance have been made: that it is adrenaline-like (Harris, 1952a, 1955) or noradrenaline itself (Vogt, 1955); that both cholinergic and adrenergic agents are involved (Markee, Sawyer and Hollinshead, 1948); that it is a histamine-like agent (Harris, Jacobsohn and Kahlson, 1952) though all of these are considered unlikely (Guillemin and Fortier, 1953; Guillemin, 1955). Recently Slusher and Roberts (1954) reported the presence of two materials in extracts of bovine hypothalamus which appeared to be capable of stimulating pituitary release of ACTH: one factor was a water soluble, non-protein substance present in all parts of the brain; the second factor was present only in the posterior hypothalamus and was a lipid or a lipo-protein in nature. There have been other reports in the literature (e.g. Hume and Wittenstein, 1950) of a similar nature and further work is required before a final conclusion can be reached, due to inherent difficulties in testing such extracts.

Any theory which assigns hypothalamic neurohumours as

56

regulators of anterior pituitary function faces the further complication of interdigitating with the hypothesis regarding the control of the secretion of antidiuretic hormone. This hormone, although at one time generally thought to be elaborated by the posterior lobe of the pituitary, is now considered to be made in the supraoptic and paraventricular nuclei of the hypothalamus and passed down as or by neurosecretory granules to be stored in the neurohypophysis (see Scharrer and Scharrer, 1954). Thus the hypothalamus may be regarded, on these theories, as being the major controlling organ of the pituitary in all its aspects. Mirsky, Stein and Paulisch (1954) carry the integration even further by suggesting that the antidiuretic hormone serves as the neurohumour responsible for activation of the adenohypophysis. These suggestions underline one difficulty of the problem in that, if the six or more adenohypophysial hormones together with the posterior lobe hormones can show independent secretory rates, then the hypothalamic neurohumour must have a differential capacity or there must be a number of these substances equivalent to that of the pituitary secretions. The 'hypothalamic' theory has not escaped criticism on other grounds (Zuckerman, 1952, 1955). That the hypothalamus is of importance is not denied, for it must clearly be in the pathway which transmits such external stimuli as light to the pituitary; but it is considered, for example, that the integrity of the hypophysial portal system is not a prerequisite for normal anterior lobe functioning.

The method by which the secretions of the pituitary are controlled both as to rate and amount is under active investigation in many places and the theories outlined above will naturally undergo rapid modifications. There seems little doubt that both the level of circulating corticosteroids and the hypothalamus are important factors in the control of ACTH secretion. It may be, then, that the hypothalamus is normally one partner in the dual control of adenohypophysial secretions reacting to nervous system stimuli and perhaps to an 'inherent rhythm' to dominate and modify, from time to time, the basic controlling mechanism of pituitary/target-organ balance.

The Adrenal Cortex

(VI) FUNCTIONS OF THE ADRENAL CORTEX

General

The first step in exploring the function of a gland is to remove it experimentally and observe the consequences. Even so it is difficult to distinguish between primary and secondary effects. Adrenalectomy became a standard experimental method only when aseptic surgery allowed animal operations to be conducted without infection masking the results and when the importance of adrenal accessory tissue was recognized.

Ingle (1944) has made a very useful list of the effects of adrenalectomy:

Digestive	loss of appetite		
	slowed absorption		
	nausea and vomiting		
	diarrhoea		
Circulatory	physical	(i)	haemoconcentration
		(ii)	decreased blood pressure
		(iii)	decreased blood flow
	chemical	(i)	decreased sodium, chloride, bicarbonate and glucose in serum
		(ii)	increased potassium and non-protein nitrogen in serum
Tissue	physical		muscular asthenia
	chemical	(i)	decreased sodium in muscle
		(ii)	increased potassium and water in muscle
		(iii)	decreased glycogen stores in liver and muscle
Renal	increased excretion of sodium, chloride, and bicarbonate		
	decreased excretion of potassium and nitrogen		
Growth	cessation of growth. Loss of weight		
Resistance	decreased resistance to all forms of stress		
	death results in untreated animals		

Functions of the Adrenal Cortex

In the untreated adrenalectomized animal death quickly supervenes. Cowie (1949) examined in detail the influence of age and sex on the life span of adrenalectomized rats. He showed that the survival time after adrenalectomy of both sexes increases with age of the rat at the time of operation, and that, in post-pubertal rats, females outlive males after this operation. In the male, there was a regular increase in the life span, from 7·9 to 16·2 days after adrenalectomy, at ages between three weeks and four months, with the survival period remaining relatively constant in each age group. In the female there was a similar trend, the survival time after operation increasing from 8·2 to 19·2 days over the same age range. Similar figures for survival times after adrenalectomy in other species have been obtained. Of nineteen British Saanens goats (eighteen male castrated kids, 2–10 months old, and one adult female) adrenalectomized by Cowie and Stewart (1949), seventeen were dead within a week, the remaining two died on the 16th to 18th day after operation. Greep, Knobil, Hofmann and Jones (1952) found that the average survival time for four adrenalectomized monkeys was 13 days. Untreated adrenalectomized cats (Rogoff and Stewart, 1929) lived 10·8 days on the average, and for the dog the same authors (1926a, b) found a survival time of about 7 days. These figures serve as a general indication of survival times after adrenalectomy. It may, however, be much shorter, due to shock, especially operative trauma, or much longer, due to adrenal accessory tissue.

Although the removal of the adrenal glands is followed by a characteristic series of events, a statement of the exact function of the adrenocortical hormones is still impossible though the lines on which they must act are becoming clearer. It should be noted that adrenocortical hormones are part, not only of the balanced hormonal system of the body, but also of the whole set of integrated effectors of homeostasis. The classical approach of endocrinology which notes the consequences of removal of a gland and the subsequent effects of the gland extract do not therefore necessarily reveal the physiological role of the hormones involved. Not only does the mere fact of experimentation interfere with the normally acting mechanisms but also it seems to be generally true that most, if not all, naturally occurring substances will manifest some purely pharmacological action when administered in larger than physiological amounts (Ingle, 1948). The question of the physiological amount of hormone in itself has two aspects. On the one hand, the hormone can cause characteristic alteration of metabolism measurable, for example, by an end product specific to its action; and, on the other, in smaller and indeed minimal quantities, it may permit other reactions to go on: it may have a 'permissive' quality (Ingle, 1952 *a*, *b*). We may suppose that hormones liberated into the blood pass to all parts of the body and the specific effect of one hormone is demonstrated by its entry into one set of enzyme systems in a cell. The action of hormones on enzyme systems falls into four possible categories (Dorfman, 1952): (i) by changing tissue-enzyme concentrations; (ii) by functioning as a component of an enzyme system; (iii) by accelerating or inhibiting an enzyme system; or (iv) by direct or indirect effects on accelerators and/or inhibitors of enzyme systems. Following this thought we may suppose that firstly adrenocortical hormones are helpful though not essential catalysts in all cellular metabolic activities, and secondly that they are essential for the normal working of some cells or groups of cells. Adrenocortical hormones do not initiate processes but influence them.

Two major aspects of the role of adrenocortical hormones can be considered, namely water and salt-electrolyte metabolism and carbohydrate metabolism. These two aspects are by no means divorced from each other and are probably more closely related than can at present be defined. For, to name but one aspect of the problem, physiological processes must ultimately be expressed in terms of metabolism of the cell and in this unit the flux of ionic exchange and potential plays a vital role in energy-releasing activities just as much as changes in carbohydrate metabolism must reflect back on ionic balance.

Carbohydrate metabolism

A relationship between the adrenal cortex and carbohydrate metabolism was noted at the turn of the century when Bierry and Malloizel (1908) and Porges (1909, 1910) reported the occurrence of hypoglycaemia in adrenalectomized dogs and in Addison's disease. Later work on many animals gave more information on the low liver glycogen and low blood sugar levels following removal of the adrenal glands (table 6; Kuriyama, 1918; Bøggild, 1925; Britton and Silvette, 1931, 1932 *a*, *b*; Britton, Silvette and Klein, 1938; Rogoff and Stewart, 1926 *a*, *b*; Swingle, 1927; Hartman, MacArthur, Gunn, Hartman and MacDonald, 1927; Artundo, 1927; Houssay and Artundo, 1929; Cori and Cori, 1929). The altered carbohydrate levels in the absence of adrenocortical hormones must be considered in the light of the suggested pathways of carbohydrate metabolism and of the influence of other factors thereupon. Not all the ways in which the three major classes of foodstuffs, carbohydrates, fats and proteins are utilized by the body are known. Some of the reactions, however, by which energy is made available have been shown (text-fig. 9).

TABLE 6. *Carbohydrate levels of normal and adrenalectomized rats and mice (from Noble, 1950)*

Fasting (hr.)	Glycogen (mg./100 g.)		Blood glucose (mg./100 g.)
	Liver	Muscle	
	Mice		
0	2840	435	—
0	*2177*	*479*	—
6	171	332	—
6	*86*	*238*	—
12	252	298	—
12	*102*	*269*	—
24	346	228	—
24	*44*	*158*	—
	Rats		
0	1780	590	124
0	*2310*	*533*	*97*
24	25	520	72
24	*16*	*432*	*54*
48	233	507	80
48	*70*	*358*	*30*

Values for adrenalectomized animals are in italics.

The adrenocortical hormones constitute but one factor which enters into the intricate network of these processes. Insulin secreted by the β cells of the pancreatic islets of

The Adrenal Cortex

Langerhans is the hormone predominantly active in maintaining a constant blood-sugar level. Insulin works conjointly with several other hormones:

(i) The adrenocortical hormones (especially the glucocorticoids by definition).

(ii) Hormones of the anterior lobe of the pituitary. Of these, the actions of adrenotrophin are mediated completely (or nearly completely?) by the adrenal cortex. Growth hormone has a great influence on carbohydrate metabolism and this hormone may be the 'diabetogenic hormone' (Young, 1952) or they may be separate entities (Raben and Westermeyer, 1952), though this seems doubtful (Reid, 1953; Houssay and Rodriguez, 1953).

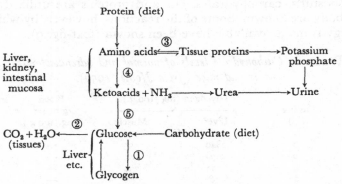

Text-fig. 9. Possible sites (1–5) of action of adrenal cortical hormones on protein and carbohydrate metabolism (from Long, 1953).

(iii) The hyperglycaemic factor or glucagon (Sutherland, 1950, 1951), ascribed to the α cells of the islets of Langerhans.

(iv) Adrenaline and noradrenaline (epinephrine and norepinephrine), the secretions of the adrenal medulla.

(v) The secretion of the thyroid gland.

Absence of insulin causes an increase in blood-sugar level (hyperglycaemia), the appearance of sugar (glycosuria) and of ketone bodies (ketonuria) in the urine; symptoms which characterize the human disease, diabetes mellitus. Absence of adrenocortical hormones causes, if food is withheld, hypoglycaemia and marked reduction of liver glycogen which does

Functions of the Adrenal Cortex

not rise with continued fasting as it does in the intact animal
(Long, 1953). The diabetic animal clearly offered interesting
possibilities for the display of adrenal influences which were
examined, in the first place, by Hartman and Brownell (1934)
and Long and Lukens (1936). The latter found that after
adrenalectomy of the cat with pancreatic diabetes, the high
blood-sugar levels fell to a normal or hypoglycaemic value,
there was a marked reduction in glycosuria and ketonuria
and, in addition, a significant decline in urine nitrogen excretion. Adrenalectomy of the rat with phloridzin diabetes
(Evans, 1936) or with diabetes following partial (95 % the
maximum in the rat) pancreatectomy (Long, Katzin and Fry,
1940) gave similar results. This amelioration of diabetes is
similar to that obtained by hypophysectomy of the diabetic
animal, i.e. the Houssay animal which is pancreatectomized
and hypophysectomized. The changes following adrenalectomy of the diabetic animal can be reversed by injection
of adrenocortical hormones (Long, Katzin and Fry, 1940;
Thorn, Koepf, Lewis and Olsen, 1940; Evans, 1941; Ingle
and Thorn, 1941). Long et al. (1940) found not only that
injection of adrenocortical extract or of glucocorticoids would
re-establish the diabetic symptoms but also that, in fasting,
adrenalectomized or hypophysectomized animals, adrenocortical hormone injection produced an increase in blood-
sugar level together with a deposition of glycogen in the liver
(table 7). There was, however, no deposition of glycogen in the
muscle with this treatment nor do fasting muscle glycogen values
change greatly after adrenalectomy (tables 6 and 7; Russell and
Bennett, 1936; Russell, 1939; Bennett and Perkins, 1945).

A further significant finding was that, following the injection
of adrenocortical hormones, urinary nitrogen excretion was
enhanced. Similar findings were obtained in fed rats; administration of cortical extract or adrenal steroids could produce
hyperglycaemia, glycosuria and deposition of glycogen in this
case not only in the liver but also in the muscle. Later work
has shown that steroids differ in their capacity to effect deposition of liver glycogen in adrenalectomized rats and mice and
in order of effectiveness in this respect they are hydrocortisone (compound F), cortisone (compound E), corticosterone

63

The Adrenal Cortex

TABLE 7. *Effect of cortical extract and adrenal steroids on glycogen (from Noble, 1950)*

Notes and treatment	Glycogen (mg./100 g.)	
	Liver	Muscle
Normal mice		
24 hr. fasted controls	346	288
Cortical extract, 0·25 ml./hr., 10–12 times	2986	223
Corticosterone, 0·025–0·5 mg., 8–10 times	1890	—
Dehydrocorticosterone, 0·25–0·5 mg., 8 times	2260	—
Fed controls	2840	435
Cortical extract, 0·25 ml./hr., 8–24 times	9196	1014
Adrenalectomized mice		
24 hr. fasted controls	44	158
Cortical extract, 0·25 ml./hr., 12 times	2370	182

(compound B), dehydrocorticosterone (compound A) and deoxy-corticosterone (Venning, Kazmin and Bell, 1946; table 4). The influence of the adrenal steroids on the carbohydrate metabolism of normal rats was shown by Ingle (1941) who force-fed normal animals a high carbohydrate diet (which had a normal calorific value). On injection with 11-dehydro-17-hydroxycorticosterone the rats showed marked hyperglycaemia and glycosuria. These effects have also been produced by corticosterone and 17-hydroxycorticosterone (Ingle, 1948) and with cortisone acetate (Lazarow and Berman, 1950). Such effects are called adrenal steroid diabetes and it is noteworthy that the presence of glucose in the urine is accompanied by increased urinary nitrogen. Adrenal steroid diabetes differs from pancreatic diabetes in that injections of insulin will not lower the blood-sugar level—it is insulin-resistant (Ingle, Sheppard, Evans and Kuizenga, 1945). The production of adrenal steroid diabetes as such may be an overdosage phenomenon, but the condition is due to physiological actions of adrenocortical hormones, though magnified. Furthermore, injections of ACTH can produce similar symptoms of hyperglycaemia, glycosuria and increased urinary nitrogen, indicating that such a rate and amount of steroid secretion is within the capacity of the adrenal cortex (Ingle, Winter, Li and Evans, 1945; Ingle, 1948).

Two conclusions are reached from these types of experiment: (i) because the increased production of carbohydrate by glucocorticoids is accompanied by a concomitant increase in urinary nitrogen, it was postulated that gluconeogenesis occurred with protein catabolism as the source of the newly formed carbohydrate (Long, Katzin and Fry, 1940); (ii) the amount of increased carbohydrate production was greater than could be accounted for by protein breakdown when the urinary non-protein nitrogen was used as an index of this gluconeogenesis (Ingle and Thorn, 1941; Ingle, 1948). For this reason, therefore, it was postulated that, in addition to their capacity to stimulate gluconeogenesis, the adrenocortical hormones also inhibit the utilization of carbohydrate by the tissues. Soskin and Levine (1946) do not support the suggestion that 'carbohydrate oxidation' is suppressed by adrenocortical hormones and, denying that any change occurs in the rate of sugar uptake by the peripheral tissue in the presence of increased amounts of adrenal steroids, consider that alterations in carbohydrate metabolism can best be explained by stimulated gluconeogenesis alone. The recent trend in opinion, however, favours the hypothesis that insulin facilitates peripheral utilization of glucose and adrenocortical hormones inhibit it. Thus recently Welt, Stetten, Ingle and Morley (1952), by injecting ^{14}C glucose at a constant rate and comparing the specific activity of administered and excreted glucose, estimated the glucose production from non-carbohydrate precursors. They found that treatment with cortisone resulted in a sevenfold increase in gluconeogenesis with no great change in glucose oxidation in anaesthetized but definite impairment in active unanaesthetized animals.

Should the hypothesis be true that adrenocortical hormones inhibit peripheral utilization, then these hormones can be considered 'anti-insulin' in effect. This possibility can be examined either by using the whole animal or isolated tissues. In the former type of experiment, de Bodo, Sinkoff, Kurtz, Lane and Kiang (1953) demonstrated that hypophysectomized dogs which manifest extreme sensitivity to insulin together with an abnormal secondary hypoglycaemia during the glucose tolerance test, gave normal or nearly normal reactions after a slight adrenaline hyperglycaemia when either ACTH,

The Adrenal Cortex

cortisone or hydrocortisone had been injected. Complete substitution is not achieved by the injections of these hormones and the authors conclude that the absence of anterior lobe hormones other than ACTH is an important factor in the abnormal carbohydrate metabolism of the hypophysectomized animal. Results from studies *in vitro* throw more light on the actual reactions which are affected by cortical hormones.

Work by Cori and his colleagues (Price, Cori and Colowick, 1945; Colowick, Cori and Slein, 1947; Cori, 1946) suggested that hexokinase (glucokinase) activity is stimulated by insulin and inhibited by adrenocortical and anterior pituitary hormones. It was shown that, although a commerical extract (Upjohn) of the adrenal cortex was by itself without effect, it increased the inhibitory action of anterior pituitary extracts on a purified hexokinase preparation. Analysing the hexokinase activity of crude extracts of muscles from alloxan diabetic rats (where there should be a preponderance of the pituitary factor over insulin because of the destruction of the islet tissue by alloxan), it was found that addition of adrenal cortex extract alone produced inhibition of the enzymic activity. This inhibitory effect of the adrenal cortex depended on the conditions of the experiment, and could be removed by the addition of insulin and re-imposed by further additions of cortical extract.

Krahl (1951), by further *in vitro* experiments, has confirmed in the case of striated muscle that part of the insulin-reversible inhibition of glucose uptake is produced by products of the pituitary and the adrenal glands, though other factors also are operative. The pituitary factor involved may well be growth hormone. Thus the glucose uptake of diaphragms taken from rats more than 10 days after hypophysectomy is higher than normal (Krahl and Park, 1948; Villee and Hastings, 1949; Park and Daughaday, 1949). After adrenalectomy, Krahl and Cori (1947) found glucose uptake of the diaphragms of the operated rats to be slightly increased above normal and Villee and Hastings (1949) found larger increases, though in neither case was the change sufficient to account for the increase in glucose uptake which follows hypophysectomy. However, insulin still stimulates glucose uptake in the dia-

66

phragm of the hypophysectomized-adrenalectomized animal so that although the absence of the pituitary and adrenal hormones removed an inhibition, the reaction still requires insulin for maximal efficiency (Krahl and Park, 1948; Perlmutter and Greep, 1948). It may be that, not only does insulin stimulate the hexokinase system but also has an additional effect on one or more reactions involved in the further oxidation of pyruvate which is antagonized by the adrenocortical hormones (Villee, White and Hastings, 1952). It is puzzling to find, on the other hand, that corticosteroids inhibit glycogen formation in the rat diaphragm (Verzar and Wenner, 1948; Bartlett, Wick and MacKay, 1949), and Verzar (1952) resolves this by supposing it to be due to stimulation of phosphorylase activity.

The isolated liver has also been employed in studies *in vitro* although by no means to the same extent as diaphragm preparations. In the isolated liver, adrenal hormones may prevent glycogenolysis (Seckel, 1940) or, when perfused with glucose, actually accelerate glycogen formation: these findings are more in keeping with those given above for the whole animal. Of course, as Krahl (1951) points out, direct balance experiments on uptake and utilization of glucose by liver are not possible, because glucose transformed to glucose-6-phosphate by the hexokinase reaction may be immediately returned to the medium as glucose by the glucose-6-phosphatase which is present in liver but not in muscle. The use of ^{14}C helps to overcome these difficulties; and Teng, Sinex, Deane and Hastings (1952) found that rates of formation of glycogen and other constituents of liver (fats, amino acids) from ^{14}C-labelled glucose by liver slices were decreased in diabetes and increased after adrenalectomy.

The actual mechansim involved in the inhibition of the hexokinase reaction is not known. Hexokinase requires the presence of sulphydryl groups for activity (Barron, 1951). Bacila and Barron (1954) demonstrated that cortical hormones inhibit glycolysis by inhibition of hexokinase and they added the further interesting suggestion that these hormones combine with the —SH groups to effect this inactivation.

Protein metabolism. In addition to their suggested action in retarding the release and promoting the formation of glycogen and in inhibiting glucose utilization, the adrenocortical hormones affect protein metabolism. Is the action of corticoids centred on whole protein in the cell or upon amino acids? Engel (1951) used bilaterally nephrectomized rats to answer this question. He found that there was a regular increase in urea nitrogen accumulation on treatment with ACTH and adrenal cortical extract and this, he considered, could be due to five possibilities: (i) a decrease in protein anabolism; (ii) an increase in protein catabolism; (iii) increased deamination of amino acids; (iv) an increased rate of conversion of ammonia to urea; and (v) secondarily, from a decreased rate of utilization of carbohydrate. Cases (i) and (ii) are distinguishable by the measurement of net nitrogen change; and Engel dismissed possibility (iv) on Hoberman's (1950) evidence that the rate of this step is far too rapid to permit serious consideration of a hormonal control over the urea cycle of Krebs. Engel marshalled evidence against possibility (iii) that deamination was the central locus of action of adrenocortical hormones. He showed that suppression of the stimulatory effect of steroid hormones on urea accumulation could be obtained by giving amino acids together with cortical extract.

Engel supported the view, then, that the locus of action of adrenocortical hormones was protein degradation and synthesis, and showed that, in contrast to what happens when amino acids are given, the injection of albumen into the rat preparations treated with cortical extract resulted in a significantly greater rate of urea formation than when either cortical extract or albumen was given separately. If protein breakdown or the ability to incorporate amino acids into protein were the site of action of the adrenocortical hormones, it would be expected that changes could be detected in plasma or tissue amino acid levels of intact animals or in the rate of amino acid accumulation in the plasma of the liverless (hepatectomized) animal under the influence of these hormones. Engel brings together evidence supporting this and recently Bondy, Ingle and Meeks (1954) have reinvestigated the problem. After surgical removal of the liver, amino acids released from the tissue proteins cannot be deaminated and

68

therefore accumulate in the plasma. The rate of rise of the plasma amino acids may therefore be considered a rough indication of the net rate of protein breakdown. When adrenal-ectomized animals are eviscerated (that is not only the liver but the other abdominal viscera also removed), the accumulation of plasma amino acids is depressed (Ingle, Prestrud and Nezamis, 1948). Exogenous steroids (adrenal cortical extract, cortisone, and hydrocortisone) given to eviscerated rats adrenalectomized 24 hr. previously, produced an accelerated rise of plasma amino nitrogen, showing that the cortical hormones have their effect on protein in the absence of the liver. Furthermore, these experiments of Bondy *et al.* do not support the hypothesis that adrenal steroids, at least as typified by hydrocortisone, must be metabolized by the liver in some way before they are physiologically active.

Finally, Hoberman (1950) and Clark (1950), using the amino acid glycine labelled with ^{15}N, showed that when cortisone is administered to rats there is an increased excretion of the isotope, indicating that, in the presence of considerable endogenous protein breakdown, the exogenously administered isotope cannot be used for protein synthesis but is itself excreted. We may conclude that not only do the cortico-steroids cause a catabolism of protein with gluconeogenesis but also prevent the anabolism of protein. A rider must be added to such a generalization; for although the catabolic effect is usually dominant, anabolism may not be prevented and may be enhanced. This appears to depend on the amounts of circulating steroids and the metabolic state of the animal. The action of corticoids both in protein and carbohydrate metabolism and in water and salt-electrolyte metabolism are often contradictory. The apparent confusion may be due to their being homeostatic agents so that their effect depends on the direction of the deviation from normal metabolism (Sayers, 1950)

Fat metabolism

The adrenocortical hormones must enter into fat metabolism, but the way in which they do this is even less precisely defined than their effects on carbohydrate and protein metabolism. The ruminants are resistant to the hypoglycaemic action of

The Adrenal Cortex

insulin in the adult (Reid, 1951) and apparently use fatty acids in energy-producing metabolism so that it would be of great interest to know the role of corticoids in such animals. Certainly 17-hydroxycorticosterone is secreted by the calf (Balfour, 1953). The rate of lipogenesis, like that of carbohydrate utilization, is regulated by the balance of hormones which are secreted by the pancreas on the one hand and the pituitary and adrenals on the other. Thus, fat synthesis is almost abolished by pancreatectomy, i.e. in the absence of insulin, while fat synthesis becomes normal or nearly normal in two main types of preparation, (i) in the Houssay animal that is hypophysectomized, pancreatectomized and (ii) in the adrenalectomized-pancreatectomized animal (Lukens, 1953).

Welt and Wilhelmi (1950) estimated the uptake of deuterium in the liver and body fat of rats to measure the effects of insulin and the principal contra-insulin hormones, growth and adrenotrophic hormones, the latter working through the adrenocortical hormones. If we examine the curves showing the rate of incorporation of deuterium into fatty acids during a period of 8 days, it can be seen that there is a striking increase of fat synthesis after adrenalectomy, and a retarded or inhibited synthesis when normal rats were treated with either growth hormone or ACTH (Brady, Lukens and Gurin, 1951). It seems, as Lukens comments, that treatment with growth hormone caused a progressive decline in the amount of carcase fatty acids similar to that described by previous workers (Lee and Ayres, 1936) and that ACTH has a similar effect. On the other hand, despite increased fat synthesis in the adrenalectomized rats 'the amount of body fat was nearly constant during the period of observation'. From this it is concluded that though inhibition of fat synthesis by adrenocortical hormones can be demonstrated, it is possible that paradoxically they also promote lipogenesis. In Cushing's syndrome, in which adrenal steroid secretion is very high, the increased deposition of fat may reflect the generally increased metabolism in these cases and the rate of fat synthesis may depend on the active carbohydrate metabolism. Furthermore, other unknown factors must be involved, as the fat in Cushing's syndrome is not distributed on the normal pattern.

70

Functions of the Adrenal Cortex

Muscle Work

A very characteristic finding in adrenalectomized animals and in patients with adrenal insufficiency is that of muscular weakness and quick muscular fatigue. Both Ingle (1944) and Everse and de Fremery (1932) developed this 'fatiguability' of muscle into a work test. The Ingle work test involved faradic stimulation of the gastrocnemius muscle of adrenalectomized rats and the injection of adrenal extracts. This gave a method by which steroids could be bio-assayed by observing the decreased rate of 'fatigue'. The method has proved to be a sensitive one and there is a close parallelism between the relative potencies of adrenocortical extracts and steroids in their ability to relieve the fatigue of muscle from adrenalectomized animals and their capacity to effect glycogen deposition in the liver (Ingle and Nezamis, 1948; Pabst, Sheppard and Kuizenga, 1947). This fatigue is related to the hypoglycaemia of adrenalectomy so that it might be presumed that lack of glucose means that the muscle cannot perform as much work. Certainly one intravenous injection of a solution of glucose in the 'fatigued' adrenalectomized rat gives a temporary improvement in the rate of work (Ingle and Lukens, 1941). On the other hand, continuous intravenous administration of glucose to the adrenalectomized animal does not significantly improve work performance (Ingle and Nezamis, 1948). There is no lack of available carbohydrate nor of the capacity to assimilate glucose during the time that the muscle is failing.

It may mean that this is evidence of circulatory failure, a real problem in the 'fatigued' adrenalectomized rat (Hales, Haslerud and Ingle, 1935; Ingle and Lukens, 1941), interfering thereby with energy liberation (Swingle and Remington, 1944). The administration of adrenocortical extracts and 11-oxysteroids can sustain optimal ability to work in the Ingle work test. This presents a paradox, for we have discussed above how one of the main sites of action of corticoids is the inhibition of the hexokinase system with consequent decrease in carbohydrate utilization. How then, as Ingle and Nezamis say, can hormones which seem to inhibit the utilization of

71

The Adrenal Cortex

carbohydrate sustain an optimum ability to work? This echoes the feeling of Soskin and Levine (1946). The corticoids make available carbohydrate from the tissue proteins and would thereby prevent hypoglycaemia and its consequences on the nervous system; but this does not apply in this case, as administered glucose maintains normal blood levels, with no alteration of fatigue. Ingle and Nezamis suggest that in physiological amounts the cortical hormones probably do not limit the ability of the animal to attain a peak of carbohydrate oxidation at high and continued levels of energy output. The feeling nevertheless remains that the corticoids must enter in a positive way to help cell metabolism and this is in agreement with Verzar's opinion that 'general experience points to some fundamental process in which the corticoids influence the utilization of glucose as a main source of energy production (Verzar, 1952).

Water and salt-electrolyte metabolism

Another main group of metabolic activities over which the adrenocortical hormones exert an influence is water and salt-electrolyte metabolism. That there were disturbed relationships in this aspect of physiological processes had not gone

TABLE 8. *The effects of adrenalectomy on plasma potassium and sodium concentrations in different species*

	Normal (mEq./l.)		Adrenalectomized (mEq./l.)	
	Na	K	Na	K
Dog*	155·9	3·5	139·6	9·1
Cat†	177	3·9	138	5·92
Goat‡	150·5	6·72	133·9	9·46
Rat§	143·62±0·58	4·40±0·08	134·04±2·46	5·76±0·15
Monkey‖	148·5 ±0·99	4·53±0·08	138	7
Man¶	143·2	4·0	135·5	5·4

* From Noble (1950).
† From Hegnauer and Robinson (1936).
‡ From Cowie and Stewart (1949) (their figures averaged).
§ From Christianson and Chester Jones (unpublished).
‖ From Greep, Knobil, Hofmann and Jones (1952), the figures for the adrenalectomized monkey (no. 17) read off the graph.
¶ From Sherwood Jones (unpublished)—untreated Addisonian.

unnoticed in the early work in this field (Soddu, 1898; Banting and Gairns, 1926; Lucas, 1926; Marine and Baumann, 1927; Baumann and Kurland, 1927; Rogoff and Stewart, 1928). It was, however, Loeb (1932, 1933) and his associates (Loeb, Atchley, Benedict and Leland, 1933), and two other groups of workers, Swingle, Pfiffner, Vars, Bott and Parkins (1933), and Harrop, Soffer, Ellsworth and Trescher (1933), who focused attention on the disturbance of salt-electrolyte metabolism as a major consequence of adrenocortical insufficiency in the Addisonian and in the adrenalectomized animal. The literature is now considerable and helpful reviews will be found in Kendall (1948), Noble (1950), Justin, Besançon, Lamotte, Lamotte-Barrillon and Barbier (1951).

TABLE 9. *Changes in plasma and muscle after adrenalectomy*
(Christianson and Chester Jones, unpublished)

Group	No. of animals	Body weight	Plasma mEq./l. Na	K	Muscle mEq./kg. wet weight Na	K	Water (%)
Normal females in oestrus	9	229·67± 3·23	144·61± 0·95	4·55± 0·12	25·82± 0·46	101·7± 0·71	76·06± 0·40
In dioestrus	9	233·6 ± 4·67	143·62± 0·58	4·40± 0·08	24·64± 0·61	100·9± 0·96	75·98± 0·15
Adrenalecto-mized 6 days	10	216·15± 4·20 236·15± 4·78*	134·04± 2·46	5·76± 0·15	18·34± 0·53	102·7± 0·53	76·5± 0·14

* Body weight before adrenalectomy.

The most common finding in the Eutheria after adrenalectomy or in adrenal insufficiency is an abnormal loss of sodium through the kidney, together with a decline in the concentration of plasma sodium (tables 8 and 9); the kidney is unable to limit sodium excretion when normal or subnormal quantities of salt are ingested (Loeb, 1941; Roemmelt, Sartorius and Pitts, 1949). Chloride is also lost but not relatively as rapidly as the sodium. Muntwyler, Mellors and Mautz (1940) found that the ratio of serum sodium loss to serum chloride loss was roughly 1·6 : 1; for the rat the ratio is almost 3 : 1 (Buell and

Turner, 1941); for the white-faced monkey, 1·7 : 1 (Britton, Kline and Silvette, 1938). Although this is the general finding, it is not always true, for Harrison and Darrow (1938) reported a 1 : 1 ratio in their rats, and Greep, Knobil, Hofmann and Jones (1952) actually found the ratio reversed in the male monkey (*Macaca mulatta*), namely 0·6 : 1. Concomitant with these changes, after adrenalectomy, the concentration of potassium in the plasma rises (tables 8 and 9) and there is a reduction, compared to normal, of the rate of loss through the kidney: the kidney is unable to excrete potassium adequately (Chester Jones, 1956 *a*, *b*). Other chemical constituents of the blood also change. There is an increase in the concentration of blood non-protein nitrogen and urea, and of uric acid, creatinine, sulphate and phosphate, and a fall in glucose, bicarbonate and in total base and total acid (table 10) (Noble, 1950). Amino acid nitrogen decreases in the fasting, adrenalectomized rat (Friedberg and Greenberg, 1947), but in the rhesus monkey, Greep *et al.* (1952), found an initial increase followed by an approach to normal levels as adrenal insufficiency progressed. There is decreased renal ammonia production (Sartorius, Calhoon and Pitts, 1953). The volume of the plasma decreases and there is an increase in the viscosity of the blood, in erythrocyte number and in haemoglobin and haematocrit (p.c.v.) values. The blood pressure falls, leading

TABLE 10. *Summary of data for normal and adrenalectomized rat plasma (from Conway and Hingerty, 1946)*

	Normal (mmol./l.)	Adrenalectomized (mmol./l.)
pH	7·32 ± 0·06	7·14 ± 0·07
K	5·91 ± 0·18	9·16 ± 0·40
Na	138 ± 1·8	128 ± 3·0
Ca	3·11 ± 0·12	2·82 ± 0·10
Mg	1·46 ± 0·06	2·03 ± 0·09
Cl	110 ± 0·9	104 ± 1·0
HCO_3 (as CO_2)	22·4 ± 0·5	20·1 ± 0·3
Inorganic P	2·17 ± 0·05	2·72 ± 0·21
Total P	3·89 ± 0·15	4·54 ± 0·20
Urea	6·7 ± 0·8	8·3 ± 1·0
Total molar concentration taking normal value as 100	100 ± 5·8	86 ± 5·4
Water content (g./100 g.)	92·4 ± 0·1	92·3 ± 0·3

74

to a hypotension which is very characteristic of the adrenal-ectomized animal. The capillaries of such animals become atonic and dilated causing a pooling of the blood with stasis (Noble, 1950). There is bradycardia, a fall in body temperature and a lower respiratory quotient.

TABLE 11. *Summary of data for normal and adrenalectomized rat muscle taken from Conway and Hingerty (1946). Details of the alteration of concentrations of hexose esters are omitted*

	Normal (mmol./kg.)	Adrenalectomized (mmol./kg.)
Potassium	101 ±1·9	118·5 ±1·3
Sodium	27·1 ±1·1	19·7 ±0·9
Calcium	1·56±0·04	1·54±0·06
Magnesium	11·0 ±0·3	12·4 ±0·4
Chloride	16·2 ±1·5	14·2 ±1·3
Total P	82·9 ±0·9	79·5 ±0·8
Acid-soluble-P	61·1 ±0·8	60·7 ±1·6
Phosphocreatine-P	24·4 ±0·4	27·7 ±0·4
Adenosine triphosphate-P	19·5 ±0·8	19·9 ±0·5
Total hexose monophosphate-P	10·04±0·87	5·67±0·46
Water content (%)	76·8 ±0·3	77·4 ±0·3

The haemo-concentration has a dual origin. In the first place, the increased excretion of sodium following removal of the adrenals necessarily carries water with it so that there can be an actual diuresis, the adrenalectomized animal coming into negative water balance. In the second place, there is a shift of water within the body itself. This is manifested by an increase in intracellular water at the expense of the extra-cellular and intercellular compartments. The peripheral tissues therefore reflect these changes. Skeletal muscle, a tissue commonly examined in the adrenalectomized animal, shows a decline in the total amount of sodium and chloride compared to normal values (tables 9 and 11). The decline in sodium concentration is due to the general drainage of sodium from the body, and it may well be also that a small part of the sodium lost is donated by the cell itself (Cole, 1953*a*, *b*, *c*). Tissue potassium values may show an increase after adrenal-ectomy, and the increase, when present, is due to the increased concentration of potassium in the intercellular fluid and to the increased potassium content of the cells themselves. Analysis of the whole muscle, however, may both reveal and mask the processes occurring, for the increase in cell-water content may offset the increase in cell-potassium values when expressed in mEq./kg. wet weight, as is normally the case (table 9).

It must be remembered, too, that the consequences of adrenalectomy are a function of time, it is a progressive process, though not necessarily a smooth one; erratic episodes are very likely to occur depending, among other things, on the diet and the frequency of taking food, which in turn may rest on the extent of the developing asthenia and possible attendant psychosomatic symptoms. Water accumulates in the cells as it is lost from the intercellular fluid; but for some time after adrenalectomy, figures for the water content of muscle may show no increase over normal. Thus Christianson and Chester Jones (unpublished) found that untreated adrenalectomized female rats 6 days after adrenalectomy had normal muscle water content, while Chester Jones (1956 b) found, 8 days after operation, an increase in muscle water (tables 9 and 12). Similarly potassium muscle concentrations may not be significantly elevated until adrenal insufficiency is well advanced, though an increased plasma potassium level is a constant finding. Changes in other substances have not been as thoroughly investigated as the salt-electrolytes. Conway and Hingerty (1946), however, found that in the rat 3 to 6 days after adrenalectomy there was an increase in magnesium, phosphocreatine and glucose-1-phosphate in skeletal muscle, and a decrease in glucose-6-phosphate and fructose-1 : 6-diphosphate; in the plasma there was an increase in magnesium, urea, total phosphate, inorganic phosphate and a decrease in bicarbonate and calcium and probably plasma pH (tables 10 and 11).

One of the most striking facts, then, about the adrenalectomized animal is the sodium loss, and it was Loeb and his associates (*loc. cit.*) who, finding patients with Addison's disease had lower than normal concentrations of blood sodium and chloride, demonstrated the marked clinical improvement obtained when an adequate amount of sodium chloride was added to the diet. The adrenalectomized rat, particularly, can be kept in good health for many months if it is given 1 % sodium chloride instead of tap-water to drink. In general, adrenalectomized animals, especially dogs, do better when a mixture of sodium salts, the chloride and bicarbonate or citrate, are given, thus allowing for the greater excretion of sodium over chloride and thereby preventing acidosis (Harrop, Soffer, Nicholson and Strauss, 1934; Allers, 1935). Interestingly enough, the adrenalectomized rat will choose saline rather than water to drink after adrenalectomy and will drink increasing quantities *pari passu* with time after the operation to combat the loss of water associated with the continuous increasing sodium excretion (Richter, 1936; Chester Jones and Spalding, 1954). The other factor necessary in successful salt therapy is to keep the intake of potassium as low as possible (Allers and Kendall, 1937; Wilder, Kendall, Snell, Kepler, Rynearson and Adams, 1937). Certainly the adrenalectomized rat, and to a varying extent other animals (including man) investigated, can be kept in good health with normal amounts of blood and tissue constituents by keeping the sodium intake up and that of potassium down. In the salt-maintained rat, too, the rate of deposition of glycogen in the liver is normal (Long, Katzin and Fry, 1940; Andersen, Herring and Joseph, 1940), as is the rate of absorption of glucose from the gut (Deuel, Hallman, Murray and Samuels, 1937; Althausen, Andersen and Stockholm, 1939). On the other hand, although in good health, the salt-maintained rat cannot withstand any form of stress as well as a normal animal, and the muscles are easily fatigued. Nevertheless, this therapy does substitute surprisingly well for the absence of the adrenocortical hormones. To conclude that the consequences of adrenal insufficiency stem from sodium loss would, however, be unjustified. To consider this point further, some of the mechanisms at work must be briefly reviewed.

The conservation of sodium is one of the functions of the kidney and in the normal animal sodium can be retained in the face of starvation or low food intake. The physiology of the kidney is a complex subject (Homer Smith, 1951), but as far as the hormones are concerned we must take cognizance of certain aspects (text-fig. 10). It is estimated that in man, the 180 litres of plasma which are filtered and reabsorbed during 24 hr. contain 25,000 mEq. of sodium, largely as bicarbonate and chloride; a deviation of as little as 1 % from the amount of sodium normally filtered and reabsorbed would mean the gain or loss of 11 g. of sodium and accompanying anions per day, a not insignificant amount (Selkurt, 1954). It is believed (Smith, 1951; see Selkurt, 1954) that the protein-free plasma is filtered through the glomerulus into the proximal tubule where there is an active reabsorption of sodium, other solutes and water following this cation; this is 'obligatory' reabsorption and accounts for about 80–90 % of the filtrate which

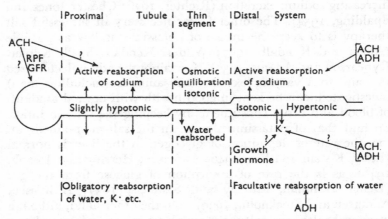

Text-fig. 10. Possible sites of action of hormones on water and salt-electrolyte excretion in the nephron (based on Smith, 1951; Gaunt and Birnie, 1951; Selkurt, 1954). ACH, adrenocortical hormones; ADH, anti-diuretic hormone of the posterior lobe of the pituitary; RPF, renal plasma flow; GFR, glomerular filtration rate.

entered the proximal tubule. The remaining 10–20 % of the filtrate arrives at the thin segment isosmotic or slightly hypotonic to the original filtrate due to the more rapid reabsorption of solutes proximally. The thin segment is considered the region where isosmotic equilibration is restored, possibly with the osmotic pressure of the plasma proteins of the peri-tubular capillaries (Selkurt, 1954). Further changes in the filtrate reaching the distal part of the nephron depend on the physiological needs of the animal—the reabsorption here is therefore termed 'facultative'. In the distal tubule both sodium and water are reabsorbed by processes which seem independent and the urine becomes more concentrated. Selkurt (1954) points out that this theory of Homer Smith and his associates is not necessarily true for all conditions of nephron activity and it may be that the amount excreted is a changing fraction of that filtered, rather than a constant fraction (proximal segment) plus a constant amount (distal segment). Nevertheless it serves as a valuable hypothesis for the discussion of the influence of hormones on the kidney.

The activity of the kidney depends on many factors other than hormones even as regards mineral metabolism (cf. Ingle, Meeks and Humphrey, 1953). The filtrate reflects the composition of the blood and its volume depends on the amount of blood reaching the glomerulus per unit time (the renal plasma flow) which in turn depends on the blood pressure, as does, among other things, the amount of plasma filtered per unit time, the glomerular filtration rate (GFR). The diet is important, influencing as it does the composition of blood and tissues (Cole, 1953*a*). In the absence of the adrenocortical hormones, there is hypotension and haemo-concentration and, therefore, decreased renal plasma flow and glomerular filtration rate. Even in adrenal insufficiency it is supposed that obligatory reabsorption still takes place in the proximal tubule so that maximum sodium loss could only be 15 % of that in the filtrate. It is possible, however, though not experimentally demonstrated, that the adrenocortical hormones influence the enzyme or carrier systems concerned in the active reabsorption mechanisms in the proximal tubule so that proportionally less sodium is reabsorbed with consequent less obligatory reabsorption of attendant solutes. It is clear, however, that in the absence of the adrenal gland the sodium reabsorptive mechanisms in the distal tubule do not operate in a normal way. So we can say that the corticosteroids act upon the distal tubule cells and effect the reabsorption of sodium facultatively.

But there is another factor at work (neglecting factors such as ionic balance, and rate of passage of ions past the tubular cells), namely the antidiuretic hormone, secreted by the posterior lobe of the pituitary (the neurohypophysis) in conjunction with vasopressin. The antidiuretic hormone is thought

The Adrenal Cortex

to act also at the level of the distal tubule, effecting the reabsorption of water liberated by the distal reabsorption of sodium in the normal animal and in addition effecting urine concentration at the same sites or possibly more distal ones in the nephron. The action of the antidiuretic hormone (ADH) on the reabsorption of water is unequivocal and ADH is secreted, possibly through the mediation of osmoreceptors, under the influence of varying 'effective' osmotic concentrations of the blood (Verney, 1947; Heller, 1956). ADH has, however, other properties assigned to it, one of which causes increased excretion of sodium (and chloride)—a natriuretic property. It has been difficult to show whether or not this action of ADH on sodium occurs within the same physiological range as that which effects the reabsorption of water (Selkurt, 1954). The possibility at least exists of an antagonism of action at the distal tubule level between ADH as a 'sodium excretor' and the adrenocortical hormones as 'sodium retainers'. Silvette and Britton originally developed the idea of adrenocortical/posterior-pituitary antagonism (Silvette and Britton, 1936; Britton and Silvette, 1937a, b; Silvette and Britton, 1938a, b; Corey, Silvette and Britton, 1939; Corey and Britton, 1941). The work of Gaunt and his associates, of Lloyd and of Sartorius and Pitts extended and modified the original concept so that the sequelae of adrenalectomy are considered to be occasioned negatively by the deficiency of adrenocortical hormones and positively by the water retaining and natriuretic properties of ADH (Gaunt, 1951; Roemmelt, Sartorius and Pitts, 1949; Sartorius and Roberts, 1949; Lloyd, 1952). One part of the argument was that antidiuretic substances accumulated in the blood of adrenalectomized animals owing to the lack of opposition from corticosteroids, but this now seems unlikely (Ames and van Dyke, 1952; Ginsburg, 1954).

Nevertheless, the possibility that there is a dual hormonal control on water and salt-electrolyte metabolism has been strengthened by investigations on the rat (Chester Jones, 1953; 1956a, b; Chester Jones and Wright, 1954a, b), extending earlier ones on the cat (Winter, Gross and Ingram, 1938; Ingram and Winter, 1938; Winter, Ingram, Gross and Sattler, 1941). Diabetes insipidus was induced in rats by removal of the neurohypophysis. Such 'diabetic' (in the sense of diabetes insipidus, having no connexion with diabetes mellitus) animals drink and excrete about five times the normal quantity of water but they remain in sodium and potassium balance (text-fig. 11). It was argued that if the effects of adrenalectomy on water and salt-electrolyte metabolism were solely due to the absence of adrenocortical hormones, there should be no

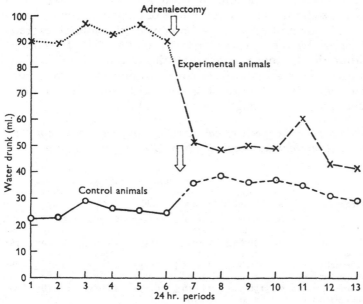

Text-fig. 11. The water intake of normal rats and rats with permanent diabetes insipidus before and after adrenalectomy. Experimental animals: x ... x posterior lobe of the pituitary removed, resulting in diabetes insipidus; x – – – x after adrenalectomy. Control animals: O——O before adrenalectomy; O - - - - O after adrenalectomy.

differences between animals without adrenals but with the posterior lobe (i.e. with ADH) and ones without adrenals and without ADH. It was found, however, that 8 days after adrenalectomy, diabetic rats had plasma sodium values within the normal range while the control adrenalectomized animals had values characteristically low (table 12). Similarly, the depletion of sodium in the muscle was not as great in the diabetic adrenalectomized rats as in the control adrenalectomized animals. On the other hand, the diabetic adrenalectomized animal shows pronounced elevation of plasma potassium concentration, though not so marked as in the control adrenalectomized animal. Similarly, the muscle potassium concentration of the former group was not as high as in the latter (table 12). It seems that the rat, in the absence of both the adrenocortical and the antidiuretic hormones, was able to conserve sodium and this in the face of the continued excretion of a large amount of water—about twice the volume excreted per 24 hr. by the normal animal (text-fig. 11). Injection of antidiuretic hormone into diabetic adrenalectomized rats depressed the plasma sodium concentration to the low control adrenalectomized figures, together with some elevation of plasma potassium though not to control levels.

TABLE 12. *The effects of adrenalectomy on female rats with and without diabetes insipidus (from Chester Jones, 1956b)*

Group	n	Body weight	Plasma mEq./l.		Muscle mEq./kg. wet weight		% water
			Na	K	Na	K	
Controls	12	190·63± 4·94	148·36± 4·22	4·96± 0·22	23·39± 0·31	100·26± 0·99	75·43± 0·33
Diabetes insipidus for 2–3 months	9	176·85± 2·61	146·40± 1·81	4·93± 0·19	25·95± 1·48	99·26± 1·63	75·78± 0·29
Controls adrenalectomized for 8 days	12	189·88± 5·90	131·85± 2·70	7·84± 0·13	18·33± 1·49	110·08± 1·92	76·60± 0·39
Diabetes insipidus adrenalectomized for 8 days	10	196·50± 5·50	146·16± 3·10	6·03± 0·18	19·68± 0·60	105·18± 1·82	76·00± 0·67

The conclusion seems to be justified that ADH does play a part in helping to promote the sodium loss characteristic of the adrenalectomized animal. In the normal nephron, then, we can envisage the adrenocortical hormones affecting the enzyme systems so as to cause sodium reabsorption, and the antidiuretic hormone inhibiting this action, perhaps by depressing the sodium carrier system (mitochondria?). As the adrenalectomized animal shows loss of water, in spite of the postulated activity of ADH, it may be supposed that the increased amount of sodium in the filtrate as it proceeds down the distal tubule carries with it an obligatory amount of water and water reabsorption under the influence of ADH cannot occur because of the osmotic forces involved.

Potassium excretion is complicated by the co-existence in its renal transport of both secretory and reabsorptive processes (Berliner, 1952). The evidence seems to suggest that the adrenocortical hormones facilitate potassium excretion and in their absence the normal copious secretion of potassium in the distal tubule is prevented. Conway and Hingerty (1953) have suggested that adrenalectomy may cause increased reabsorption of potassium in the renal tubules and that this tends quantitatively to reduce sodium reabsorption. Roberts and Pitts (1952), however, found that tubular secretion of H˙ or K˙ in exchange for Na˙ in the dog was not directly influenced by adrenocortical hormones but depended on the phosphate load presented to the tubules. Probably two distinct transfer mechanisms for Na˙ and K˙ are involved. Thus it is possible for an increase of K˙ excretion to occur with DOC without alteration of Na˙ reabsorption. Antidiuretic hormone may well have some effect on potassium excretion and may influence the cation exchange mechanism via the hydrogen ion, whereby it is postulated that potassium is secreted (Chester Jones, 1956a). Among the multifarious factors which determine the ratio of the amounts of sodium and potassium excreted, it should be noted that growth hormone can cause potassium retention, though whether this is a normally operating mechanism is still to be determined (Whitney, Bennett and Li, 1952). Furthermore, various experimental procedures can produce in normal animals depression of plasma sodium without marked change in potassium and, on the other hand, using the animal with regenerated adrenal cortex after enucleation, rats with disturbed water and potassium metabolism but with normal plasma sodium values or with normal water and sodium metabolism but with elevated plasma potassium can be produced (Chester Jones and Spalding, 1954). In the vast array of homeostatic effectors, hormones play an important part, but in salt-electrolyte metabolism not necessarily a dominant one (Hays, 1952).

In addition, several studies have been concerned with the possible interrelationship of changes of potassium and of

The Adrenal Cortex

protein and carbohydrate metabolism under the influence of adrenocortical hormones. Ingestion of large amounts of KCl prevents the increase in diabetes expected when ACTH is injected into diabetic rats (Glafkides, Bennett and George, 1952). Potassium chloride has produced a similar effect in the ACTH- or cortisone-treated diabetic human being inasmuch as the anticipated increase in insulin requirement did not occur (Kinsell *et al.* 1954). The suggestion (Whitney and Bennett,

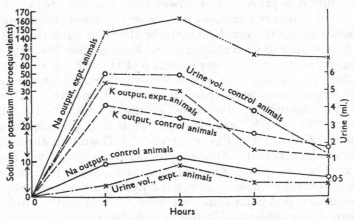

Text-fig. 12. The output of urine, sodium and potassium of normal and adrenalectomized male rats over a 4-hr. period, after administration of water by stomach tube (the volume of water administered was on the basis of 3 ml./100 cm.² surface area; this meant that the amount of water given was (in round figures) 11 ml. at hr. − 1 and 11 ml. at hr. 0 per rat). Note poor excretion of water by adrenalectomized animals but their high output of sodium.

The experimental animals were adrenalectomized, and maintained on tap-water; food was removed 24 hr. before water-loading.

1952; Bennett, Liddle and Bentnick, 1953) that large doses of potassium chloride might prevent protein breakdown and the weight loss induced by the injections of cortisone or of ACTH has not, however, been confirmed (Rupp, Paschkis and Cantarow, 1955).

One further aspect of the disturbed metabolism of the adrenalectomized animal must be described. The patterns of excretion of both the normal and the adrenalectomized animal are different when it eats and drinks *ad libitum* than when it is

84

presented with the stress of a water load. Thus, when the human subject ingests a large volume of water or when a laboratory animal is given a water load by stomach tube, the administered water is excreted rapidly. It is a characteristic of adrenal insufficiency that ingested water cannot be excreted at a rapid rate and this is the basis of a test for Addison's disease (text-fig. 12; Robinson, Power and Kepler, 1941). Excessive loads of water lead to water 'intoxication', convulsions and death, and this occurs more quickly, and with a smaller volume of water, in adrenalectomized than in normal animals; adrenocortical hormones provide protection (Gaunt, 1944*a*, *b*, 1951). While water given to the adrenalectomized animal is excreted very slowly, the amount of sodium excretion is high compared with normal (text-fig. 12) and potassium excretion usually reduced, though the GFR may be reduced to one fourth of normal values (Chester Jones, 1956*a*, *b*). This failure to excrete water rapidly in adrenal insufficiency is probably not merely due to the unopposed action of ADH but also because rapid water excretion demands the positive action of adrenocortical hormones as typified by cortisone and cortisol. Thus the need to excrete a water load or to eliminate large quantities of water regularly as in the animal with diabetes insipidus calls forth increased cortical activity (Chester Jones and Wright, 1954*b*). Injection of cortisone and cortisol into the adrenalectomized animal or the Addisonian improves the capacity to excrete water loads at a normal rate, an action beyond that of improving the glomerular filtration rate.

The presentation of loads of sodium chloride to the adrenalectomized animals brings problems of renal physiology somewhat outside the scope of this section. It should be noted, however, that an effective and rapid diuresis can be achieved in an adrenalectomized animal by giving sodium chloride of the correct percentage and this changes with the time after operation (text-fig. 13; Drill and Bristol, 1951; Hays, 1952). Potassium chloride loads given to adrenalectomized rats often result in death; before the onset of terminal symptoms, however, much less potassium is excreted than in the normal animal under similar conditions, though there is a considerable outpouring of sodium with little water (Chester Jones, 1956*a*).

The Adrenal Cortex

The results of the absence of adrenocortical hormones have been discussed, but which corticosteroids are responsible for the symptoms have still to be noted. Hartman and his colleagues separated a sodium-retaining factor from the cortex (Hartman, Spoor and Lewis, 1939; Hartman and Spoor, 1940)

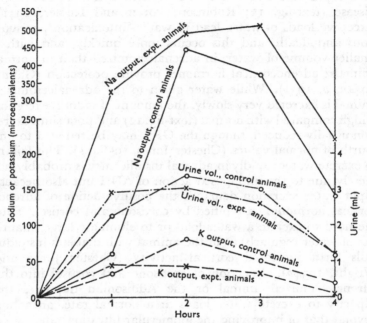

Text-fig. 13. The output of water, sodium and potassium of normal and adrenalectomized male rats over a 4-hr. period, after administration of a 1·2 % solution sodium chloride (volume as in text-fig. 12). Note that under these particular conditions the excretion pattern of the adrenalectomized animals was similar to that of the controls. The increased potassium excretion, though no potassium was given, is also of interest.

The experimental animals were adrenalectomized, and maintained on tapwater; food was removed 24 hr. before water-loading.

and the amorphous fraction has been known for some time to have a potent influence on sodium retention (Kendall, 1948). Cortisone itself affects water and mineral metabolism (table 4). First reports showed that cortisone administration resulted in an increased renal secretion of sodium (Gaunt, Birnie and

Functions of the Adrenal Cortex

Eversole, 1949; Thorn, Engel and Lewis, 1941); but sodium retention can occur in the normal, and does so regularly in the adrenalectomized, animal and the Addisonian (McEwen, Bunion, Baldwin, Kuttner, Appel and Kaltman, 1950; Perera, Pines, Hamilton and Vislocky, 1949; Sprague, Power, Mason, Albert, Mathieson, Hench, Kendall, Slocumb and Polley, 1950; Roberts and Pitts, 1952). It was felt at one time that cortisone and cortisol had sufficient influence on water and salt-electrolyte metabolism to obviate the need of postulating any specific additional steroid. Deoxycorticosterone was held to mirror the action of those adrenocortical hormones which influenced mineral metabolism, as it has effective sodium retaining properties; this retention was often accompanied by water retention. This compound is rather pharmacological than physiological in its actions and in continued administration produces a 'diabetes insipidus'-like syndrome (Ferrebee, Parker, Carnes, Gerity, Atchley and Loeb, 1941). On the basis that DOC was similar to a true corticosteroid in action and that cortisone and similar steroids had a variable influence on sodium excretion, Woodbury, Cheng, Sayers and Goodman (1950) suggested that corticosteroids tended to 'normalize' the concentration of sodium in the plasma irrespective of the direction of deviation from normal.

Recently aldosterone representing most, if not all, the activity of the amorphous fraction has been isolated (p. 37) and found to have a great influence on mineral metabolism (table 4). Aldosterone inhibits sodium excretion and promotes sodium retention and in this respect is 25 to 120 times as active as DOC in comparable amounts (Desaulles, Tripod and Schuler, 1953; Gross and Gysel, 1954; Simpson and Tait, 1955). Swingle, Maxwell, Ben, Baker, LeBrie and Eisler (1954) have examined the effect of aldosterone on the serum electrolytes of adrenalectomized dogs and compared it with that of other steroids. These authors show that the method of administration is important; they found that the minimum maintenance dose of aldosterone was 10 μg. per adrenalectomized dog per day when administered in 10 % alcohol in divided doses twice daily and that the daily dose for complete maintenance varied between 6·2 and 12·5 μg. per day (dogs

weighed between 13 and 20 Kg.). The least amount of DOC
which was necessary to maintain the adrenalectomized dog
was 125–250µg. per day. Thus they found that aldosterone is
12 to 25 times more potent than DOC when tested under the
conditions of their experiment. In contrast, the minimum
maintenance dose of aqueous-alcohol solutions of cortisone and
cortisol was 5000µg. per adrenalectomized dog per day, 500
times the maintenance requirement for aldosterone. The
properties of aldosterone are being actively investigated by
many workers and many of them have yet to be clarified,
especially at high dosage levels. Clinically it has been shown
to be very effective in the Addisonian (Mach, Fabre, Duckert,
Borth and Ducommun, 1954). The potency of aldosterone in
facilitation of potassium excretion is apparently not as great
relatively as its effect on sodium retention, Desaulles *et al.*
(1953) giving it a factor of five times the effectiveness of DOC.
Swingle *et al.* (1954) make the interesting point that it took
more aldosterone to maintain blood pressure and normal serum
potassium levels in the adrenalectomized dog than it did to
maintain normal serum sodium levels. Aldosterone, so far as
can be judged at present, does not facilitate water excretion as
does cortisone, so that adrenalectomized dogs or Addisonians
adequately maintained on aldosterone still cannot excrete
water at a normal rate (Gross and Gysel, 1954; Mach *et al.*
1954; Kekwick and Pawan, 1954). It does not produce the
'diabetes insipidus'-like effect of DOC or other pathological
side-effects of the latter compound (Gross, Loustalot and
Meier, 1955).

Aldosterone exerts its effects, on sodium metabolism at
least, in very small quantities as compared with other known
adrenal steroids (table 4). Thus, Simpson and Tait (1955)
found in man that in the pooled peripheral blood of twenty
normal males, the aldosterone content was of the order of
0·08µg. per 100 ml. whole blood and that of hydrocortisone
2·5µg. There is presumably 25 to 60 times as much cortisol as
aldosterone circulating normally. It would have been convenient, perhaps, to find that aldosterone dealt with the
problems of mineral metabolism (even if not water metabolism)
and that we had at last a compound solely with mineralo-

corticoid attributes. This is not the case. Aldosterone shares with cortisone and cortisol several properties as revealed by tests commonly employed to evaluate adrenocortical hormones (Dorfman, 1953; Gaunt, 1955). Thus aldosterone has one-third the activity of cortisone in restoring liver glycogen values in starved adrenalectomized mice in the Venning test (Venning *et al.* 1946), and in this test is 30 times as active as DCA (Schuler, Desaulles and Meier, 1954); aldosterone is about one half to one fourth as active as cortisone in reducing the number of circulating eosinophils in the Speirs and Meyer (1949, 1951) test, while DOC has no effect (Speirs, Simpson and Tait, 1954; Gaunt, 1955); aldosterone influenced the permeability of rat isolated diaphragm to ^{24}Na and ^{42}K in qualitatively the same way as did cortisol (Flückiger and Verzar, 1954). There is a difference of degree but generally not one of kind in the influence of aldosterone and other adrenal steroids on some enzymes such as alkaline and acid phosphatase, lipase, and esterase in testes, intestine and kidney (Loustalot and Meier, 1954).

It seems, despite aldosterone's powerful effect on sodium metabolism, that any rigid differentiation of the known adreno-cortical hormones into glucocorticoids and mineralocorticoids is unjustified. All the processes of the body are articulated one with another, and carbohydrate metabolism, mineral metabolism and water metabolism must inevitably be inter-dependent (Zuckerman, 1952). We do not know yet if aldosterone is always secreted in small amounts, thereby limiting it more strictly to its mineral activity, nor how its effects are integrated with those of compounds B and F which are normally secreted in large enough quantities to have some effect on mineral metabolism. It is possible, too, that aldosterone may be secreted in response to different evocators from those which stimulate the secretion of the other adrenal steroids (Axelrad, Johnson and Luetscher, 1954; Singer and Stack-Dunne, 1955), though nevertheless it has a basic relationship with ACTH (Renzi, Gilman and Gaunt, 1954), and the anterior pituitary does influence the rate of its secretion, if not as markedly as it does other corticosteroids.

The adrenocortical hormones necessarily act at the cellular

level. For the most part, cell membranes are permeable to sodium, potassium and chloride ions, although the concentration differences inside/outside are considerable. Replacement of cell potassium with sodium is achieved *in vivo* only after prolonged feeding with a low-potassium, high-sodium diet (Conway and Hingerty, 1948). A change of cell permeability may well be, therefore, a significant factor in changes of electrolyte distribution after adrenalectomy, though this is not the only mechanism at work (Ussing, 1954). The high intracellular potassium of rat muscle after adrenalectomy has been ascribed to passive inward diffusion of K due to the presence of intracellular organic anions, formed during metabolism, and which do not diffuse outwards (Boyle and Conway, 1941). As regards sodium, it is thought that there is a specific extrusion system and it has been suggested that this is based on a reduction-oxidation system located in the cell membrane with affinity for a particular cation, in this case sodium (Conway, 1947, 1949, 1951, 1952, 1954, 1956). According to this hypothesis the system proceeds cyclically and depends on the redox potential outside the cell. In what way this system is modified by the adrenocortical hormones is not known. However, it appears that the normal intracellular concentrations of Na and K in brain and kidney cortex *in vitro* are associated with rapid 'turnover rates' of the ions and this demands a continuous supply of energy from cell metabolism (Krebs, Eggleston and Terner, 1951; Davies and Galston, 1951). Even apart from sodium extrusion, active uptake of K from extracellular fluids has been postulated to account for maintenance of normal intracellular concentrations. These mechanisms can be altered by cortical hormones; sodium extrusion from cold-stored blood cells which have been transferred to a temperature of 37° C. is largely inhibited by deoxycorticosterone glycoside (Sherwood Jones, 1955). Moreover, the turnover rate of ^{42}K in erythrocytes is reduced by adrenocortical steroids (Streeter and Solomon, 1954). It is clear, however, that the effects of adrenocortical steroids on electrolyte transfer at the cell level are not known. We must look for the action of these hormones on mitochondria, for example—which may well be involved in renal tubular transport of water and cations (Whittam and

Functions of the Adrenal Cortex

Davies, 1954; Davies, 1954)—and on the enzyme systems. Dehydrogenase may take part in the renal excretion of Na and K as these depend on tubular secretion of H and NH_4. It is of interest to find therefore, that, using histochemical methods, Bourne and Malaty (1953) found that adrenalectomy reduced tubular dehydrogenase, treatment with cortisone partially restored it and DCA rendered it almost normal.

(VII) RELATIONSHIP OF THE CORTEX TO PERIPHERAL TISSUE

General

The adrenocortical hormones have a widespread inhibitory effect on various manifestations of growth (Asling, Reinhardt and Li, 1951; Baker, 1950; Evans, Simpson and Li, 1943; Ingle, Sheppard, Oberle and Kuizenga, 1946) and this involves their protein catabolic effects, the suppression of cell division and their capacity to cause disintegration of some types of cell. Overall somatic growth is suppressed in normal rats after treatment with large doses of adrenal cortical extract (Ingle, Higgins and Kendall, 1938) and with corticosterone and cortone (Wells and Kendall, 1940b; Ingle, Prestrud and Rice, 1950). Similarly, injections of ACTH which stimulate the adrenal to produce corticoids bring about suppression of growth (Moon, 1937; Evans, Simpson and Li, 1943; Ingle, Prestrud, Li and Evans, 1947). In the hypophysectomized rat, ACTH has been shown to inhibit the growth produced by growth hormone (Marx, Simpson, Li and Evans, 1943). Other signs of the interference of corticoids with protein metabolism are shown by retardation of chondrogenesis and osteogenesis and atrophy of red bone marrow following ACTH injections (Becks, Simpson, Li and Evans, 1944; Baker and Ingle, 1948). The application of the adrenal hormones to the skin may cause it to atrophy and the hair to be lost (Baker, Ingle, Li and Evans, 1948; Baker and Whitaker, 1948). Another very obvious effect of corticoids in sufficient dosage is the delay or prevention of normal wound healing, and none of the fundamental inflammatory tissue elements, such as fibroblasts, inflammatory cells and lymphocytes, appear at the site of injury (Baker and Whitaker, 1950; Creditor, Bevans, Mundy and Ragan, 1950; Ragan, Howes, Plotz, Meyer and Blunt, 1949; Bishop, 1954).

Not only, then, under the influence of adrenal hormones in excess, is there breakdown of protein from muscles, connective tissue elements, skin and blood vessels, but also from the matrix of the bone, leading to osteoporosis and to delay in the healing of fractures (Blunt, Plotz, Lattes, Howes, Meyer and Ragan, 1950; Bishop, 1954). As is so frequently the case, these effects of corticoids depend on dose and are not consistently in the direction of inhibition of growth. For example, the force-fed adrenalectomized rat, given 1 to 2 ml. of adrenal cortical extract daily, showed decreased output of urinary nonprotein nitrogen compared to saline treated controls, or those receiving larger amounts of hormone (Ingle and Prestrud, 1949). In addition to the amount of corticoids, various other factors such as diet and general metabolic state are operative in determining their metabolic effect. This is particularly evident in man where the existence of disease may well alter the consequences of corticoid administration, the protein made available being re-used by the body so that nitrogen excretion does not alter (Bishop, 1954). In other clinical cases, negative nitrogen balances have not appeared with large doses of ACTH or cortisone (Conn, 1948; Elkington, Hunt, Godfrey, McCrory, Rogerson and Stokes, 1949; Forsham, Thorn, Prunty and Hills, 1948; Sprague, Mason and Power, 1951). Browne (1951; 1952) has suggested that the fundamental process involved is a 'mobilization' or 'loosening' of protein by corticoids, the products then available being excreted or used by the body according to the needs of the moment. 'Mobilization' presumably means hydrolysis of the cell protein giving an increased amount of amino acids in the blood.

Cartilage and bone

Growth of cartilage and bone occurs at the ends of the long bones. Bone itself is in a dynamic state with constant solution and deposition of bony tissue, processes regulated by the calcium and phosphate ratios and by the parathyroid gland hormone. High secretion of adrenocortical hormones, as produced experimentally by ACTH injections, interferes with these normal processes so that the epiphysial cartilage of adult rats is thinner, proliferation of the cartilage cells reduced, as shown by the reduction in height of the cell columns and the atrophic condition of the cells themselves; erosion of the cartilaginous matrix seems to be impaired (Baker, 1950). Increased corticosteroid production, too, reduces the number of active osteoblasts, indicating a retardation in new bone formation. Clinically continued secretion of large amounts of adrenocortical hormones, as in Cushing's syndrome, often leads to osteoporosis—in twenty-two of the forty cases considered by Soffer, Eisenberg, Iannaccone and Gabrilove (1955)—characterized by decalcification of the bone. This process can be arrested by treatment which suppresses the cortical activity, for example, among other palliatives, by irradiation of the pituitary when there is over-secretion of ACTH by a basophilic adenoma or by partial adrenalectomy when a cortical tumour is the prime cause of excessive corticosteroid production (Browne, Beck, Dyrenfurth, Giroud, Hawthorne, Johnson, MacKenzie and Venning, 1955).

Skin and hair

In the skin there are three epithelial structures which exhibit significant cellular proliferation, namely hair bulbs, epidermis and sebaceous glands. In the rat after adrenalectomy, hair grows at a more rapid rate (Baker, 1951). Although oestrogen is also an inhibitor of hair growth, it fails to prevent the acceleration which follows adrenal removal (Baker and Whitaker, 1949; Ingle and Baker, 1951). Injection of ACTH into the rat can cause hair growth to cease almost completely (Baker, 1950); and the same effect can be obtained by local application of corticosteroids to the skin when the inhibition is manifest only in the area of treatment (Baker, 1951). The hair follicles are arrested in the epithelial bud stage, but growth is resumed after cessation of treatment with reinstatement of the normal patterns of growth. Prolonged treatment with corticosteroids results in a generalized atrophy of the skin: the epidermis is thinned and the hair follicles and sebaceous glands are atrophic; the dermis is much thinner and

93

exhibits a loss of discreteness in the outline of the collagenous fibres, which appear to be fused together; the number of fibroblasts, mast cells and other cellular components of connective tissue are reduced (Baker, 1951). Similar effects are seen in Cushing's syndrome.

Lymphatic tissue and blood cells

Adrenal insufficiency, either in the Addisonian or in the adrenalectomized animal, can be associated with a hyperplasia of lymphoid tissue, including enlargement of the thymus, and some increase in the number of circulating lymphocytes—lymphocytosis (Yoffey, 1950; Dougherty, 1952). Administration of ACTH or of corticosteroids or their endogenous increase by stress and disease results in a generalized reduction in the amount of lymphoid tissue in the body. The thymus gland may be reduced to a thin strand of lymphoid tissue embedded in fat (Wells and Kendall, 1940a; Baker, Ingle and Li, 1951). There frequently is a reduction in the number of circulating lymphocytes—lymphocytopaenia—together with an increase in neutrophils. In fact, a lymphocytopaenia has been taken as indicative of a high secretion rate of corticosteroids. One theory suggested that the reduction of the number of lymphocytes in stress, after ACTH administration and so on, was achieved by their lysis with consequent release of gamma-globulins. Thus the defence mechanism of the body was enriched by increase in anti-bodies under the influence of adrenocortical hormones (Dougherty and White, 1944, 1945, 1947). This does not seem to be the case (Noble, 1950; Yoffey, 1950) and the hypothesis has been withdrawn (Long, 1951). Preparations of extracts of the adrenal cortex and the corticoteroids do not induce lymphocytolysis *in vitro*, except for the anomalous case of the commercial adrenal extract, 'Lipo-adrenal Cortex' (Miller, 1954). It seems that induced lymphocytopaenia may be part of the general manifestation of inhibition of cell division by excess corticosteroids.

The eosinophilic leucocytes are of interest from the point of view of the adrenal cortex in that the numbers of these cells, of unknown origin and function, seem to depend on the amount of circulating adrenocortical hormones (Dalton and

Relationship of the Cortex to Peripheral Tissue

Selye, 1939). From the original observation, two tests have been developed; they are empirical, as is so often the case with corticosteroid estimation. First, a method employed clinically to evaluate the adrenocortical competence of a patient depends on the fact that the eosinophil cell count normally declines markedly after a known dose of ACTH over a given test period in the normal subject, while in the Addisonian, for example, such a decline is slight or absent (Thorn, Forsham, Prunty and Hills, 1948). The efficacy of the test is sometimes marred by other factors contributing to eosinophil cell number, such as the existence of an allergy. Secondly, the eosinophil count in mice has been used successfully as a biological test for the potency of various adrenocortical steroids (Speirs and Meyer, 1949, 1950a, b, 1951), and dose/response curves can be obtained of statistical validity (Rosemberg, Cornfield, Bates and Anderson, 1954).

The success of the Speirs method turned in the first place on finding a strain of mice, the C_{57}, which is particularly sensitive—handling alone of the male can produce a drop of as much as 80 % in the eosinophil count in 4 hr. In principle, the method uses adrenalectomized mice of this strain maintained with salt, and the assay is based on a decrease in the number of circulating eosinophils during a 3-hr. period following subcutaneous injection of a corticosteroid. Non-specific eosinophil count depression can occur but, interestingly enough, a single injection of adrenaline renders the eosinophils refractory to further injections of this substance or of toxic materials for 4 hr., so that corticosteroids under test can then be injected and the resultant eosinopaenia can be attributed specifically to them. Speirs (1953) reviewed his experience of the activity of adrenaline in the test and correlated adrenaline-induced eosinopaenia in the adrenalectomized mouse with the presence of adrenal accessory tissue, which was found as small, oval, yellowish masses of cortical tissue along the vena cava and renal vessels. Furthermore, he noted Hummel's observation that accessory adrenals were found in 49·6 % of 3173 mice representing nine strains. Speirs suggests that adrenaline acts directly on the accessory tissue to produce corticosteroids, a mechanism which, if confirmed, is of considerable theoretical importance.

95

Padawer and Gordon (1952) point out that the effect of corticosteroids on eosinophils may fall into three categories: (i) an inhibiton of the manufacture or release of these cells from bone marrow, (ii) a re-distribution of eosinophils to other organs, (iii) an increased destruction of eosinophils. While Padawer and Gordon favour the last suggestion, experiments *in vitro* do not confirm any direct effect of glucocorticoids on eosinophils (Essellier, Jeanneret, Kopp and Morandi, 1954). The first hypothesis seems to be more in line with the general mitotic inhibitory effect of adrenocortical hormones.

The formation of red cells is affected by the adrenocortical hormones, but they are only one of many agents affecting erythropoiesis (Gordon, 1954). The salt-maintained adrenalectomized rat shows an anaemia characterized by reduction in red cell counts, haemoglobin concentration and haematocrit values; red cell fragilities are also reduced (Gordon, Piliero and Landau, 1951; Gordon, 1954). In the bone marrow there is a reduced concentration of erythroid cells together with an increase in myeloid elements. All white cell elements appear to participate with the most striking increase noted in the numbers of immature neutrophilic and total eosinophilic cells (*loc. cit.*). The anaemia obtaining in the adrenalectomized rat has been characterized as of the hyperchromic macrocytic variety, while that of the Addisonian appears to be normochromic and normocytic (*loc. cit.*). The majority of the formed blood cell values in the adrenalectomized animal display their most significant deviations from normal at approximately two to three weeks following the operation and then return gradually to the control levels. This latter tendency may reflect the development of cortical rests, a constant danger in long-term adrenalectomized animals (Gordon *et al.* 1951).

Connective tissue

Fibroelastic connective tissue is made up of three essential components, namely the fibroblast, the fibre (the majority composed of collagen), and a mucoid ground substance (Ingle and Baker, 1953*a*). In addition, in this tissue, are found neutrophilic leucocytes, lymphocytes and monocytes from the blood, macrophages and mast cells. The ground substance is made up of two compounds, chondroitin sulphate, the formation of which is inhibited by cortisone, and hyaluronic acid. The latter component is modified by the enzyme hyaluronidase, so that the spread of substances through the connective tissue is facilitated. Hayes and Baker (1951) found that when the animal was pre-treated with corticosteroids (as adrenal extract), then the spreading of the test substance, haemoglobin, was increased, in extent and in rate, over that of the controls. The authors considered this effect to be due to the general action of adrenocortical hormones in the modification of skin. On the other hand, without pre-treatment, an acute experiment with one large injection of adrenal extract showed a decrease in the spreading reaction, and this was interpreted as a direct inhibitory effect of the adrenocortical hormones as typified by cortisone on hyaluronidase.

Tissue when damaged shows an inflammatory reaction and the adrenocortical hormones suppress all stages of this response. Thus, on damaging skin, the passage of leucocytes through the capillary wall into the injured area, the appearance of mononuclear and mast cells and the formation of fibrin and oedema are all retarded by cortisone, and there is a reduction in capillary permeability (Michael and Whorton, 1951; Ingle and Baker, 1953a). Wound-healing in the presence of excess corticosteroids is delayed as new fibroblast formation is suppressed, collagen fibres appear only slowly, the growth of new capillaries is retarded, and granulation tissue fails to form at normal rates. The wounds eventually do heal, even in the presence of excess cortical hormone, but the tensile strength of the scars may be reduced due to the reduction of the essential binding element, the collagen fibre (Ingle and Baker, 1953a). The connective tissue reaction to cortisone can be seen by implantation of a pellet of this steroid into orbital connective tissue. In these cases, the appearance of epithelioid macrophages, giant cells and the formation of new fibroblasts and macrophages were suppressed in the neighbourhood of the implanted pellet (Baker, 1954).

In man, collagen undergoes a specific change in rheumatoid arthritis, lupus erythematosus, scleroderma, periarteritis nodosa, dermatomyositis and rheumatic fever, and on this basis

The Adrenal Cortex

Klemperer (1950) grouped these diseases together as 'collagen diseases'. It was in the treatment of rheumatoid arthritis that cortisone and hydrocortisone came into marked clinical prominence (Hench, Kendall, Slocumb and Polley, 1949). The disease is undoubtedly alleviated by corticosteroid treatment, primarily through its anti-inflammatory action, but the effect is a pharmacological one and does not touch the unknown root cause of the disturbance. Bishop (1954) points out (i) that in order to avoid serious side effects, dosage of cortisone must be kept low, (ii) that treatment is likely to be disappointing in severe cases, and (iii) that long-standing remissions are not to be expected. Cortisone has been tried in many diseases with varying results; in many it acts to give temporary relief and sometimes it is more a triumph of euphoria than of systemic betterment. The Merck and Co. publication *The Cortisone Investigator* gives abstracts of the world literature on the clinical use of adrenocortical preparations, and surveys are given in Ingle and Baker (1953a), Bishop (1954), among many others. Corticosteroids and corticotrophin are useful, for example, in certain allergic conditions, in status asthmaticus, giving temporary relief in difficult cases, which may then be treated by other methods, in inflammatory diseases of the eye, and in rheumatoid arthritis (Bayliss, 1955).

Mammary tissue

Cowie and Folley (1947), using male and female gonadectomized weanling rats, found that when any significant morphological changes were observed in mammary tissue following adrenalectomy they were always in the direction of regression. On the other hand, while Cowie and Folley provide the fullest examination of the position, their findings do not agree with those of earlier workers. Butcher (1939) showed that adrenalectomy of underfed intact and spayed female rats caused the mammary glands to grow more rapidly, to increase the number of bud-like growths along the course and at the ends of the ducts and to increase the area covered by the gland. Similarly, Reeder and Leonard (1944), using immature male rats, found that adrenalectomy was followed by increase in the number of lateral buds or short branches along the duct system, though the actual area of the gland did not change. Little work has been done to demonstrate the effects of injected adrenal steroids on mammary growth. Mammary growth has been shown to be enhanced by DCA (Heuverswyn, Folley and Gardner, 1939; Speert, 1940; Mixner and Turner, 1942; Nelson, Gaunt and Schweizer, 1943). However, investigation of the role of identified corticosteroids in mammary tissue development has not so far been made. It seems probable that the growth stimulation of this tissue lies primarily with the sex steroids and that, as Cowie and Folley (1947) point out, the adrenal cortex plays no outstanding or essential role in the normal development of the mammary gland.

The importance of the adrenal cortex in hormonal control of lactation is not in doubt though its exact role is uncertain. The corticosteroids join with lactogenic hormone itself (prolactin), growth hormone, thyroid secretion and insulin in the complex control of milk production and it is not appropriate to consider these aspects here (see Folley, 1952, 1953, 1955). It should be noted, however, that adrenalectomy causes a marked depression of lactation which can be restored to normal by adequate substitution therapy (Cowie and Folley, 1948). The combination cortisone-deoxycorticosterone is very effective in this respect (Folley, 1953). Recently Cowie and Tindal (1955) have shown that small doses of chlorohydrocortisone alone gave virtually complete maintenance of lactation in the adrenalectomized rat.

(viii) ADRENAL-GONAD RELATIONSHIPS
General

The secretions of the adrenal cortex enter, as we have seen, into essential combination with the manifold activities of the body and no less with the processes of reproduction. Excellent reviews of the vast literature are available (Parkes, 1945; Burrows, 1949; Courrier, Baclesse and Marois, 1953; Zuckerman, 1953), and I shall confine myself here to a consideration of some of the problems of the relationship of the adrenal cortex with the gonads and with gonadotrophins. While these two glands perform distinct physiological roles, yet they maintain a close interdependence—expected though not inevitable from their common coelomic epithelial origin—the nature of which is still perplexing despite assiduous investigation. As the adrenal cortex depends on pituitary adrenotrophin for normal functioning, so do the gonads depend on pituitary gonadotrophins. It is possible, therefore, to have the following theoretical interactions.

(i) ACTH acting on the adrenal cortex to produce not only corticosteroids but also sex hormones (androgens, oestrogens and progesterone).

(ii) ACTH acting on the gonads to produce corticosteroids.

(iii) Gonadotrophins acting on the gonads to produce not only gonadal sex hormones but also corticosteroids.

(iv) Gonadotrophins acting on the adrenal cortex to produce sex hormones.

(v) Gonadal sex hormones entering into the metabolism of corticosteroids in the adrenal cortex and into the metabolism of the trophic hormones in the pituitary.

(vi) Cortical sex hormones entering into the metabolism of sex hormones in the gonads and into the metabolism of the trophic hormones in the pituitary.

(vii) Corticosteroids entering into the metabolism of the gonadal sex hormones and into the metabolism of the trophic hormones in the pituitary.

(viii) Gonadal corticosteroids entering into the metabolism of corticosteroids in the adrenal cortex and into the metabolism of the trophic hormones in the pituitary.

Not all these possibilities have been realized either clinically or experimentally. Clearly the problem, in the first place, is one of semantics and definition. Zuckerman (1953) has very properly pointed out that the fundamental consideration is that of biological function, and a hormone can be defined by the nature of its action on the target organ. But we must also recognize that a corticosteroid, to quote one typical example, can be either a steroid which has the action of adrenocortical hormones together with the steroidal configuration associated with these hormones, or a steroid which has physiological action on target organs which we associate with corticosteroid action, or a substance which has the steroidal configuration of the known adrenocortical hormones. At its simplest, we still require chemical information that we are dealing with steroids, as glycyrrhetinic acid mimics DOC to some extent in the Addisonian (Bayliss, 1955). Furthermore, because steroids do conform to the same general configuration, and the sex and adrenocortical hormones particularly so, it is not surprising to find an overlap of action, especially within the limitations of experimental conditions. Beyond this again there is the role of the liver in producing steroid metabolites which, by this very change, not only lose the physiological properties of their parent molecule but may gain different ones.

Gonadectomy removes the prime source of sex hormones from the body and allows gonadotrophin both to be stored in the anterior lobe of the pituitary and to be secreted copiously (Greep and Chester Jones, 1950*a*, *b*). Should, therefore, the adrenal cortex be a noteworthy secretor of sex hormones, the gonadectomized eutherian would reveal it, particularly by the continued maintenance of secondary sexual characters and other target organs of the accepted sex hormones, but in the vast majority of cases it does not. An exception is provided by the possibility, at least, that the *immature* rat adrenal can produce small amounts of material of androgen-like action, reflected in the continued differentiation of the prostate after gonadectomy (Price, 1936; Burrill and Greene, 1939; Howard, 1941). It has not been conclusively shown, however, that early differentiation of the rat prostate is dependent upon androgenic stimulation (Gersh and Grollman, 1939; Parkes, 1945). There is no doubt that the adrenal cortex has the capacity to secrete sex hormones and this is amply demonstrated by certain diseases in man (see below). In the mouse, too, gonadectomy of the newborn of certain strains, particularly the *dba*, allows adrenal tumours to arise which produce excessive secretions of androgens and oestrogens (Woolley, 1950). Reichstein demonstrated the presence of 11-oxygenated C-19 steroids in adrenal tissue (Reichstein and Shoppee, 1943) and isolated adrenosterone and 11-hydroxyisoandrosterone. A proportion of the metabolites of androgens normally found in the urine are thought to be of adrenal origin and androgens have been found in adrenal venous effluent (Gassner, Nelson, Reich, Rapala and Samuels, 1951). Furthermore, there is evidence in the literature that the adrenal hypertrophy induced by the administration of a pituitary extract rich in ACTH and prolactin caused development of the prostate, as seen by size increase and histological stimulation, and to a lesser extent of the seminal vesicles in the *castrated* rat (Davidson and Moon, 1936; cf. Nelson, 1941). There is nothing inherently improbable in these results—simulating a possible mechanism of abnormal adrenal androgen production in Cushing's disease and the adrenogenital syndrome, which is essentially the over-stimulation in the adrenal cortex of the enzymic reactions along pathways which, although unfamiliar to it, are normal in the other steroid-producing tissue, the testis—but they have not been confirmed (Li, 1953 *b*).

The absence of the gonads, though not normally associated with a change in adrenal function, does result in a change in size of the adrenal cortex, particularly in those species with post-pubertal adrenal sexual dimorphism (p. 9 above). Castration of the rat, for example, results in an increase of adrenal weight, tending towards that of the female (text-fig. 14). Spaying of the female is followed by a decline in adrenal weight, although there may be an initial transient increase (Greep and Chester Jones, 1950 *b*; Carter, 1954; text-fig. 14). The extent of the changes depends on the age of the animal at gonadectomy and on the duration of the condition. Many of the consequent histological changes seem to turn on the zona fasciculata, particularly the inner region; in the castrated male this thickens, in the spayed female it declines and the whole zona fasciculata may become smaller. Gonadal hormones reverse these changes, so that injection of testosterone in the castrate results in a diminution of adrenocortical size as it does in the normal male and female with appropriate dosages (text-fig. 14); oestrogens similarly effect an increase of adrenocortical size in the spayed female and in the normal and castrated male animal.

Sex hormones may well act directly on the adrenal cortex. Testosterone, for example, has cortex-maintaining properties in the hypophysectomized rat (Cutuly *et al.* 1938; Leonard, 1944; Zizine, Simpson and Evans, 1950), though this is not an unequivocal finding (Walsh *et al.* 1934; Nelson and Merckel, 1938; Zalesky *et al.* 1941; Rennels *et al.* 1953). In the mouse, too, testosterone has an undoubted direct effect on part of the adrenal cortex (see below). On the other hand, oestrogens have no apparent effect on the cortex of the hypophysectomized animal (Bourne and Zuckerman, 1940). The keystone of the changes consequent on the removal or exogenous addition of the homologous or heterologous sex hormones is the anterior lobe of the pituitary. Normally the adenohypophysis has a fairly constant number of cells and this, coupled with the presence of three main types of cell for the secretion of several hormones, carries with it the possibility of metabolic competition. So that increased secretion (with or without increased storage) of one hormone alters directly (Friedgood, 1946) or reciprocally (Selye, 1947) the secretion of other hormones. There can be, too, a total increased secretion of many or all adenohypohpysial hormones from one stimulus (Zuckerman, 1953).

In its simplest terms, in considering adrenal-gonad relationships, it may be supposed that there is an interplay between adrenotrophin (ACTH) and gonadotrophin (FSH and LH) storage and secretion. If the data of Greep and Chester Jones (1950*b*) on pituitary secretion and storage of gonadotrophins and on adrenal size are considered together, then a possible scheme can be outlined (text-fig. 14). The normal female pituitary contains little detectable stored gonadotrophin, and ACTH, FSH and LH are, of course, secreted in normal amounts. Spaying results in the storage of FSH and LH, though the pituitary weight does not change compared with that of the normal female. The storage of gonadotrophin can be regarded as diverting cells from the manufacture of ACTH and, together with a decline in the storage (Nowell, 1953), there is a diminished secretion of ACTH. This is reflected in the narrow zona fasciculata of spayed females, particularly well shown in Carter's material and in the drop in weight of the gland (Chester Jones, 1955). On the other hand, castration in the male is accompanied by increased storage of FSH and LH (as in the female) but also by increase in pituitary size. It may be that gonadotrophin is stored by enlargement of cells only and that no cells are taken away from ACTH production so that the zona fasciculata of the castrate male is well maintained. Furthermore, the pituitary of the castrate secretes both LH and FSH copiously, while the spayed female pituitary secretes relatively much less (Greep and Chester Jones, *loc. cit.*). The castrate adrenal cortex, particularly the inner zona fasciculata, has the opportunity, at least, of responding to gonadotrophin (see below for the mouse) which would cause a thickening of this zone. There is, however, no direct evidence for an effect of gonadotrophin on the rat cortex.

It is an interesting property of testosterone that it favours the storage of FSH in the pituitary. Thus testosterone reduces the size of the pituitary and LH secretion and this in turn will be reflected in a decrease of the inner zona fasciculata; the continued and increased storage of FSH could, especially with high doses, lead to a diminished secretion of ACTH. The overall picture would then be a smaller adrenal cortex. On the other hand, oestrogens seem to act, at least over short periods, by aiding the secretion not only of gonadotrophins but also of ACTH (Mandl and Zuckerman, 1952; Gemzell, 1952). It is possible that, in normal males, injected oestrogen liberates the stored FSH, allows more cells to produce ACTH and the consequent adrenal enlargement is aided by the reduction of testicular androgen allowing inner zona fasciculata to increase in size. A similar pattern, *mutatis mutandis*, may be true of gonadectomized animals injected with oestrogens.

In normal females only high doses of oestrogens are clearly effective in reducing adrenal weight (Chester Jones, 1955). The latter is sometimes accompanied by reduced body weight and some of the confusion in the literature results from the examination of either absolute adrenal weights or those expressed in terms of 100 g. body weight; other differences arise from age and strain of animal, dose and type of oestrogen used and period of injection. Many workers have reported adrenal enlargement after oestrogen injection (Selye *et al.* 1935; Ellison and Burch, 1936; Noble, 1938; 1939; Morrell and Hart, 1941*a*; von Haam, Hammel, Rardin and Schoene, 1941; Mark and Biskind, 1941; Janes and Nelson, 1942; Vogt, 1945), while others found no change with the same treatment

Adrenal-gonad Relationships

(Leonard *et al.* 1931; Selye *et al.* 1935; Freudenberger and Clausen, 1937 *a*, *b*; Bacsich and Folley, 1939; Selye, 1940; Mellish *et al.* 1940; Morrell and Hart, 1941 *b*), and a minority an adrenal weight reduction after oéstrogen treatment (Clausen and Freudenberger, 1939; Zondek, 1936). It seems that the normal female shows a variety of reactions to exogenous oestrogens; doses over short periods could stimulate general pituitary secretion and allow adrenal enlargement, or interfere hardly at all with the mechanisms already operated by endogenous oestrogens or, in high doses over long periods, have a pharmacological effect of general depression of pituitary secretion manifested in extreme cases by tumour formation.

TABLE 13. *The effects of gonadectomy and testosterone propionate* (T.P.) *injection on the adenohypophysial and adrenal weights of the golden hamster aged about three months. The table also illustrates the sex difference in adrenal weights which, in the case of the hamster, is contrary to that of the majority of eutherians* (Holmes, 1955)

		Body weight	Adrenal weight		Anterior pituitary weight	
	n	(g.)	(mg.)	(mg. %)	(mg.)	(mg. %)
Unmated females	6	56·1± 2·73	8·66± 1·04	15·46± 0·20	1·75± 0·19	3·18± 0·12
Spayed for 21 days	6	74·1± 6·37	9·93± 1·21	13·46± 0·82	2·84± 0·08	4·37± 0·42
Spayed and T.P. for 21 days	4	60·7± 1·71	13·80± 0·87	22·80± 1·97	2·26± 0·20	3·82± 0·17
Males	6	59·0± 2·00	11·00± 0·46	18·71± 0·43	1·64± 0·09	2·77± 0·04
Castrated for 21 days	6	82·2± 7·03	11·82± 1·22	13·09± 0·25	3·39± 0·09	4·27± 0·41
Castrated and T.P. for 21 days	4	85·5± 3·80	19·87± 1·38	23·42± 2·42	2·39± 0·14	2·82± 0·91

The golden hamster presents an unexplained paradox (table 13). It is doubtful whether spaying has much effect on adrenal size (table 13, cf. Keyes, 1949; Wexler, 1952). Castration tends to result in adrenal weight decrease and exogenous testosterone unequivocally causes an increase in adrenal weight in both males and females. The very early maturity of the male hamster, about 16 days of age, must be accompanied by significant gonadotrophin secretion in an immature soma.

The Adrenal Cortex

The basic ACTH/FSH/LH pattern may be thereby set differently and it is interesting to note that there is no sexual dimorphism of the pituitary in the hamster as there is in the rat. In the hamster, testosterone in moderate doses may effect the release and increased secretion of ACTH, a phenomenon only seen in the rat with very high amounts (Selye, 1941).

In the female animal there is a close relationship between the adrenals and the oestrous and menstrual cycles (Zuckerman, 1941, 1953; Mandl, 1951; Parkes, 1945). Mandl has recently confirmed Kostitch and Telebakovitch's (1929) observation that cyclical changes take place in the vaginal smears of spayed rats. The most attractive explanation would be that such continued cyclical changes were consequent on oestrogens of adrenal origin, presumably secreted cyclically (without going as far as assigning this to an adrenotrophic or gonadotrophic effect). There is no direct evidence for this view, though it is consonant with Bourne and Zuckerman's (1940) finding that vaginal cycles in spayed animals maintained on constant threshold dose of injected oestrogen occur in parallel with cyclical changes in the size of the adrenals. It might be implied, as Mandl suggests, that the periodicity is primarily dependent on the ovaries rather than the pituitary, and that, after ovariectomy, the adrenal cortex is the governing factor. It seems somewhat peculiar, however, that the presumed adrenal oestrogen does not affect the uterus, an organ naturally sensitive to the presence of such compounds, and that the adrenal cortex should possess the capacity for precise cyclical secretory patterns intrinsic to itself in the light of the gland's great dependence on pituitary adrenotrophin. Theoretically, the continued cyclical secretion of gonadotrophins acting upon the adrenal cortex could be invoked but, as noted above, there is no evidence for this.

The ovary itself requires the presence of adrenocortical hormones for the normal functioning of its own metabolic processes and for those which it influences peripherally. The Addisonian, for example, shows, especially in long-term cases, follicular atresia and decline in secondary sexual characters. The untreated adrenalectomized rat has irregular oestrous cycles, with at best a short oestrus (Chester Jones, unpublished).

Adrenal-gonad Relationships

When maintained with salt, however, the rat may have regular cycles for some time (Parkes, 1945; Mandl, 1954), although Swingle, Fedor, Barlow, Collins and Perlmutt (1951) found pseudopregnancy supervening in these circumstances. Mandl (1954), confirming and extending observations in the literature, has found that the decline in the size of the ovaries of the adult rat after adrenalectomy is due to an impairment of their sensitivity to FSH, rather than a decreased production of this gonadotrophin by the pituitary. Further, while the immature adrenalectomized rat does not fully respond to human chorionic gonadotrophin (utilized as being generally similar in action to pituitary luteinizing hormone), the adult rat is capable of a full response. On the other hand, the latter animal continues to be less sensitive to pregnant mare serum (used as being generally similar to pituitary follicle-stimulating hormone). Mandl concludes that, as the ovarian responses become normal with cortisone treatment, in the normal animal corticosteroids mediate at the cellular level in ovarian physiology—a working hypothesis in line with general thought.

The only real experimental evidence that adrenotrophin can act upon the gonads to produce corticosteroids comes from the work of Clayton and Prunty (1951). These authors found that ACTH administered to adrenalectomized mice produced a substance of ovarian origin which inhibited the formation of granulation tissue. It is considered that this effect is specific to corticosteroids (Billingham, Krohn and Medawar, 1951 a, b; Morgan, 1951; Sparrow, 1953; Krohn, 1955). Though Clayton and Prunty weave their finding into normally occurring physiological reactions, the experiment seems to demonstrate the possibilities of steroid-producing tissue rather than their normal function. If progesterone is an intermediary in corticosteroid formation in the cortex, then its biogenetic oxidation to a corticosteroid configuration in luteal tissue under abnormal conditions is not difficult to envisage.

In pregnancy there is generally an increase in adrenocortical secretion (Gemzell, 1953; Venning, 1946; Tobian, 1949; Bishop, 1954), and this can be associated with amelioration of the symptoms of rheumatoid arthritis and with the increased survival of homografts (Heslop, Krohn and Sparrow,

1954). It is possible for the Addisonian to become pregnant with improvement of the symptoms, though the stress of the condition, particularly parturition, may precipitate crises (Thorn *et al.* 1949; Samuels, Evans and McKelvey, 1943; Jailer and Knowlton, 1950). In the experimental animal, adrenalectomy may not interfere with the successful termination of pregnancy, though there is frequent maternal death or foetal reabsorption, especially in those not maintained with salt (Houssay, 1945; Walaas and Walaas, 1944; Christianson and Chester Jones, unpublished). Nevertheless, progesterone seems to have the capacity of prolonging the life of adrenalectomized animals, without salt therapy, and this is true both of the pregnant and pseudopregnant animal (e.g. Swingle, Parkins, Taylor, Hays and Morrell, 1937; Collings, 1941; Amour and Amour, 1939; Gaunt and Hays, 1938; Greene, Wells and Ivy, 1939; Emery and Schwabe, 1936; Schwabe and Emery, 1939; among others).

Progesterone then has one capacity equivalent to those of corticosteroids, but this is an example of overlap between adrenal and gonad secretions in function without confusion of identity. Though it is agreed that progesterone does possess this life-maintaining capacity, wherein its action lies is not clear. It would be expected that it had a sodium-retaining activity (Emery and Greco, 1940). Christianson and Chester Jones (unpublished) have found, however, that the adrenalectomized pregnant rat and the adrenalectomized pseudopregnant rat (with deciduoma) still show a decline in plasma sodium concentration to the low levels characteristic of adrenalectomized non-pregnant animals. This raises the question of the capacity for secretion of the adrenals in the foetuses. We have found (*loc. cit.*) that adrenalectomy of the pregnant rat is followed by a decline in foetal body weight and an increase in foetal absolute adrenal weight which, as it is coupled with a low plasma sodium concentration, with elevated potassium, suggests that the foetal adrenal does not secrete even minimally in these circumstances (cf. Moore, 1950; Jost, 1953). There is the further implication that in normal pregnancy in the rat the foetal adrenal cortex does not secrete.

Adrenal-gonad Relationships

The X zone

The mouse presents the only unequivocal case where gonadotrophins have been shown to have a direct effect on the adrenal cortex, although the possibility as a useful hypothesis has always been attractive (Albright, 1943). The mouse adrenal cortex possesses a juxta-medullary zone comprising cells with acidophilic cytoplasm and prominent basophilic nuclei, varying in expression according to age and sex (Pl. II, fig. 6). The cells do not show sudanophilia when undegenerated, nor do they react positively to the other histochemical tests which have been used to estimate cortical function, except that they do contain ascorbic acid. This zone is generally referred to as the X zone (Howard, 1927), but it has gone under other names ('transient cortex', Whitehead, 1933; 'boundary zone', Waring, 1935; 'interlocking zone', Roaf, 1935; 'androgenic zone', Grollman, 1936; 'sexual zone', Cano Monasterio, 1946; Botella Llusia and Cano Monasterio, 1950). The X zone has been confused many times with the zona reticularis but the two zones are entirely different. The X zone was originally described in the mouse adrenal cortex by Tamura (1926) and Masui and Tamura (1926), who thought it was the zona reticularis and thus initiated a confusion which has persisted spasmodically to the present day. Howard (1927) named the layer 'X zone', and Deanesly (1928) gave us clearly the details of the history of this layer of cortical cells. Since that time this intriguing histological entity has attracted much speculation, and many workers (references cited above; others to be found in Waring, 1942; Cano Monasterio, 1946; Chester Jones, 1948, 1949*a*, *b*, 1950, 1952; Greep and Chester Jones, 1950*a*; Howard and Benua, 1950; Benua and Howard, 1950; Zuckerman, 1953). Although the X zone cells may, perhaps, be identified earlier (Waring, 1935), they form a clearly recognizable juxta-medullary area at about 14 days of age in the mouse (Howard, 1930). The zone then increases quickly in size, as does the rest of the cortex.

The subsequent history of the zone is illustrated in text-fig. 15. In the male, the X zone collapses by the direct action of the androgens produced at maturity (Deanesly and Parkes,

1937; Chester Jones, 1949*b*). This collapse is achieved by pycnosis of the nuclei and shrinkage of cell cytoplasm, the whole settling down on the medulla and the connective tissue trabeculae forming a perimedullary capsule. Thus, the young adult male mouse from about 40 days of age onwards has no

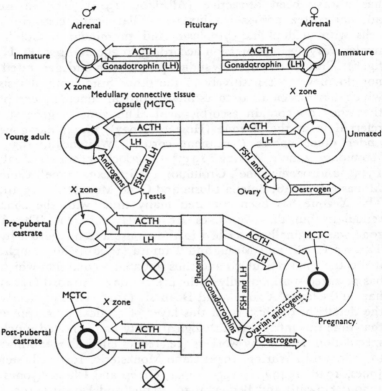

Text-fig. 15. Diagram illustrating the various changes in the adrenal cortex throughout the life of the mouse (from Chester Jones, 1955).

X zone but a connective tissue capsule surrounding the medulla. Castration before puberty removes endogenous androgens and the enhanced gonadotrophin secretion, which had been shown to be the trophic hormone, probably the LH fraction (Chester Jones, 1949*a*), allows the X zone to continue development. In the unmated female, the X zone does not disappear at puberty but comes to occupy up to 50 % of the adrenal cortex, as it may in the pre-pubertally castrated male. The extent of the X zone and the time after puberty at which it may degenerate in the unmated female depend very much on the strain of mouse. In the majority, the X zone can remain undegenerated up to three months of age and gradually in middle life the zone collapses, perhaps under the influence of small amounts of ovarian androgens (Chester Jones, 1949a, 1952). The collapse of the zone in the female is generally a 'fatty degeneration'. The cytoplasm becomes full of large lipid droplets. Cells about the middle of the zone are usually the first to show this degeneration, and in routine preparations with haematoxylin and eosin where the fatty cell contents are washed out the tracery effect gradually embraces the whole of the zone. The nuclei show pycnosis, and the degenerating zone collapses against the medulla, the X zone cells adjacent to the latter being the last to degenerate. As in the male, a medullary connective tissue capsule is formed. Peripherally to the X zone, a zona reticularis can be seen, little marked up to puberty but thereafter quite prominent, forming a sharp boundary between itself and the undegenerated X zone. This gradual process of X zone degeneration in the unmated female is a very rapid one during first pregnancy. During days seven to twelve of pregnancy, the X zone disappears rapidly by the action of ovarian androgens produced under the influence of placental gonadotrophin (Chester Jones, 1952), and a medullary connective tissue capsule is formed.

Post-pubertal castration of the adult male mouse gives a very clear example of the capacity of gonadotrophins to act upon the cells of the inner zona fasciculata to evoke a wide secondary X zone (Chester Jones, 1949a, 1955). Due to the growth of this latter zone, the whole gland increases in weight, as does the rat gland in like circumstances (see above). The mouse therefore demonstrates a potentiality in one species which may be present in others. There is no evidence that the X zone secretes anything, certainly not sex hormones (Grollman, 1936; Gersh and Grollman, 1939; Howard and Gengradom, 1940; Chester Jones, 1949a, b, 1950). The X zone does contain silver-nitrate-reducing substances, which may be ascorbic acid, but this is the only so-called histochemical property it shares with the permanent cortex, for the undegenerated X zone does not contain lipid droplets. Nor are the droplets of degeneration similar to the normal secretory droplets of the mouse zona fasciculata. It may be, just as ACTH can have two or more facets in promoting cell growth and hormone secretion, that gonadotrophin can maintain the X zone without biological indication of hormone activity. While the X zone has its most typical representation in the mouse, it does occur in other species (the cat: Davies, 1937; Lobban, 1952; the rabbit: Roaf, 1935). Holmes (1955) found that the nulliparous female of the golden hamster possessed a small X zone while the male did not at any age. The absence of the zone in the male hamster adrenal may be correlated with its early maturity (Chester Jones, 1955).

The foetal or transient cortex in man

The problem of the human foetal or transient adrenal cortex or X zone (Pl. II, fig. 10) is bound up with the general problem of the possibility of foetal endocrine secretion and its relationship with maternal hormones (text-fig. 16). The literature has recently been reviewed (Lanman, 1953; Baar, 1954). The embryonic adrenal, once formed, consists of a rim of cells, referred to as the 'permanent' cortex, while 90 % of the gland is made up of trabeculae of cells with acidophilic cytoplasm and basophilic nuclei—the transient or foetal cortex (Pl. II, fig. 10). The tissue is well supplied with blood sinuses and embedded in it are nests of sympatho-chromaffin cells which will give rise to the medulla. The foetal cortex commences to degenerate once the child is born or the involution may commence a little before full-term. At the same time, the permanent cortex grows and takes on the zonation characteristic of the adult gland. The degeneration of the foetal cortex once started goes on rapidly so that it usually has gone by six months of age, although it has been reported as taking up to eighteen months to disappear. The involution takes place by pycnosis of the nuclei and fatty degeneration, the zone shaking down, as in the mouse, so that the reticulum forms a connective tissue capsule around the medulla.

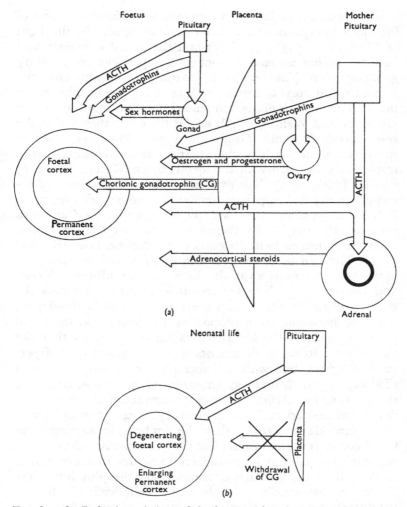

Text-fig. 16. Endocrine relations of the human adrenal cortex. (a) Showing some of the possible influences, none of which may be operative, on the developing human adrenal; (b) illustrating the enlargement of the permanent adrenal cortex after birth, with the concomitant degeneration of the foetal cortex. It is suggested that the withdrawal of chorionic gonadotrophin is the prime factor in this degeneration. There is, in fact, no evidence for hormonal influence on, nor secretion by, the human foetal cortex (from Chester Jones, 1955).

The Adrenal Cortex

The foetal cortex has been supposed to perform a variety of functions though none has been demonstrated. In the light of Albright's (1943) hypothesis and the actual demonstration that the mouse adrenal X zone was dependent on pituitary gonadotrophin (Chester Jones, 1949a), the theory that the human foetal cortex, having at least a superficial likeness to the X zone, was similarly influenced, became more attractive (Gardner and Walton, 1954), as did the older theory that the zone was androgenic (Grollman, 1936). However, there is still no evidence. It is possible that human chorionic gonadotrophin is concerned with the foetal cortex (Rotter, 1949a, b; Chester Jones, 1955). The predominance of chorionic gonadotrophin in the Anthropoidea matches its unique full expression in the embryo of this sub-class (the evidence for the Carnivora is weak (Hill, 1937)). Furthermore, chorionic gonadotrophin is at its maximum early in pregnancy and declines later (Bruner, 1951). This can be correlated with the observed degeneration of the foetal cortex at a variable interval from full-term (Keene and Hewer, 1927). The anencephalic monster, in which the brain is virtually absent and the pituitary poorly developed, has often been quoted in support of the theory that the foetal pituitary controls the foetal cortex on the grounds that the embryonic adrenal in these cases shows a normal type of permanent cortex but with the foetal cortex poorly expressed (Tähkä, 1951). Moreover, Angevine (1938) considered that there was no correlation between the amount of pituitary tissue developed and adrenal size in anencephalics. It would seem, too, from Meyer (1912) that the embryonic anencephalic foetal cortex is normal until after the fifth month and the small neonatal adrenal may be considered to be due to degeneration *in utero*. If we suppose that chorionic gonadotrophin influences the first conglomeration of coelomic epithelial cells which form the foetal cortex, the permanent cortex arising a little later (Keene and Hewer, 1927; Uotila, 1940), it would follow that the anencephalic has varying amounts of foetal cortex because of lessened chorionic gonadotrophin secretion by the placenta. On this hypothesis, failure of the usual development of brain and adrenal in anencephaly are not cause and effect, but arise from a common cause, namely placental abnormality. This hypothesis is, clearly, only a tentative one. Both in man and in the mouse, the foetal cortex and the X zone occupy, at one stage or another, such a large proportion of the adrenal gland that one feels that these aggregations of cells must be associated with some function; if it be so, we still await the demonstration.

(IX) NATURAL ABNORMALITIES OF FUNCTION

Adrenocortical insufficiency

Thorn and Forsham (1950) give the following clinical syndromes characterized by adrenal deficiency:

I. Acute adrenocortical insufficiency:

1. following acute adrenal cortical injury: infection, trauma, haemorrhage, thrombosis;

2. following removal of adrenal tissue;

3. precipitated by infection, trauma or surgery in patients with diminished adrenal cortical reserve.

II. Chronic adrenocortical insufficiency:

1. without pigmentation;

2. with pigmentation—Addison's disease;

3. as a part of a pluriglandular deficiency syndrome.

(*a*) *Waterhouse-Friderichsen syndrome.* In the first category, the Waterhouse-Friderichsen syndrome is of great interest. It can be divided into two types: the first is found in the newborn baby and most commonly follows difficult and prolonged labour, with an incidence as high as 1 % reported in babies dying at or shortly after birth (Thorn and Forsham, 1950). The second type can occur at any age, though 90 % of the cases have occurred in children under the age of nine years (Bishop, 1954). It results from severe overwhelming infections principally of meningococci or of staphylococci. The first symptoms are those associated with such infections and the onset of the Waterhouse-Friderichsen syndrome itself is heralded by cyanosis and the appearance of small spots formed by the effusion of blood (petechiae). Coma follows and death ensues in 12 to 24 hr. after onset.

In all cases the adrenals show considerable areas of blood

constituting massive and extensive haemorrhages with cortical cell degeneration. This condition is probably due to the excessive and continued demand for corticosteroids with the concomitant outpouring of ACTH operating beyond the capacity of the cortex to respond resulting in cortical collapse. Apparently injection of adrenocortical preparations in large enough amounts and early enough, when coupled with measures to alleviate the shock condition and the infection, will occasionally bring the patient through. The syndrome in the newborn, however, invariably leads to death.

(*b*) *Addison's disease.* In the secondary category of adrenocortical insufficiency, Addison's disease is of paramount importance and has been the subject of an enormous literature. Thorn, Forsham and Emerson (1949) give the following signs and symptoms of the disease in their order of importance:

1. weakness and easy fatiguability;
2. abnormal and increased pigmentation;
3. weight loss and dehydration;
4. hypotension and small heart size;
5. anorexia, nausea, vomiting and diarrhoea;
6. hypoglycaemic manifestations;
7. dizziness and syncopal attacks;
8. nervous and mental symptoms;
9. changes in gonadal function and secondary sex characteristics.

Other characteristics of Addison's disease have been noted earlier in the various sections on the function of the adrenal. In brief, among other things, the serum sodium and chloride concentration is decreased with elevation of serum potassium and blood non-protein nitrogen levels, there is haemo-concentration, low basal metabolic rate, small heart size, and increase in the number of circulating lymphocytes, and failure of the normal diuretic response to ingestion of water.

The disease is slow and insidious in development (Guttman, 1930; Rowntree and Snell, 1931) and is, therefore, not mimicked exactly by adrenalectomy of laboratory animals, though a less rapid induction of adrenal insufficiency has been obtained by venous obstruction (e.g. Elman and Rothman, 1924). Most cases arise—about 70 % according to Simpson

(1948)—from destruction of the adrenal by tuberculosis. Of the remainder, although a few appear from adrenal involvement in such diseases as syphilis, amyloidosis and scleroderma, the majority are due to a progressive atrophy of the adrenal cortex for no obvious reasons. It might be thought that adrenocortical degeneration in these non-tuberculous cases would be due to a failure in ACTH secretion but this generally does not appear to be the case. Thorn and Forsham (1950) suggest that this idiopathic atrophy may arise when the individual has had to cope with severe infection or the like resulting in some cortical degeneration. Normally the surviving healthy tissue would regenerate but, rarely, the excessive ACTH secretion might cause progressive cortical collapse on the principle of an 'over-work' phenomenon.

The treatment of the Addisonian depends on substitution for the corticosteroid deficiency by giving deoxycorticosterone, together with cortisone and supplementary sodium chloride. Individuals so treated can live fairly normal lives so long as infection and stresses of various types are guarded against, for in these cases the normal replacement therapy is not adequate to cope with the increased demand. With the advances in the preparation of more potent adrenal steroids, the progressive improvement in the life expectancy of the Addisonian may well be continued and enhanced (Thorn, Forsham and Emerson, 1949).

(c) *Simmond's disease.* Simmond's disease falls into place in the category of chronic adrenocortical insufficiency. This disease, however, is accompanied by widespread deficiency in other endocrine glands because it springs from a general failure of the secretions of the anterior lobe of the pituitary which is sometimes referred to as panhypopituitarism (Albright, 1943; Farquharson, 1950). Simmond's disease is now linked with the name of Sheehan because of his observations on cases arising from post-partum necrosis of the anterior lobe of the pituitary, one of the common causes of the disorder (Sheehan, 1939). For our purposes it may be briefly noted that patients with Simmond's disease are often slow mentally and physically, asthenic, with a marked decline in basal metabolic rate, anaemia, gonadal deficiency, a tendency to hypoglycaemia,

The Adrenal Cortex

increased insulin sensitivity; water metabolism is disturbed but blood sodium and potassium values are normal. Cachexia is not a necessary accompaniment of anterior-lobe deficiency. This has been brought out in the hypophysectomized rat by Shaw and Greep (1949). These workers showed that operated animals, on high calorific diets, could actually increase their weight above the starting body weight instead of losing weight or barely maintaining it as is frequently found.

Adrenocortical excess

(a) *Cushing's syndrome*. In man a continued excessive secretion of adrenocortical steroids gives a set of symptoms known as Cushing's syndrome (Bishop, 1954; Soffer, Eisenberg, Iannaccone and Gabrilove, 1955; Browne, Beck, Dyrenfurth, Giroud, Hawthorne, Johnson, Mackenzie and Venning, 1955). The disease is generally associated with abnormal secretion of ACTH giving adrenal hyperplasia. This arises from pituitary malfunction due to such lesions as basophil (one of the commonest types), eosinophil and chromophobe adenomas, and carcinomas. It may, however, arise from disturbance of the adrenals themselves by tumours, both carcinomas and benign adenomas. Fat is laid down in the cheeks and neck so that the face is 'round like a full moon' (Bishop). The distribution of fat in the rest of the body is peculiar, for obesity is confined to the shoulder-girdle, trunk and abdomen with none in the lower extremities. This occurs because of the excessive amounts of adrenocortical steroids in the body but the mechanism is unknown. Carbohydrate metabolism is disturbed and diabetes mellitus commonly found (p. 60). There is rarefaction of the bones leading to osteoporosis (p. 92). The skin becomes thin and the underlying capillaries and venules easily break, leading to striae, purpura and easy bruising (p. 93). The blood pressure is invariably raised (hypertension). There may be virilization and amenorrhoea in females. In some cases there is a disturbance of water and salt-electrolyte metabolism so that there may be retention of sodium and water (p. 72); the potassium level of the blood may be lowered, though Soffer and his co-workers find, when a change does occur, that it is elevated.

(b) *The adrenogenital syndrome*. In certain diseases in man the

Natural Abnormalities of Function

adrenal cortex produces large amounts of androgens. This may be associated with adrenocortical hyperplasia or with a cortical neoplasm. Enlargement of the adrenal cortex with high androgen secretion and consequent virilization produces the adrenogenital syndrome. This disease occurs in the adult, especially in the female, when masculinization occurs. Congenital hyperplasia in the child, sometimes commencing in pre-natal life, gives rise to pseudo-hermaphroditism in the female and macrogenitosomia praecox in the male (see Wilkins, Bongiovanni, Clayton, Grumbach and van Wyk, 1955). Even in those cases which show signs of masculinization at birth, there is rapid degeneration of the foetal cortex. The adrenal hyperplasia is caused by an early enlargement of the permanent cortex (Chester Jones, 1955). The large cortex of children with the adrenogenital syndrome, often three to four times the normal size, has not the usual appearance of a gland of the same size obtained from adults. Some workers have assigned the hyperplasia to an enlarged zona reticularis alone, or together with an increased zona glomerulosa (see Bishop, 1954). In my own experience, the hyperplasia is due to an increase in the size of the zona fasciculata, particularly the inner region; the zona glomerulosa is narrow. This cortical enlargement may well be due to the increased secretion of ACTH. Sydnor, Kelley, Raile, Ely and Sayers (1953) have demonstrated increased ACTH in the blood of patients with virilizing adrenal hyperplasia. It is of particular interest that the adrenogenital syndrome is often associated with adreno-cortical insufficiency, largely inadequacy in salt-electrolyte metabolism (Wilkins et al. 1955). Crises similar to those of the Addisonian may supervene and patients sometimes die in these attacks from sodium loss or dehydration or cardiac arrest and potassium retention (Bishop, 1954). This means that the production of normal corticosteroids is low; the cortical cells are producing androgens instead.

Among other schemes to account for this (see Wilkins et al. 1955), Dorfman (1955) has produced one of the most attractive (text-fig. 7). We saw earlier that the corticosteroid precursors give the C-21 steroid pregnenolone. Dorfman suggests that they also give C-19 steroids. Normally these are produced

119

only in small quantities, as a 'by-product' of corticosteroido-genesis. Along the C-19 pathway, the Δ^5-3β-hydroxy steroid, dehydroepiandrosterone, is formed from the first intermediates and the reaction is under the influence of ACTH. Dehydro-epiandrosterone, with the enzyme 3β-dehydrogenase, can form Δ^4-androstene-3,17-dione and this, with 11β-hydroxylase, is converted into 11β-hydroxy-Δ^4-androstene-3,17-dione. These three are adrenal androgens. In the adrenogenital syndrome, the increased production of ACTH results in the increased formation of C-21 and C-19 steroids. There is, however, Dorf-man suggests, a defect in the 21-hydroxylating system. There-fore, although pregnenolone and progesterone are produced in increased amounts, the corticosteroids are not. Decreased levels of circulating corticosteroids allow increased secretion of ACTH (see p. 55). The alleviation of the symptoms of the adrenogenital syndrome by cortisone administration, shown by Wilkins *et al.* (1955) would be due, on this hypothesis, to the suppression of ACTH secretion by the cortisone. Decrease in ACTH secretion would be followed by a decrease in C-19 production by the cortex. Furthermore, a varying degree of inhibition of the 21-hydroxylating system would give differing levels of corticosteroid secretion. This would show, with com-plete inhibition, as the 'salt-losing' type; with partial inhibi-tion, only the virilism would show—there would be enough corticosteroids produced for normal needs.

The diseases of adaptation

Experimentally produced enhancement of adrenocortical secretion and size mimics recurrent phenomena of an animal's day-to-day physiology. In response to diverse stimuli, the body's need for adrenocortical hormones is increased. Those stimuli which are potentially dangerous, that is, causing a 'harmful state of disturbed homeostasis within the body' (Ingle and Baker, 1953 *b*), such as cold, heat, burning, trauma, infection, anaesthetics and drugs, are 'stressor' agents. The whole phenomenon is known as 'stress'. The secretory activity of the adrenal cortex is increased during every type of stress so far investigated, depending in the first place on the increased secretion of ACTH.

Natural Abnormalities of Function

Selye (1947) has put forward the theory that long-continued exposure to stress results in non-specific, systemic reactions of the body which he calls the *general adaptation syndrome*. This he divides into three stages and defines as follows:

(i) *The 'alarm' reaction*, which is defined as the sum of all non-specific, systemic phenomena elicited by sudden exposure to stimuli to which the organism is quantitatively or qualitatively not adapted. There is active corticoid secretion with adrenal enlargement as a manifestation of active defence against 'shock'.

(ii) *The 'stage of resistance'*, which is defined as the sum of all non-specific, systemic reactions elicited by prolonged exposure to stimuli to which the organism has acquired adaptation. It is characterized by an increased resistance especially to the particular agent to which the body has been exposed and this is usually accompanied by a marked decrease in resistance to other types of stress.

(iii) *The 'stage of exhaustion'*, which representst he sum of all non-specific, systemic reactions which ultimately develop as the result of very prolonged exposure to stimuli to which adaptation has been developed but could no longer be maintained.

As I understand Selye's hypothesis, the first two stages are those frequently occurring, followed generally by return to normality. In extreme cases of stress the stage of exhaustion will be entered, and will be accompanied by an upset in general metabolism through an imposed abnormality of the hypophysial-adrenocortical relationship. On this concept, then, the Waterhouse-Friderichsen syndrome will represent the 'stage of exhaustion'. Likewise in the aetiology of idiopathic Addison's disease, it could be considered, from Thorn and Forsham's (1950) theory given above, before the onset of symptoms a stage of exhaustion must have been endured from which the adrenal cortex never recovered.

Selye himself extends the idea of the 'stage of exhaustion' considering that it can lead to the 'diseases of adaptation'. He suggests that in these circumstances the adrenal cortex secretes either an excess or an imbalance of its hormones which plays a role in the aetiology of certain diseases such as hypertension, arteriosclerosis, perarteritis nodosa, rheumatic

The Adrenal Cortex

fever, arthritis, nephritis, nephrosclerosis, gastro-intestinal ulcers, among others (Ingle and Baker, 1953 b). Selye's conception of the importance of stress in the aetiology of disease in man has stimulated many workers in their ideas and investigations. There is a feeling, however, that the primary causative agents of these diseases are still not known and any hormonal upsets are secondary manifestations of the root causes.

EXPLANATION OF PLATES

PLATE I

Fig. 1. Part of the adrenal gland of a normal adult female rat ($\times$ 100; Bouin, H. and E.). Outer connective tissue capsule at the top of the photograph, medulla at the bottom.

Fig. 2. As fig. 1 ($\times$ 55; formalin, sudan black). Zona glomerulosa heavily stained, zona fasciculata moderately and the zona reticularis sparsely. No sudanophobic zone visible; this occurs more usually in the male.

Fig. 3. As fig. 1 ($\times$ 1200). Cells of the zona glomerulosa.

Fig. 4. As fig. 1 ($\times$ 1200). Cells of the zona fasciculata.

Fig. 5. As fig. 1 ($\times$ 1200). Cells of the zona reticularis.

PLATE II

Fig. 6. Part of the adrenal gland of a young adult female mouse ($\times$ 100; Bouin, H. and E.). Outer connective tissue capsule, zona glomerulosa, zona fasciculata, no obvious zona reticularis, X zone, medulla.

Fig. 7. Part of the adrenal of an adult male mouse aged about nine months, hypophysectomized for 100 days (as fig. 6). The cortex is narrow. Externally there are healthy cells of the zona glomerulosa; from these there is a gradient to degenerated cells with pycnotic nuclei.

Fig. 8. As fig. 7 (formalin, sudan black). The external part has not taken the stain, the internal degenerating part is heavily sudanophilic.

Fig. 9. Part of the adrenal gland of an adult male rat hypophysectomized for 50 days ($\times$ 55; formalin, sudan IV). In contrast to fig. 8, the external part is heavily sudanophilic, the inner region is not.

Fig. 10. Part of the adrenal gland of a full-term baby ($\times$ 85; Bouin, H. and E.), showing the narrow rim of permanent cortex and (underneath) a portion of the extensive foetal cortex.

PLATE I

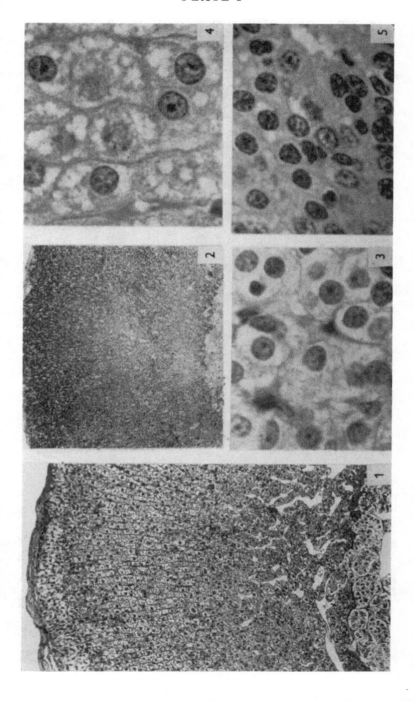

PLATE II

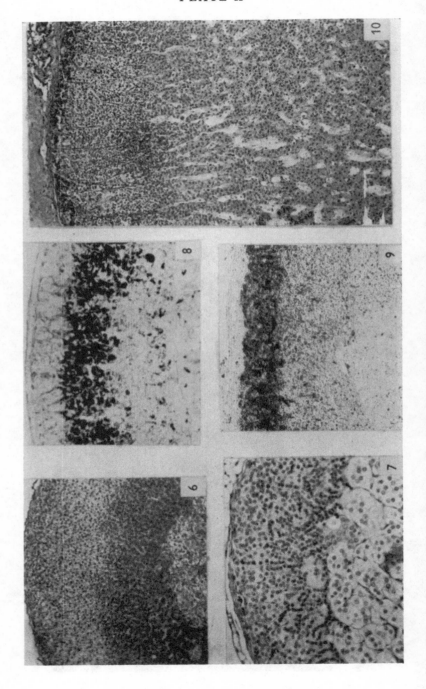

CHAPTER II

PISCES

(i) ELASMOBRANCHII

RETZIUS (1819) noted the resemblance between the adreno-cortical tissue of elasmobranchs and that of birds. Later, several workers (Nagel, 1836; Stannius, 1846; Leydig, 1851, 1852, 1853) described this organ in the cartilaginous fish. Balfour (1878) suggested the term 'interrenal' for the presumed adrenocortical homologue in *Scyllium* and he confined the name 'suprarenal' (Leydig, *loc. cit.*) to the chromaffin bodies. 'Suprarenal' has, of course, been used to denote the whole adrenal of higher forms and there exists, therefore, the possibility of confusion. I think that the synonym 'interrenal' may usefully be retained for the gland in bony and cartilaginous

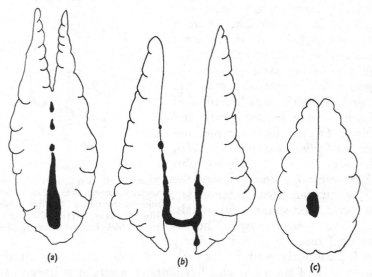

(a) (b) (c)

Text-fig. 17. Variation in form of the elasmobranch interrenal in relation to the kidney. (*a*) Rod-shaped type; (*b*) Horse-shoe shaped type; (*c*) Concentrated type (see text) (from Vincent, 1897*a*; Kisch, 1928; Dittus, 1941).

fish, though 'adrenocortical tissue' equally applies, as it does throughout the vertebrates. On the other hand 'suprarenal' should not be used, rather 'chromaffin tissue' or 'chromaffin bodies'.

Thorough investigation into the variations in form and the histological appearance of the elasmobranch interrenal was made in the last years of the nineteenth century by Diamare (1896), Vincent (1896, 1897a, b and see 1924) and Giacomini (1898). From these authors and from later work (Grynfellt, 1904a; Kisch, 1928; Dittus, 1937, 1941) the interrenal can be put anatomically into three main groups (text-fig. 17).

1. *The rod-shaped type*. The interrenal is rod-shaped, wide posteriorly, tapering off anteriorly, and lies in the mid-line between the kidneys, below the dorsal aorta and against the cardinal vein. Anteriorly in the same plane small groups of interrenal tissue are found. Dogfish (text-fig. 18) and sharks belong in this group and Dittus (1941), for example, describes *Scyllium canicula*, *S. cutulus*, *Mustelus laevis*, *M. vulgaris*, *Carcharias glaucus*, *Galeus canis* (and the holocephalous *Chimaera monstrosa*), as having rod-shaped interrenals.

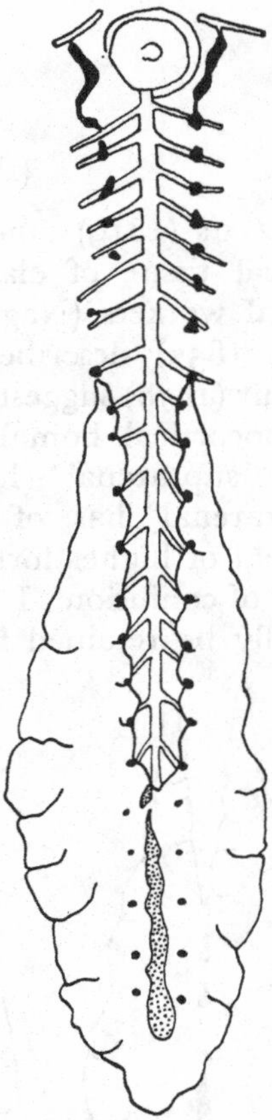

Text-fig. 18. Ventral view of the aorta, kidneys, chromaffin bodies, and interrenal of *Scyllium catulus* (from Vincent, 1897a). Chromaffin tissue, black; cortical tissue, stippled.

2. *The horse-shoe type*. The interrenal tissue is joined medially and posteriorly and is applied therefrom anteriorly to the inner side of the right and left kidneys where it is irregularly developed. In general, the interrenal has an asymmetrical appearance, in that the interrenal of the right side extends

forward further than that of the left (Leydig, 1852; Dittus, 1941). In this group are found many of the rays, Dittus (1941) noting *Raja asterias, R. batis, R. clavata* and *R. laeviraja oxyrhinus,* though the interrenal tissue in *R. clavata* is more usually broken up into small groups (Grynfellt, 1904*a*).

3. *The concentrated type.* The interrenal is confined in the majority of cases to an elongate oval body lying dorsally on the posterior part of the left kidney near the mid-line. In young animals the gland may be more ventral. The genus *Torpedo* is characterized by this concentrated type of interrenal, particularly *T. ocellata* and *T. marmorata,* although in *Raja eglanteria* (Hartman, Lewis, Brownell, Angerer and Shelden, 1944) among the rays it is of similar type. In *Torpedo,* in a minority of cases, the interrenal may be paired or, if single, may lie on the right kidney (Kisch, 1928). Dittus (1941), however, found that out of 145 fish (*T. ocellata* and *T. marmorata*) 134 had a single interrenal on the left kidney, five a single one on the right kidney and six had a pair, one on each.

The actual size of the elasmobranch interrenal naturally varies considerably, though there is a lack of precise data. Kisch (1928) gives some measurements of *Torpedo* interrenals; thus a female *T. ocellata* weighing 1430 g. possesses an interrenal measuring 17 mm. long by 8 mm. wide, and a female *T. marmorata,* weight 1530 g., a gland 10 mm. by 5 mm. Hartman, Shelden and Green (1943) weighed the interrenals of 135 animals representing ten species. Those with large enough numbers in the groups are reproduced in table 14.

The interrenal may be creamy white, yellow or brown in colour, depending principally on the presence of lipids, lipochromes and the state of activity of the gland (Vincent, 1897*b*; Fraser, 1929; Fancello, 1937; Pitotti, 1938; Dittus, 1937, 1941, among others). In general the gland is surrounded by a connective tissue capsule thinner in young than in old animals where it is often well developed. The interrenal cells are arranged in irregular, disorientated cords and groups of them are surrounded more or less completely by fine connective tissue (Pl. III, figs. 11 and 12). Capillaries and sinuses between the lobules are frequently abundant. There are no 'lumina' in the gland in well-fixed material (Dittus, 1941). Each cell is

The Adrenal Cortex

round or polygonal, with a prominent nucleus with strongly
basophilic chromatin and one or more nucleoli (fig. 12). The
eosinophilic cytoplasm is vacuolated, after routine histological
methods, due to the washing out of the fat droplets.

TABLE 14. *Body and interrenal weights of three species of
elasmobranchs (from Hartman, Shelden and Green, 1943)*

	n	Body weight (g.)	Interrenal weight (mg.)	(mg./100 g. body weight)
Raja erinacea				
Male	17	562± 9·6	23± 1·25	3·6
Female	56	532± 9·5	18± 0·12	3·4
Mustelis canis				
Male	9	2564±209·4	243±33·9	9·2
Female	6	2933±626·0	194±43·7	6·6
Squalus acanthias				
Females*	20	2987±151·3	171±14·4	5·7

* With embryos typically weighing 30 g. each.

This lipid has many of the properties displayed by that of
the eutherian adrenal cortex: it occurs in cytoplasmic droplets
of varying size, it can appear birefringent, it takes up osmic
acid and sudan colours, it gives reactions indicating the presence
of cholesterol esters, lecithin and non-saturated glycerides and
has a positive plasmal and phenyl-hydrazine reaction (Ramalho,
1917, 1921; Fraser, 1929; Aboim, 1939, 1944, 1946). These
methods do not indicate the adrenocortical hormones in
elasmobranchs any more than they do in mammals (Deane
and Seligman, 1953), but the variation and properties of
adrenal lipid may sometimes with caution be correlated with
adrenal activity. Furthermore, the reactions are at least
indicative that a similar set of chemical processes is going on
in the cortex of both lower and higher forms.

With this similarity it is not surprising then that workers
have correlated variation in the histological appearance of the
elasmobranch interrenal with changes in functional state as
many have done, *mutatis mutandis*, in mammals. Fancello
(1937), for example, considered that with maturity the dogfish (*Scyllium canicula*) interrenal became more lobulated,

together with an increase in the amount of lipid and of cell cytoplasm, and in the size of the nucleus, though no actual measurements were taken. Pitotti (1938) described the changes in the interrenal of *Torpedo* and of *Trygon* concomitant with sexual maturity and with 'pregnancy'. The most extensive work in this field has been done by Dittus (1941) who also investigated the torpedo, *T. ocellata* and *T. marmorata*. These animals are viviparous, pregnancy in the former lasting seven months and in the latter, ten months.

Dittus described four groups of these animals:

1. *The interrenal of torpedo embryos shortly before birth.* Distinct lobules made up of cords of cells clearly demarcated from each other; abundant capillaries and sinuses. The nucleus has a 'normal' appearance, the cytoplasm is granulated with little or no lipid. It is suggested that the granules are formed from nucleoli substance explosively extruded from the nucleus. There is active cell division, both mitotic and 'amitotic'; about 10 % of the cells have two or more nuclei.

2. *The interrenal of torpedo shortly after birth.* Three to four days after birth the interrenal has a compressed appearance, the lobules are close together and big sinuses are not seen. Cell division is much diminished. Obvious connective tissue fibres run through the gland, more or less in parallel lines. The four animals in this group had interrenals of this characteristic appearance. The cell nuclei were particularly basophilic with many nucleoli, and the cytoplasm had marked granulation, with some lipid.

3. *The interrenal of young animals with immature gonads.* The interrenal has a compact appearance with the lobules, although present, less discernible. In somewhat older animals (35 to 100 g.) there is a peripheral zone of cells with the cords of cells tending to lie parallel to the outer connective tissue capsule of the gland. The cytoplasmic granulation in this outer zone is more marked than that of the inner but in neither is it so conspicuous as the previous two stages. Lipid is present in both large and small droplets.

4. *The interrenal of animals during maturition of the gonads and reproduction.* In animals with eggs up to 0·5 cm. in diameter there seems to be some enlargement of the lobules with well-

developed connective tissue and increased vascularization. The cytoplasm has many granules. Animals with eggs up to 3 cm. in diameter and gravid animals possessed interrenals with even richer blood supply, marked basophilia of the nuclei, and increase in the extrusion of nuclear particles to form abundant granulation. There is little cytoplasmic lipid. Dittus believes that lipid is plentiful before the formation of granules, but after their formation the fat disappears; that is, the lipid is an essential forerunner of the interrenal secretion which is produced from the granules. Thus, the extent of cytoplasmic granulation can be correlated with the activity of the interrenal. Probably further work is required before we can say that the mature dogfish (Fancello, 1937) and the mature and gravid torpedo (Dittus, 1941) have particularly active interrenals.

The interrenal of *Torpedo* depends for its normal histological appearance on the pituitary for, 8 days after hypophysectomy, all the lobules have collapsed and the characteristic appearance is lost. The nucleus has a poorly developed chromatin network, with few or no nucleoli, there are no cytoplasmic granules; lipid is present but not so extensive as in the gland of immature animals (Dittus, 1941). It is interesting to note that Dittus does not describe actual cell degeneration with pycnosis of the nuclei; without the pituitary the interrenal goes into 'vegetative' form which can be changed to give an active appearance by the injection of mammalian ACTH (Dittus, 1941). The general correspondence, as far as it goes, between the findings for the eutherian and the elasmobranch is further exemplified by the consequences of interrenalectomy (Biedl, 1913; Kisch, 1928; Dittus, 1937). The proviso must be made, however, that normal elasmobranchs, especially torpedos, do not keep well in captivity and sham-operated animals lived only 4 to 20 days (Kisch, 1928). Kisch found that *T. ocellata* lived only a half to a third as long as *T. marmorata* after interrenalectomy, the former 26 hr. to $3\frac{1}{4}$ days and the latter $3\frac{1}{4}$ to 6 days, and Dittus (1937) has similar figures. In all cases, interrenalectomy was followed by a change in colour, a greying due to contraction of the melanophores, failing respiration, lassitude, muscular contraction with consequent opisthotonus, and finally death. The injection of mammalian adrenocortical extract into interrenalectomized fish produced improvements in the symptoms (Dittus, 1937).

It might be supposed, by analogy with the mammal, that interrenalectomy would be followed by a disturbance of the salt-electrolyte metabolism. Only Hartman, Lewis, Brownell, Angerer and Shelden (1944) have investigated this aspect of the problem and they came to the conclusion that 'there was no evidence that the interrenal plays any role in electrolyte metabolism' (p. 237). Chester Jones (1956a) considered, however, that their data indicated in fact some electrolytic change consequent on interrenalectomy (table 15). Their control figures (a control fish is a captured one placed on a board for blood extraction and therefore not necessarily normal) for the skate (*Raja erinacea*) show a very high serum sodium level and a high potassium level compared with the values in many vertebrate classes (cf. table 8). While in the sham-operated animals there was a tendency for both plasma sodium and potassium values to rise, the increase above the control figures is not statistically significant. Interrenalectomy whether the blood was taken from fish that were still active or sluggish after the operation, was followed by a rise in the level of plasma

TABLE 15. *Plasma Na and K changes after interrenalectomy in the skate,* Raja erinacea *(recalculated from Hartman, Lewis, Brownell, Angerer and Shelden, 1944)*

		Plasma		
		Na		K
	n	mEq./l.	n	mEq./l.
Sham-operated controls immediately prior to operation	8	$257 \cdot 42 \pm 4 \cdot 43$	7	$6 \cdot 22 \pm 0 \cdot 62$
2·5–13 days after operation (average 7½ days)		$261 \cdot 37 \pm 6 \cdot 52$		$8 \cdot 02 \pm 0 \cdot 84$
Active 'interrenalectomized' immediately prior to operation	15	$246 \cdot 60 \pm 3 \cdot 53$	15	$7 \cdot 80 \pm 0 \cdot 50$
2–15 days after operation (average 7 days)		$260 \cdot 40 \pm 3 \cdot 89$		$10 \cdot 52 \pm 0 \cdot 72$
'Markedly insufficient' interrenalectomized immediately prior to operation	9	$253 \cdot 66 \pm 4 \cdot 53$	9	$6 \cdot 75 \pm 0 \cdot 49$
2–10 days after operation (average 5 days)		$286 \cdot 11 \pm 6 \cdot 51$		$9 \cdot 72 \pm 0 \cdot 76$
All control figures prior to operation	32	$251 \cdot 25 \pm 2 \cdot 48$	31	$7 \cdot 14 \pm 0 \cdot 32$
All interrenalectomized figures	24	$265 \cdot 41 \pm 3 \cdot 96$	24	$10 \cdot 22 \pm 0 \cdot 40$

potassium. This increase is significant when judged against either the unoperated or sham-operated controls. The rise in the value of serum sodium is of doubtful significance for, although the increase after interrenalectomy is marked when compared with the unoperated animals, little difference is shown from that of the sham-operated controls. The whole problem needs re-investigation but there is an indication that removal of the interrenal in the skate is followed by an increased plasma potassium value as adrenalectomy normally is in eutherians. Elasmobranchs are, of course, practically all marine and have the property of actively re-absorbing urea from the renal tubules and thus raising the osmotic pressure of the blood. Water is thus absorbed osmotically from sea water through the gills and oral membranes and actual ingestion is not required (Smith, 1951). These facts must influence the electrolyte picture profoundly apart from the presence or absence of hormones. It might be supposed, however, that the elasmobranch living in an environment where sodium concentration is about 470 mEq./l. would have little difficulty, even after interrenalectomy, in keeping up the plasma sodium level to about 260 mEq./l. On the other hand, a typical figure for sea-water potassium is 9·96 mEq./l. so that, if interrenal hormones facilitate potassium excretion, interrenalectomy might well result in increased plasma concentration towards the level found in the external medium.

There is little evidence of the exact nature of the interrenal hormones. Crude extracts of elasmobranch interrenals have been used. On the one hand, Cleghorn (1932) could not prevent the development of insufficiency symptoms in adrenalectomized cats by injection of such extracts. On the other hand, with similar preparations, Grollman, Firor and Grollman (1934) maintained normal growth in young adrenalectomized rats. Evidence of a less direct nature comes from Santa (1940) who equated selachian interrenal extract with deoxycorticosterone acetate (DCA) on the basis of similar ability to contract carp melanophores. Whatever the interrenal hormones are found to be by direct biochemical analysis, the evidence now available from histological appearance, from the reactions of the cytoplasmic lipid, from the results after interrenalectomy, hypophysectomy, and the injection of adrenocortical extract, points to secretions not vastly dissimilar from those produced by the eutherian adrenal cortex and to a similar adenohypophysial-adrenocortical relationship.

(II) TELEOSTEI

There has been some doubt over the representative of the adrenal cortex in teleosts but the consensus now confirms the interrenal of the head-kidney in this role and not the corpuscles of Stannius (see below). Giacomini, in numerous papers over many years, laid the foundations of our knowledge on the anatomy and histology of the teleost adrenal cortex (Giacomini, 1902 *a*, 1905, 1908, 1909 *a*, *b*, 1910, 1911 *a*, *b*, *c*, *d*, 1912, 1920, 1921, 1922, 1928, 1933). The 'anterior interrenal' was so named by Giacomini (1908) in contrast to the corpuscles of Stannius which were the 'posterior interrenal'; this latter name is no longer used. The adrenocortical homologue, the interrenal as it can simply be called, lies in the pronephros which in the adult fish is generally represented by lymphoid kidney, though the head-kidney of *Fundulus* like that of other cyprinodonts has numerous glomeruli (Krauter, 1952; Pickford, 1953 *a*, *b*). The interrenal itself is formed into groups of varying size and distribution, depending on the genus, but usually it lies about the cardinal veins and their tributaries.

Baecker (1928) brought together information on the distribution of interrenal and chromaffin tissue in different species. In the twelve species he examined, the interrenal was confined to the head-kidney. In seven of the twelve, the tissue was equally distributed in the two halves of the kidney, in four cases it was predominantly in the right, and in one case (*Rhodeus amarus*) solely confined to the right half of the head-kidney. The amount of interrenal tissue varies considerably not only between species (text-fig. 19) but also between individuals of the one species. This may be correlated with reproduction and other factors on which, in teleosts, research has hardly been started. In salmonids (Giacomini, 1911 *a*, *b*), areas of interrenal tissue are scattered randomly throughout the lymphoid tissue of the head-kidney, with the bigger islets

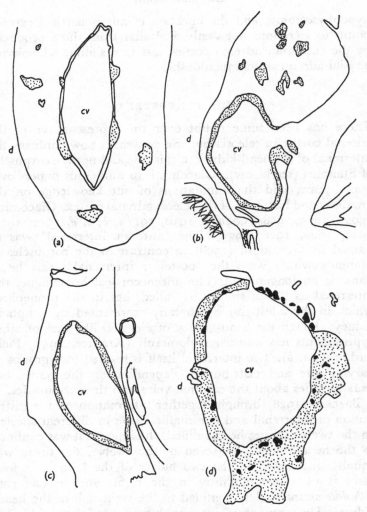

Text-fig. 19. This illustrates the increasing concentration of cortical tissue around the cardinal veins in various teleost species. One example shows that this concentration can be accompanied by a marked intermixing of chromaffin tissue. (a) *Salmo* species. T.S. head-kidney, left side; (b) *Carassius carassius*. T.S. head-kidney, right side (drawn from microphotograph (Baecker, 1928)); (c) *Rhodeus amarus* as b; (d) *Perca fluviatilis* as b.

Stippled, cortical tissue; black, chromaffin tissue; *cv*, cardinal vein; *d*, dorsal side of the head-kidney.

found more towards the middle (text-fig. 19*a*). The tissue is not related directly to the cardinal veins and Baecker thinks this can be regarded as the primitive type of distribution. Although Baecker differentiates the pike type (*Esox lucius*) from the salmonids, the arrangement of its interrenal tissue, lying in round balls scattered in the head-kidney, is essentially similar. Other types have the common basis of being associated with the cardinal veins and their tributaries. Thus, in the cyprinodont, *Leuciscus cephalus* (and perhaps in *Cyprinus carpio* and *Nuria danrica*), the interrenal tissue is several layers thick surrounding the cardinal veins and its branches in the head-kidney. In the sprat, *Clupea sprattus*, the interrenal tissue lies around the proximal part of the lumen of the right cardinal vein and also around the bigger venous branches in both halves of the head-kidney. In the stickleback, *Gasterosteus aculeatus*, the interrenal surrounds the cardinal veins and also sends fingers of cells into the surrounding lymphatic tissue and this occurs to some degree in the sprat too. Another variant is given by the cyprinodonts, *Carassius carassius* (text-fig. 19*b*) and *Gobio fluviatilis*, where the interrenal surrounds the veins but also occurs as isolated islets here and there in the head-kidney. Other fish possess more concentrated interrenals, and this applies especially to *Rhodeus amarus*, *Perca fluviatilis* (text-fig. 19*c* and *d*) and *Acerina schraetzer* and *Cottus gobio* where the tissue is applied to the cardinal veins and forms a compact mass ventrally thereto. Chromaffin tissue in all these forms lies either near the interrenal tissue, particularly in the walls of the cardinal veins, or scattered as single cells or groups of cells in the interrenal tissue. Thus, in salmonids, there is no actual intermingling of chromaffin and interrenal tissue, while in the perch most of the chromaffin tissue lies intermingled with the interrenal (text-fig. 19*d*).

The cells of the interrenal form themselves into layers, or, when in more rounded masses, seem for the most part to be made up of cords with no constant orientation but may be likened to a jumbled ball of string (text-fig. 20). A connective tissue framework is absent in many interrenals, such as those of the salmonids (Pl. III, figs. 13 and 14), although in others there is a reticulum, for example *Acerina schraetzer* (Baecker,

1928) and fish such as the eels have glands which are bounded by a capsule (Giacomini, 1911 *a*, 1912). Each cell, after routine histological methods, shows a large, round vesicular basophilic nucleus with acidophilic cytoplasm which in the minority of known cases may be granulated (fig. 14). In the salmonid interrenal (Spalding, unpublished) no lipid droplets can be seen, the sudan stains colour the cytoplasm diffusely and faintly; cholesterol cannot be demonstrated, the plasmal reaction is negative, there is no birefringent material; phospholipins are not shown after Baker's (1946) haematein method. The absence of lipid droplets by the interrenal cells and their unreactivity to many histochemical tests has been found in most teleosts (Comolli, 1913; Ramalho, 1921, 1923; Baecker, 1928; Diamare, 1933, 1934; Rasquin, 1951), although Comolli (1913) found black granules after osmic acid in *Anguilla vulgaris*, and Ramalho (1921) described, in *Labrax*, sudan-staining globules which did not react, however, to nile blue sulphate, osmic acid or after Fischler's method. Ascorbic acid (and other silver-nitrate-reducing substances) cannot be demonstrated histologically in teleost adrenocortical cells (Rasquin, 1951; Pickford, 1953*b*; Spalding, unpublished). Using chemical methods (Roe and Kuether, 1943) Hatey (1952) found that sections of the kidney of eel and rainbow trout containing interrenal had more ascorbic acid than did those comprising the more caudal renal portions. Later, Fontaine and Hatey (1954*b*) obtained similar findings in other teleosts. The teleost interrenal, therefore, does not possess, apart from ascorbic acid, many of the properties thought to be characteristic of the eutherian adrenocortical cell. This does not indicate necessarily that fish and mammal cortical secretions are different, for lipid droplets, reactive to the usual histochemical tests, are not necessary adjuncts of steroid hormone formation.

The function of the interrenal in teleosts is not known. Interrenalectomy has so far proved impossible because of the arrangement of the tissue in the head-kidney and about blood vessels. Indications of function, come, so far, from the results obtained by the techniques of hypophysectomy, the injection of hormones and the chemical analysis of blood. The silver eel (*Anguilla anguilla* L.) is a useful experimental animal, as it is hardy and has a particularly big aggregation of interrenal tissue in the walls of the anterior cardinal and the cephalad part of the posterior cardinal veins where it forms a thick

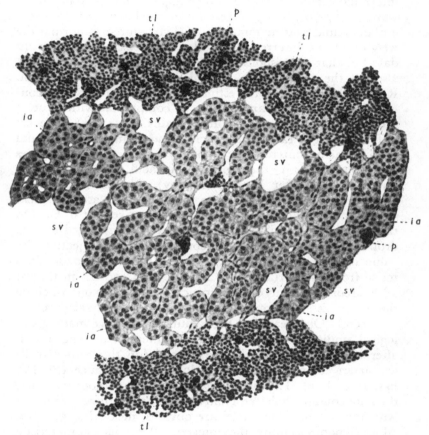

Text-fig. 20. An area of cortical tissue lying in the head-kidney of *Salmo fario* (from Giacomini, 1911 a). *ia*, cords of cortical cells; *sv*, blood sinuses; *tl*, lymphocytes; *p*, pigment.

ring 2 to 5 mm. in width and 1 to 2 cm. in length in adult eels. These compact masses can be removed and weighed. In female eels of about 150 to 250 g. the interrenal tissue (which necessarily includes the vein walls and a little other tissue) weighed 3·38 ± 0·26 mg. per 100 g. body weight. Hypophysectomy was followed by atrophy of the interrenal to about one-third of the normal weight. Six days after operation the interrenal weight was 1·23 ± 0·33 mg. % and 12 days afterwards, 1·07 ± 0·7 mg. % (Fontaine and Hatey, 1953). These authors found that mammalian ACTH restored to normal the weights of the interrenals of eels hypophysectomized for 12 days but that the response depended not only on the dose but also on the temperature at which the animals were kept. Results obtained from the estimation of the ascorbic acid content of eel interrenals are not exactly comparable with those of the rat after similar procedures (Hatey, 1954a, b). The eel differs from the rat in that there is no change in interrenal ascorbic acid content after hypophysectomy (normal figures 61·0 ± 3·91 v. 57·9 ± 6·42 μg. ascorbic acid per 100 g. interrenal 10 days after hypophysectomy). Neither does mammalian ACTH injection alter the interrenal ascorbic acid content of the hypophysectomized eel when kept at 6° C. On the other hand, at 16° C. mammalian ACTH produced a 56 % drop in interrenal ascorbic acid content 3 hr. after injection, the value returning to normal 4 hr. later. Of course, deductions made from the variations of ascorbic acid content in the rat adrenal cortex are only on an empirical basis; extension to include the teleosts is at the moment, therefore, rather hazardous.

Recently Olivereau and Fromentin (1954) have made histological studies of the interrenal of the female eel one month after hypophysectomy. In the normal fish the cortical cells are arranged in loops giving a lobular appearance (Pl. IV, figs. 17 and 18). The cytoplasm is granular, without vacuoles, does not contain cholesterol and stains only faintly and diffusely with the sudans. The nuclei are obvious, measuring 5 to 6μ. After hypophysectomy the interrenal loses its characteristic appearance, the lobules collapse and are no longer clearly visible (Pl. IV, figs. 19 and 20). The cells themselves become smaller; the diameter of normal cells is about 14 to 16μ while in operated fish the measurement is about 9 to 11μ. Some cells show signs of nuclear degeneration, while others do not. These two types are not arranged in zones but are intermixed. A similar condition prevails in the adrenal cortices of

Teleostei

Amphibia and Reptilia after hypophysectomy. Rasquin and her co-workers (Rasquin, 1951; Rasquin and Atz, 1952; Rasquin and Rosenbloom, 1954) have used a small fresh-water fish, the characin (*Astyanax mexicanus* Filippi). The interrenal in this fish is usually found as a single or double layer of cuboidal cells closely associated with the posterior cardinal veins and their ramifications. Implantation of carp pituitary, or injection of mammalian ACTH or shock by cold treatment, or, to some extent, injection of Holtfreter's saline, was followed by hypertrophy of the interrenal and increased vascularization. Mammalian ACTH was particularly effective, and 4 hr. after injection, bands of tissue next to the vein lumen comprised as many as ten to twelve cells in width compared with the normal two cell thickness. Accompanying this interrenal change was a reduction in the number of small lymphocytes in the head-kidney together with the appearance of conspicuous blood sinuses, abundant macrophages and many 'oedematous' spaces filled with a cellular granulated substance (Rasquin, 1951).

These two fresh-water fish, the eel and the characin, show pituitary-adrenocortical relationships on the overall pattern familiar in eutherians, although the lymphocytopaenia occurred in the head-kidney and the thymus was unaffected. On the other hand, this relationship is not so clear in a marine form, *Fundulus heteroclitus* L., the killifish, for the interrenal of the hypophysectomized male did not show degenerative changes nor indeed any apparent histological difference from that of the normal animal (Pickford, 1953 b). It is indicated, however, that the *Fundulus* adrenal cortex does hypertrophy after injections of fish pituitary preparations, a presumptive corticotrophic effect.

Evidence that the secretions of the teleost adrenal cortex may have a similarity to those of the mammals is given by the finding (Fontaine and Hatey, 1954a; Fontaine, 1956) of 17, 21-dihydroxy 20-ketosteroids in the plasma of salmon, a compound found in man (Bliss, Sandberg, Nelson and Eik-Nes, 1953) and in some measure indicative of cortical function. In the salmon the smolt stage (in process of migrating to the sea) has a higher plasma titre of this steroid than does the non-migratory parr. The reason for greater interrenal activity, if increased plasma 17-hydroxycorticosteroid level indicates this,

The Adrenal Cortex

is not known; correlation with the changing metabolism of the smolt and its preparation for a marine environment would presumably be involved (Fontaine, 1956). More indirect evidence is suggestive of the similarity of interrenal to mammalian adrenocortical hormones: firstly, the injection of cortisone into brown trout induces cellular degeneration in the islets of interrenal tissue (Spalding, unpublished); secondly, the hypophysectomized eel loses liver glycogen reserves and gluconeogenesis is impaired (Hatey, 1951 a, b) although these findings depend on other hormone deficiencies besides that of the glucocorticoids.

One of the important properties of mammalian adrenocortical hormones is their influence on water and salt-electrolyte metabolism and a major problem of the bony fish is osmoregulation. Especially is this true of those anadromous or catadromous fish which migrate. The teleost in fresh water must resist the lower osmotic pressure of the external medium and, *inter alia*, conserve sodium. For example, the plasma level of sodium of the spent female *Salmo salar* in December is 137·33 mEq./l. and that of potassium is 6·13 mEq./l.; the river water itself had 2·39 mEq./l. of sodium and 0·55 mEq./l. of potassium (Spalding and Chester Jones, unpublished). The fish in sea water, on the other hand, must adapt itself to a higher external osmotic pressure and it drinks, excreting the excess sodium and retaining the water. We do not know if interrenal hormones play a part in the mechanisms involved. Certainly, hormones which are known to affect salt-electrolyte metabolism in mammals also have an influence on the trout (table 16). Another suggestive finding was that the smolt

TABLE 16. *Plasma and muscle sodium and potassium levels in the normal brown trout* (Salmo trutta) *and their variation after the injection of mammalian hormones* (*from M. H. Spalding, unpublished*)

	Plasma		Muscle	
	Na	K	Na	K
	mEq./l.		mEq./kg. wet wt.	
Normal	144·24	8·93	17·20	122·01
DCA injected	115·2	6·45	26·48	102·3
Cortisone injected	109·9	6·61	25·18	103·32
ACTH injected	165·8	6·60	23·37	109·9

adapted to sea water in the laboratory showed cellular degeneration of the interrenal and the formation of 'chloride secreting cells' (Keys and Willmer, 1932) in the gills (Spalding and Chester Jones, unpublished). We do not know if the change from the necessity of conserving sodium to its active excretion is accompanied by quiescence of the interrenal which may, on analogy with mammals, secrete sodium-conserving hormones, though it is interesting, as noted earlier, that the interrenal of the killifish in salt water appears inactive (Pickford, 1953 b). If this be true, it may be that the role of the posterior lobe of the pituitary becomes the dominant one in the control of water and salt-electrolyte metabolism in marine forms while the interrenal assumes control in the fresh-water forms (Chester Jones, 1956 a).

The corpuscles of Stannius (Stannius, 1839) or the 'posterior interrenal' (Giacomini, 1908 a, b) were considered in the past to be the true interrenal of the teleost. The first reference to adrenal bodies in teleosts was made by Rathke (1827, 1828) who, in cyprinids, presumed that the head-kidney was their representative. Stannius favoured his corpuscles in this role but in 1854 preferred Rathke's suggestion to his own. After some speculation by Balfour (1882), Weldon (1884, 1885) and Grosglik (1885, 1886), Diamare (1895) was the first to assign definitely the corpuscles of Stannius as the adrenocortical homologues of teleosts. He never wavered from this view throughout the years (Diamare, 1896, 1899, 1905, 1933, 1934, 1935). Vincent, on the other hand, who agreed with Diamare for many years (Vincent, 1897 a, b, 1898 a, b, 1924), eventually considered 'the organ of Giacomini', rather than the corpuscles of Stannius, to be the teleost interrenal (Vincent and Curtis, 1927).

The corpuscles of Stannius arise as bud-like evaginations or series of evaginations from the wall of the pronephric duct (Huot, 1897, 1898, 1902; Giacomini, 1911 b, 1921; Garrett, 1942) and they are not, therefore, developmentally homologous with vertebrate adrenocortical tissue which arises from coelomic mesoderm (ch. VII). In the adult, the corpuscles are small, irregularly spherical, white, or yellowish-pink bodies which lie on or are embedded in the dorsal or ventral surfaces of the posterior region of the kidney. In most teleosts there is generally a pair of these bodies, but in some genera, especially salmonids, there may be several (Vincent, 1897a; Giacomini, 1911a) and in others, for example the cat-fish (Garrett, 1942), only one. The corpuscles are surrounded by connective tissue, trabeculae from which penetrate the tissue dividing it into small groups of cells (Pl. IV, figs. 21 and 22). In each group the cells are arranged in cords or lobules. The nuclei are smaller than those of the interrenal and more basophilic, the cytoplasm is not markedly acidophilic. Towards the centre of each group there is some cell degeneration (Pettit, 1896), and this is particularly so in older eels where an encircling layer of cells encloses a mass of degenerate tissue in many lobules (Bobin, 1948).

In the corpuscles of salmonids there is neither lipid nor bire-fringent material, nor did these organs change after the injection of mammalian ACTH and cortisone, though in the same animals the interrenals were affected (Spalding, unpublished). These data confirm similar observations on *Astyanax* by Rasquin (1951). Bobin (1948), however, recently reports the presence of lipid in the corpuscles of the eel though its nature is in some doubt as the substance took $\frac{1}{2}$ to 2 hr. to colour with sudan black, whereas adrenocortical lipid colours rapidly within a minute, and osmic acid made cytoplasmic fuchsinophil granules brown only, not black. In any case, on the basis of the presence and absence of lipid, homology with vertebrate adrenocortical tissue can only be very dubiously claimed. Experimental evidence is scanty. The early work of Pettit (1896) reports compensatory hypertrophy of the right Stannius corpuscle after removal of the left in the eel, but there is a wide variation in the size of these bodies normally in any case. Pettit further reported changes in the corpuscles after injection of pilocarpinem curare and in particular of diphtheria toxin. Vincent (1898a) of the same era also used the experimental approach and found that extirpation of the corpuscles in three eels which survived the operation had no effect on the animals. Since that time only Callamand (1943) has used methods other than histological for the resolution of the problem. She found that extracts of the Stannius corpuscles of the eel induced melano-phore reactions in *Cyprinus carpio* scales and contraction of the Rouget cells of capillaries in a like manner to mammalian cortical preparations (Giroud, Santa and Martinet, 1940).

Dipnoi

Clearly the time has come for a re-investigation of the corpuscles of Stannius, the experiments conducted in parallel with those on the interrenal. Rasquin (personal communication) has informed me that she has started such experiments and considers, from her preliminary data, that the corpuscles may in some way be concerned in osmoregulation. On the grounds of absence of reaction to ACTH and to cortisone, and of their embryological origin, I think it unlikely that the wheel will turn full circle and adrenocortical function again be specifically assigned to these enigmatic bodies.

(III) DIPNOI

There has been little work done on the Dipnoi or lung-fishes (Parker, 1892; Pettit, 1896; Giacomini, 1906; Jenkin, 1928), but recently there has been a resurgence of interest (Holmes, 1950; Gérard, 1951). The earlier workers did not find adrenocortical tissue and despite the later investigations there is still confusion about the identity of the adrenal cortex in this group. On the basis of positive reactions to Baker's haematin test and Liebermann's cholesterol test, Holmes assigned, as a hypothesis, the location of cortical tissue to the lipine cells lying principally in the epithelial coelomic connective tissue. Holmes says, 'isolated, or in small groups, they are scattered throughout this tissue and are thus often found very close to the aorta and sympathetic chain. At various points they are found in masses, particularly on the dorsal surface of the kidney and they form a substantial part of what is, anatomically, the kidney. Here they form the "lymphoid tissue" of Parker (1892). This covering to the kidney, present but thinner on its ventral surface, extends also over the dorsal surface of the gonad, but may not surround it completely. Aggregates are also found between the kidneys and, particularly in the posterior region of the body, solid masses of lipine cells lie in a position between and somewhat dorsal to the kidneys, in a position corresponding with that of the median selachian interrenal; here they are sometimes associated with adipose tissue' (text-fig. 21).

By examination of sections made from a block of tissue kindly supplied by Dr Holmes, I found that much of the lipine tissue he described bore a superficial resemblance to adrenocortical tissue but, also, it was similar in appearance to 'brown fat'. This latter kind of adipose tissue is found in the Eutheria both around abdominal organs and between the shoulder blades (Fawcett, 1948*a*, *b*; Fawcett and Chester Jones, 1949; Baker, Ingle and Li, 1950). Brown fat can be distinguished from ordinary white or yellow adipose tissue by its brown colour and lobular, gland-like appearance and among other features it is rich in phospholipid and glycogen. In addition to the tissue similar in appearance to eutherian brown fat, areas of granulocytopoiesis as described by Jordan and Speidel (1931) seem to

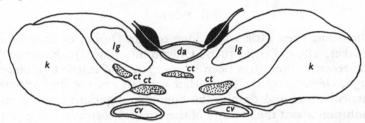

Text-fig. 21. A section through the mid-region of a lung-fish (*Protopterus*) redrawn from Holmes (1950), but with the distribution of cortical tissue more in accord with Gérard (1951). Chromaffin tissue, black; cortical tissue, stippled; *lg*, lung; *k*, kidney; *cv*, cardinal vein; *da*, dorsal aorta.

make up the bulk of the tissue considered to be adrenocortical in nature. Gérard, using the Liebermann cholesterol test as did Holmes, found only small groups of positively reacting cells by or near the kidney. The groups were confined to the ventral margin of the kidney and also occurred in isolated pockets in the walls of the venous capillaries near to their junction with the posterior cardinal veins. Birefringent crystals do not form in the cytoplasm of these cells, but they do take up the sudan stains, turn blue with nile blue sulphate and become yellow after Bennett's phenyl-hydrazine method. These reactions do not occur after treatment with fat solvents so that these presumptive adrenocortical cells contain lipid. The remainder of the tissue surrounding the abdominal viscera is considered by Gérard to be granulocytopoietic with fat intermixed. This author feels that in any case adrenocortical tissue would hardly be as abundant as that so designated by Holmes. Professor Gérard very kindly sent me his preparations and in them groups of four or five cells on the ventral side of the kidney show up very distinctly after the methods used (Pl. III, fig. 15). The difference between these cells and those surrounding the kidney is, however, only one of degree as regards reaction to sudan and nile blue sulphate when the granulocytes and fat cells stain intensely and the distinction really turns on the presence of cholesterol in the presumptive cortical tissue.

None of the reactions is, however, specific for adrenocortical cells, so the problem of the presence of this tissue in the lung-fish has not been finally resolved. As these fish have the amphibian vascular feature of an inferior vena cava together with the teleost characteristic of more dorsal cardinal veins joining the ductus Cuvieri (Young, 1950), it seems reasonable to conjecture that the location of the adrenocortical tissue is similar to that in teleost fish. Gérard's findings of pockets of tissue in the big capillaries which feed the posterior cardinals would be in accord with this view. Further, it would be expected that some cortical tissue would also be found more anteriorly by the anterior cardinal veins. It will be most interesting when the form and function of the cortex, not only in Dipnoi but also in the related coelocanth, *Latimeria*, are revealed.

(IV) CYCLOSTOMATA

There has been no recent work to my knowledge on adreno-cortical tissue in cyclostomes. The adrenals in this group were first mentioned by Rathke (1827) as being small structures close to the heart near the cardinal veins, and in 1828, after reconsideration, he put forward the suggestion that the cortical tissue was represented by the pronephros which, although degenerate, was still present as the head-kidney. These two theories were variously supported in later years, Stannius (1846) and Leydig (1852) accepting Rathke's first theory, Weldon (1884) and Kirkaldy (1894) the second, with modifications. In addition, Ecker (1846) and Pettit (1896) considered the interrenal in *Petromyzon* to be two irregular yellowish organs found between the cardinal veins and the aorta. Collinge and Vincent in 1896, however, in reviewing the position, came to the conclusion that the 'anterior interrenal' had not been discovered in the cyclostomes.

Giacomini (1902 *b*), working chiefly with *P. marinus* but also with *P. planeri* and *Ammocoetes*, found that the chromaffin tissue is found in every segment from the second branchial to nearly the end of the postanal region. This author also identified adrenocortical tissue as consisting of scattered islets, very

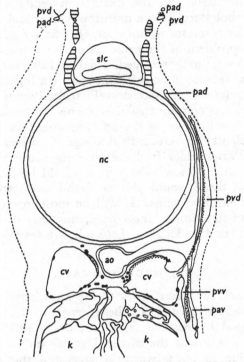

Text-fig. 22. Section through the mid-part of the trunk of *Petromyzon marinus* (redrawn from Giacomini, 1902*b*). *nc*, notochord; *cv*, cardinal veins; *ao*, aorta; *k*, kidney; *slc*, spinal cord; *pad* and *pvd*, dorsal parietal arteries and veins; *pav* and *pvv*, ventral parietal arteries and veins; small groups of cortical cells, black; chromaffin tissue, stippled.

often composed of two or three cells, which lie in the tissue round the cardinal veins on their lateral and ventral aspects (Pl. III, fig. 16), and, although not extending into the branchial region, is found posteriorly as far as the caudal chromaffin tissue; in addition, cortical tissue occurs as nodules in the pronephros and around the main kidney veins (text-fig. 22).

PLATE III

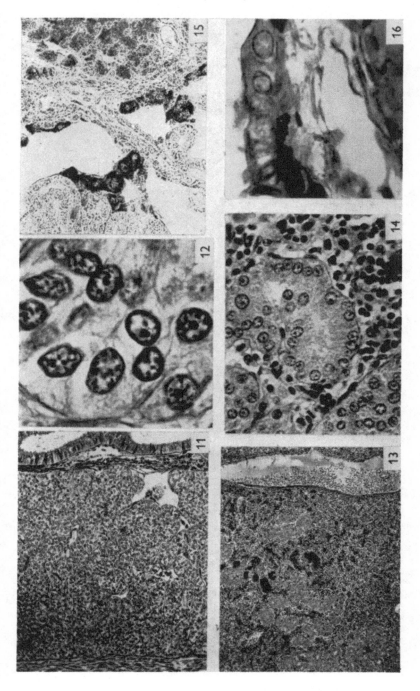

PLATE III

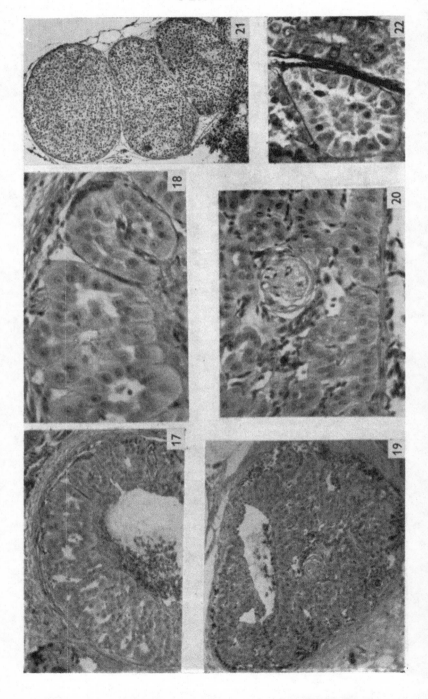

Cyclostomata

Gaskell (1912), concerned chiefly with the chromaffin tissue in *P. fluviatilis*, confirmed Giacomini's description of the distribution of cortical tissue in this species. Giacomini, however, relied, for the most part, on osmic acid preparations to differentiate cortical cells from surrounding neutral fat cells, so that in view of this and the lapse of fifty years it might be well worth while for the adrenocortical histology and physiology of cyclostomes to be reinvestigated.

EXPLANATION OF PLATES

PLATE III

Fig. 11. Part of the interrenal of the dogfish (× 85; Bouin, H. and E.). The extreme right of the picture shows some kidney tubules.

Fig. 12. As fig. 11 (× 1200). Cells of the dogfish interrenal.

Fig. 13. Part of the head-kidney of the trout (*S. trutta*) (× 50; Bouin, H. and E.). Areas of the interrenal tissue are scattered among the lymphocytes of the head-kidney.

Fig. 14. As fig. 13 (× 400). Groups of interrenal cells surrounded by lymphocytes.

Fig. 15. Part of the area in the mid-region of *Protopterus* below the kidney (× 100; formalin, sudan III). A small part of the kidney is on the left of the photograph; lying against it and by the veins are a small group of sudanophilic cells suggested by Gérard (1951) to be adrenocortical cells. To the right of the photograph is the granulopoietic tissue containing occasional pigment cells.

Fig. 16. Part of the lower wall of the right posterior cardinal vein of *Petromyzon planeri* (× 100; Bouin, H. and E.). At the very top of the photograph is the vein lumen; then a small group of suggested adrenocortical cells, underneath is some fatty tissue and the vein wall.

PLATE IV

Fig. 17. A transverse section of half the interrenal of a normal female silver eel (× 280; Bouin, H. and E.). The lumen of the anterior cardinal vein is in the centre surrounded by cortical tissue lying in the wall of the vein.

Fig. 18. As fig. 17 (× 750). The arrangement of the cords of cells into lobules is evident.

Fig. 19. A transverse section of the interrenal of a silver eel hypophysectomized for 1 month (as fig. 17). The collapsed nature of the tissue is obvious.

Fig. 20. As fig. 19 (× 750). Although the lobules and cells are smaller most of the nuclei have not degenerated.

Fig. 21. A transverse section of the corpuscle of Stannius of a salmon (× 68; Bouin, H. and E.). The group of three together is less frequently found than just one. The kidney lies at the bottom of the photograph.

Fig. 22. As fig. 21 (× 350). One small lobule of a corpuscle of Stannius showing peripheral cells round a central group. On the right of the photograph is a kidney tubule separated from the lobule by connective tissue.

CHAPTER III

AMPHIBIA

SWAMMERDAM, working in the seventeenth century, was the
first to notice the adrenals of the frog (probably *Rana esculenta*)
and referred to them as 'corpora heterogenea'. In Lloyd's
English translation (1758) the relevant passage reads: 'Beneath
the testicles and under the skin of the kidneys, there lie two
other singular and strange bodies, but I neglected duly examining them' (p. 106, part II). The foundations of our knowledge
of the micro-anatomy of the frog adrenal were then laid by a
steady stream of workers (Roesel von Rosenhof, 1758; Nagel,
1836; Rathke, 1825, 1839; Ecker, 1846; Gruby, 1842; Semon,
1891; Pettit, 1896; Stilling, 1887, 1898; Srdinko, 1898, 1900a, b;
Giacomini, 1897, 1902c; Grynfellt, 1904b).

In the Amphibia the adrenal comprises discrete or scattered
tissue lying on the mesonephros. In the Anura the adrenal
appears as a thin yellowish line on the anterior two-thirds of
the outer lateral border of the ventral surface of the kidney and
sometimes projects along the anterior renal artery. In *Rana
temporaria*, the common British species, the adrenal is situated
on the ventral surface of the kidney, its yellow colour often
contrasting with the underlying red. The exact position of the
adrenal varies from individual to individual but it usually lies
halfway between the midline and the outer edge of the kidney.
The blood supply is illustrated in text-fig. 23. Branches from
the dorsal aorta feed the gland on each side in addition to the
kidneys. The blood drains away through venules of the renal
veins which join the posterior vena cava. Venous blood from
the renal portal veins, which break up into numerous small
vessels after reaching the outer edges of the kidneys, is also
carried to the vena cava by the same route.

Adrenal weights in Amphibia are usually too difficult to
estimate. Crile and Quiring (1940), however, record six male
bullfrogs (*Rana catesbiana*) with average body weight 519·9 g.
and adrenal weight 134·2 mg.; two male horned toads (*Phryno-*

Amphibia

soma coronatum), average body weight 25 g., adrenal weight 9·3 mg.; and three females, 24·9 g. and 8·7 mg. respectively.

In the Urodela, the adrenal tissue is less coalesced than in the Anura. The adrenals of *Taricha torosa* (*Triturus torosus*), a salamander common in the United States, are two elongated multi-segmented orange to orange-yellow strands of tissue lying just within the mesonephric capsule on its ventro-medial portion; it extends the entire length of the mesonephros and usually for a variable distance cephalically along the walls of

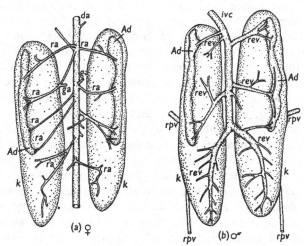

Text-fig. 23. Blood supply of the adrenal glands in the frog (*Rana temporaria*). (*a*) female arterial system; (*b*) male venous system: *da*, dorsal aorta; *Ad*, adrenal gland; *ga*, gonadal arteries; *ra*, renal arteries and those supplying the adrenal; *ivc*, inferior vena cava; *rpv*, renal portal vein; *rev*, renal efferent veins; *k*, kidney.

the vena cava (Villee, 1943; Miller, 1953). Hartman and Brownell (1949) note the distribution of the adrenal in three other species. The adrenal bodies in *Necturus* lie along the midline on top and partially embedded in the ventral side of the kidney, arranged in a double row of small bodies coalescing in some places; in *Amphiuma* they lie embedded in the ventral surface towards the midline of each kidney; in Siren they are placed in two rows between the kidneys and as a single row anterior to the kidneys in the midline. One species of the Apoda, *Ichthyophis glutinosus*, has been studied in detail by

10-2

Dittus (1936). The adrenal is here made up, as in the urodeles, of smaller and larger groups of cells. Dittus found that there were four zones of adrenal tissue: (i) small isolated islets at the anterior end of the kidney; (ii) a continuous covering of the anterior renal vein occupying about a third of the length of the kidney in adult animals; (iii) a zone (not always present) of scattered groups of cells and (iv), most posteriorly, islets of segmentally arranged tissue running to the caudad end of the kidney.

The appearance and extent of the adrenal in Amphibia depend very much on the condition and age of the animal and, particularly, on the season (cf. Stenger and Charipper, 1946; Fustinoni and Porto, 1938; Singer and Zwemer, 1934). Dittus (1936) found this to be true for *Ichthyophis*. The adrenal in *Taricha* during the breeding season (January and February) is brilliant orange and expanded, whereas three to five months after breeding, in April, May and June, the gland is duller, smaller and flatter (Miller, 1953). Similarly, in the anuran, *R. temporaria*, the adrenals of animals in the breeding season often appear a brighter orange and more elevated than those of frogs in the quiescent period (Fowler, unpublished). Holzapfel (1937) observed changes in the macroscopic appearance of the adrenals of *R. pipiens* and *R. catesbiana* throughout the year. During the spring breeding period the glands were deep yellow in colour and had a beaded appearance. In June, July and August, the adrenal consisted of inconspicuous red spots on the surface of the kidneys. As September approached, they became white or pale yellow, gradually darkening throughout October, November and December, to reach a deep yellow or orange in January and February, just before the breeding season.

The histological appearance of the adrenal cortex is very similar throughout the Amphibia (Pl. V, figs. 23, 24 and 25). In *R. temporaria* the gland comprises short, anastomosing cords of cells running more anterior-posteriorly than laterally and separated by wide blood sinuses (Radu, 1931; Fowler, unpublished). Thus, in transverse sections across the kidney, the adrenal appears as numerous round groups of cells, each surrounded by connective tissue. In sagittal sections, the cells are grouped into elongated islands and cords. The cortical cells on the inside merge with the kidney but surrounding both types of cell in this region is a varying amount of lymphoid tissue. There is no histological difference between islands lying to the inside and to the outside of the gland, that is, there is no zonation as suggested by Cater and Lever (1954) (see also ch. VIII). Cortical cells are for the most part oblong, in section, and are closely packed together within each island; they have round or oval prominent basophilic nuclei of varying size with one or two nucleoli and with a light scattering of nucleoplasmic chromatin granules. The cytoplasm, after routine methods, has numerous vacuoles. These, in frozen sections, are seen to be derived from lipid droplets which are sudanophilic and osmophilic (fig. 24); the lipid may include cholesterol and its esters, neutral fat and may form anisotropic digitonide crystals (Radu, 1931; Fowler, unpublished). The amphibian adrenal cortex, then, has the histological appearance and reactions of many, though not all, vertebrate cortical cells. Chromaffin cells occur throughout the cortex placed against cords of cells or contained within them or, frequently, forming a discrete islet of tissue on their own. Beatty (1940) noted that the chromaffin cells in *R. temporaria* occurred both singly and in islets, whereas in *R. esculenta* they were usually scattered as individual cells or, at most, in groups of two or three.

The histological appearance of the adrenal cortex probably depends on the season of the year in all Amphibia. Fowler (unpublished) finds such a relationship to exist in a general way in frogs. Certainly the rate of metabolism depends very much on season and on temperature (Krogh, 1904; Dolk and Postma, 1927). In England, *R. temporaria* can be classified into four adult types: (i) winter frogs, roughly September to December inclusive, quiescent, feeding avidly then irregularly; (ii) spawning, January to March; (iii) post-spawning, April, and may include the end of March and the beginning of May, feeding, becoming terrestrial; (iv) summer frogs, May to August inclusive, feeding, fat body laid down, gonads regenerating. The experimentalist has a further type, the captive frog which always tends towards a low rate of metabolism unless great care is taken. We keep frogs at Liverpool in a green-house on peat moss with water baths available and the colony breeds normally.

The adrenal cortex of the winter frog has both the normal undegenerated cells described above in the general account, and also cells of a second type showing varying degrees of degeneration. Usually an island shows all of one or all of the other type of cell. The cytoplasm of the second type of cell is very heavily vacuolated and the lipid droplets tend to run together; the nuclei are irregular and crenated and the chromatin granules are more obvious, the nucleoli less so. Such a picture is often associated with low or no secretory activity. In the breeding frog, type (ii) above, the islands are composed uniformly of the normal undegenerated type of cell. In types (iii) and (iv), although the majority of islands are of this form, there are scattered groups, the cells of which have crenated nuclei and markedly vacuolated cytoplasm. Tentatively it can be supposed that the adrenal cortex of the winter frog is relatively inactive, that of the spawning frog very active, and that of the post-spawning and summer frog fairly active.

Seasonal variation of the adrenal cortex is not so clear cut in all species. Indeed, Radu (1931) could not correlate adrenal activity with season in *R. temporaria* and *esculenta* but he confined himself to animals between January and September. Klose (1941) also failed to establish such a correlation in *Triturus vulgaris*. Miller (1953) feels that a distinction can be made between those amphibians, for example *T. torosus*, where the secondary sex characters disappear within one or two months after spawning and do not reappear until shortly before the onset of the next breeding season, and those, for example *T. cristatus, vulgaris* and *viridescens*, where the secondary sex characters reappear with the regrowth of the gonads and remain developed throughout the long overwintering period. The former type shows adrenocortical inactivity in the gonad quiescent period while the latter shows no adrenocortical seasonal variation as the gonads are quickly regenerated after breeding. It is doubtful if this division can be maintained for Amphibia as a whole and it does not apply in particular to *R. temporaria* in which regeneration of the gonads occurs before the winter period. Clearly the activity of the adrenal is connected with the overall metabolic rate of the animal at any one period and not solely with one manifestation of this. Differentiation of Amphibia into types would turn rather on those species which have a definite quiescent or winter period and those which maintain a fairly active life throughout the year.

Some experimental procedures result in variations in the histology of the adrenal cortex from which a rough interpretation of structure in terms of function can be obtained. Removal of the anterior lobe of the pituitary, adenohypophysectomy, or all three lobes, hypophysectomy, is followed by atrophy or some form of degeneration of the adrenal cortex in all species studied (see Houssay, 1949; Sluiter *et al.* 1949; Miller, 1953). The degeneration is characterized by malformation of the nucleus together with either a shrinkage of cell cytoplasm (e.g. *R. sylvatica* and *pipiens*: Atwell, 1935; triton: Tuchmann-Duplessis, 1945; Adams and Boyd, 1933) or the accumulation of cytoplasmic lipid (e.g. *B. arenarum*: Porto, 1940) or a mixture of both forms of cytoplasmic abnormality (e.g. *R. esculenta*: Bulliard, Maillet and Droz, 1953; *R. boylii* and *boreus* tadpoles: Smith, 1920). Hypophysectomy of larvae prevents metamorphosis chiefly due to the withdrawal of thyrotrophic stimulation of the thyroid and the adrenal cortex retains its larval position. The effects of hypophysectomy of the adult depend largely on the season of the year during which it is done.

In a resting or 'hibernation' period the general metabolism of an amphibian is at a low rate and removal of the adenohypophysis (and hence the supply of protein hormones) would not be expected to have immediate or dramatic consequences. Nor has it. *R. esculenta* hypophysectomized during the non-breeding season shows no adrenocortical changes until two months after operation and severe atrophy not until five to seven months thereafter (Sluiter *et al.* 1949). On the other hand, hypophysectomy of *T. torosa* in the breeding season is quickly followed by a thickening and darkening of the skin due to thyroid deficiency and consequent cessation of moulting. Grossly, 12 days post-operatively, the cords of cells had flattened and by the end of one month most glands were reduced to an estimated 75 % of the breeding volume and appeared smaller than the adrenal of the non-breeding animal (Miller,

1953). Histological changes were detected by day 6 postoperatively so that roughly 10% of the cortical nuclei were degenerating and by the twelfth day there was a definite decrease in volume of the gland as well as elongation, crenation and abnormal appearance of the cells. It is interesting to note that the first cells Miller found showing diminution in cytoplasmic volume were those lying most centrally in the cords. By the end of the first month after operation the adrenal cells were much reduced in size and contained osmophilic droplets slightly larger than normal. After two months all animals had markedly atrophic adrenals with a decrease in volume of approximately 50% and 80–85% abnormal nuclei. Despite continued atrophy, however, even four months after hypophysectomy, 5–10% of the adrenocortical cells still contained undegenerated nuclei. This may mean some continued adrenocortical secretion at a low level by the least affected cells (see below and ch. VIII).

The consequences of hypophysectomy are very different in the two main types of *R. temporaria*, namely 'winter' and 'summer' frogs. After adenohypophysectomy the winter frog remains apparently healthy and survives long periods, certainly up to 40 days (Smith, 1950) and for several months (van Oordt, Sluiter and van Oordt, 1951). There is some change apparent in the adrenocortical cells in that more and more islands have irregular nuclei and the cells become full of lipid. It seems that fewer and fewer cells, *pari passu* with time, are capable of secretion (Fowler, unpublished). Hypophysectomy of the summer frog is followed by death in 24–28 hr. (Fowler, unpublished) and this susceptibility is shared by *R. esculenta* but not *R. pipiens* which accepts the operation equably at any season of the year (Levinsky and Sawyer, 1953). Much of the adrenal cortex of the hypophysectomized *R. temporaria* in the short time before death undergoes a very rapid degeneration, the nuclei appearing crenated and often showing advanced pycnosis and the aggregation and coalescence of large globules of lipid in the cell cytoplasm (Pl. V, fig. 26). This degeneration extends to all islands of the cortical tissue and is therefore much greater in extent than that of the winter frog many days after operation. However, the hypophysectomized winter frog reaches the condition of the operated summer frog after five weeks (Fowler, unpublished).

The normal function of the adrenal cortex depends to a great extent, but not completely, on an adrenotrophin biologically similar to that of mammals. The extraction of an amphibian ACTH has not so far been attempted. Mammalian ACTH has a trophic effect on the amphibian cortex. On injection into summer frogs, it usually causes cortical hyperplasia, a decrease in the size of lipid droplets and an increase in nuclear size (Pl. V, fig. 27) (Fowler, unpublished). Injection of mammalian ACTH into hypophysectomized winter and summer frogs restores the histological appearance of the adrenal to normal and, additionally, in the latter prevents death. Other workers have shown similar effects in adults and larvae with injections of anterior pituitary extracts or ACTH (Smith and Smith, 1923; Atwell, 1935, 1937; Buillard, Maillet and Droz, 1953). On the other hand, Miller (1953) did not restore the adrenals of long-term hypophysectomized *T. torosa* with injections of mammalian ACTH. The normal adrenal of this species responds well to ACTH (Villee, 1943; Miller, 1953). It seems therefore, that, given enough time without ACTH, all cortical cells atrophy and the condition is eventually irreversible. The amphibian adrenal cortex can be over-stimulated with ACTH so that cell exhaustion and collapse results. This is particularly true of breeding animals when the cortex is active. Fowler found that in some animals the cortex of *R. temporaria* in the summer, given 1 i.u. ACTH (Armour) a day for 10 days, showed pycnotic nuclei and lipid accumulation. Miller (1953), using *T. torosa*, produced degenerative changes in the adrenal cortex of non-breeding animals with large doses and in breeding animals with small doses of ACTH. Cortical collapse by over-stimulation with ACTH has also been demonstrated in reptiles and mammals.

In general, therefore, as judged by variation in histological appearance, the adrenal cortex has much the same relationship to adrenotrophin in the Amphibia as it has in other classes of vertebrates. Biochemical isolation and characterization of the ACTH of representatives of each class would be most interesting.

Little information is available from the results of the injection of adrenocortical steroids, widely used in mammalian work, into Amphibia (see Atwell, 1932). Miller (1953) found that cortisone injected into breeding newts (*T. torosa*) produced no consistent effect on the histology of the adrenal. Only 50 % of the animals treated for two or more days exhibited a slight decrease in cell volume together with a tendency for a small proportion of the nuclei (about 5 %) to show darkening and diminution in size. The small inactive glands of non-breeding newts were unaffected by cortisone injections. These results may well have not been clear-cut because of the dose used (1/300 to 1·2 mg.) or the duration of the experiment. It is less likely, on present information, that the negative results were occasioned because the normal amphibian adrenal cortex does not produce cortisone-like 11-oxysteroids. Deoxycorticosterone acetate (DCA), *biologically* similar in some respects to the naturally occurring aldosterone in mammals, did have a pronounced effect in both breeding and non-breeding newts (Miller, 1953). In the former, DCA produced a definite but slight decrease in the volume of cell cytoplasm and degeneration of the nuclei in the adrenal; in the latter a reduction of cell size and in lipid content together with marked degenerative changes in the nuclei consisting of crenation and pycnosis. Fowler (unpublished) found that DCA injected into the summer frog produced marked degenerative changes with crenation of the nuclei (Pl. V, fig. 28) and increase in lipid droplet size. Both these two authors used large doses of DCA, 1 to 2 mg., and this may account for the marked changes obtained in the adrenal histology.

Another feature of the histology of the adrenal in some species is the 'summer cells'. These cells were first described by Stilling (1898) and it is now preferable to call them Stilling cells (Magath, 1915) as their appearance is not confined, as Stilling thought, to that season. Stilling cells are characterized by the very acidophilic nature of the cytoplasm which contains, for example, after routine haematoxylin and eosin staining, a large number of eosinophilic granules. These cells are found all the year round in the adrenals of some species (Grynfellt, 1904*b*; Patzelt and Kubik, 1913; Radu, 1931). Although Giacomini (1897) thought them characteristic of the genus *Rana*, they do not appear in *R. temporaria* but are particularly

Amphibia

plentiful in *esculenta*. Magath found Stilling cells in *R. pipiens* and *clamata*, and Fustinoni (1938*a*) has described them in *Bufo arenarum*. The Stilling cells are to be seen among the cortical cells and seem to be superimposed upon them (Beatty, 1940). They vary in appearance and Kucnerowicz (1935) described three principal forms in the adrenal of *R. esculenta* at the breeding season: (i) with numerous, large eosinophilic granules, spherical nucleus placed eccentrically; (ii) with larger nucleus with less chromatin and fewer acidophilic granules; (iii) small cells with little cytoplasm and few granules. Kucnerowicz correlated these variations in appearance with a secretory cycle but it is not known of what.

Indeed the nature and function of Stilling cells is still problematical. On the whole, the evidence favours the theory that they are not cortical cells but belong to the reticulo-endothelial system and are a type of acidophilic granulocyte. In *temporaria*, too, basophilic granulocytes are scattered throughout the cortex. Grynfellt (1904*b*) and Radu (1931) considered that the granules of Stilling cells would stain metachromatically and this property would bring them nearer in type to the very common basophilic granulocyte. Radu (1931) thought that the Stilling type of acidophilic granulocyte was not confined to the cortex but was to be found also in the renal blood vessels. On the other hand, Kucnerowicz (1935) thought extra-adrenal granulocytes could be distinguished from true Stilling cells by the appearance of the nucleus, by cell size and by the coloration of the granules. In view, however, of his description of the variations of the Stilling cell, the distinction is difficult to maintain. Stilling cells have, in common with cortical cells, the fact that they regress in *esculenta* after hypophysectomy (Sluiter *et al.* 1949; Bulliard, Maillet and Droz, 1953). In our *esculenta* material acidophilic granulocytes abound among the lymphoid tissue which occurs where the adrenal abuts on to the kidney and they may well be found in other parts of the body. Their waxing and waning in number and their change in appearance may be correlated with the general metabolic state of the body which would depend partly on the activity of the adrenal cortex. There is no real reason, however, to suppose that Stilling cells secrete adrenocortical hormones.

The Adrenal Cortex

The adrenal cortex has an intimate relationship with oestrogens (Padoa, 1938; Witschi, 1953; Segal, 1953). Witschi allowed tadpoles of *R. sylvatica, temporaria* and *pipiens* to live in aquarium water in which oestrogenic substances were dissolved. Treatment started at the gill-sac stage and continued until metamorphosis. A marked enlargement of the adrenal, an increase of tenfold or more in volume over controls, was obtained by high doses (1·0 to 2·5 mg. per litre). Oestrone, ethinyl-oestradiol and equilenin were effective; stilboestrol, benzoestrol and fenocyclin were not. This adrenal hyperplasia resulted in the adrenal appearing as a thick solid column lying centrally on the kidneys at metamorphosis. This compares with the appearance of the adrenals in the control animals where they lie as interconnecting islets marking the lateral borders of the caval vein and occupying the median edges of the anterior three-fourths of the mesonephros. Witschi found that the adrenal enlarges at the expense of the mesonephros, and of the primordia of the efferent ductules and gonadal medulla. He concluded that the intermediate mesoderm is pluri-potent (ch. vii). An additional observation of great interest was that, while moderate doses of oestrogens feminized the frogs, as would be expected, large doses with the concomitant adrenal hyperplasia produced masculinization. Witschi considered that either the gonadal cortex had been destroyed by the high doses, or the adrenal had been stimulated to produce androgens. This would then be a type of adrenogenital syndrome (p. 118). No adrenal enlargement is produced in hypophysectomized larvae with oestrogen (and therefore its action is mediated by the pituitary) but masculinization occurs. It seems doubtful, therefore, that the frog adrenal was induced to produce androgens.

The results obtained from the surgical removal of the adrenals are some help in attempting to elucidate the function of these glands in Amphibia. The operation is by no means a simple one because of the danger of damage to the underlying tissue, with the attendant symptoms of kidney failure masking those peculiar to adrenalectomy alone. By use of cold cautery to remove the glands and by completely controlling the experiments with sham-operated animals these difficulties can be overcome. The earliest investigation of the consequences of adrenalectomy in frogs was by Abelous and Langlois (1891 a, b) followed by Albanese (1892 a, b), Gourfein (1897), Strehl and Weiss (1901), Loewi and Gettwert (1914), Hirase (1926), Ochoa (1933), Csik and v. Ludany (1933), Wachholder and Morgenstern (1933), Moschini (1934) and Stippich (1935). In 1937, Maes made an excellent study of adrenalectomized male frogs, *R. esculenta* and *temporaria*. He used for the most part frogs in the autumn and winter, as animals operated on in the spring and summer died very rapidly. Abelous and Langlois (1891 a, b) also found this to be true and the seasonal difference has been more recently confirmed by Fustinoni (1938 b), Flexner and Grollman (1939), Clark, Brackney and Miliner (1944). Maes's adrenalectomized 'winter'-type frogs survived for long periods, depending on the temperature, before marked symptoms of adrenal insufficiency supervened (table 17). When these symptoms appeared, their duration before death in like manner depended on temperature (table 17).

TABLE 17. *Survival periods in days of adrenalectomized 'winter'-type frogs,* R. temporaria *and* esculenta, *at different temperatures (from Maes, 1937)*

	Temperature ° C.		
Survival-period in days	5	15	28
(a) before ⎫ the appearance of (b) after ⎬ marked adrenal in- ⎭ sufficiency	More than 50 5 to 21	7 to 34 2 to 5	Less than 7 1 to 2

The syndrome of adrenal insufficiency in frogs is characterized by (i) a progressive asthenia which is first manifested by a loss in spontaneous activity, by the failure of the righting reflex and by a slow corneal reflex; (ii) circulatory disturbances, slowing of the heartbeat rate, haemostasis so that reddish areas appear particularly on the feet, congestion of the viscera; the blood becoming very viscous; (iii) hypoglycaemia and the loss of glycogen reserves in the liver and muscle; (iv) oedema and salt-electrolyte disturbances. In addition, in the absence of adrenocortical hormones muscle phosphagen is lost (Moschini, 1934; Ochoa, 1933), muscle chronaxie is depressed (Hoppe and Vogel, 1951); the isolated muscle of the adrenalectomized frog has 40 % less capacity to perform work than that of the normal animal (Ochoa, 1933). The adrenalectomized frog also loses the capacity for selective reabsorption of sugars by the gut; the normal animal can reabsorb glucose and galactose from isotonic solutions two to three times faster than all other sugars (Minibeck, 1939).

TABLE 18

| | | Level of blood sugar (mg./l.) | Glycogen (g./100 g.) | |
			Liver	Muscle
	Normal	56·6 (5)	6·51 (2)	0·64 (2)
Active frogs captured less than 2 months	Sham-operated	—	6·10 (1)	0·95 (1)
	Adrenalectomized			
	(a) Before onset of asthenia	56·0 (2)	8·41 (1)	0·74 (1)
	(b) During asthenia	—	1·98 (2)	0·49 (2)
Frogs with inanition for 6 months to 1 year	Sham-operated	16·5 (2)	—	—
	Adrenalectomized			
	(a) Before onset of asthenia	20·0 (1)	—	—
	(b) During asthenia	9·0 (2)	—	—

Tables 6 and 7 of Maes (1937) combined to show the fall in the level of blood sugar and liver and muscle glycogen in the adrenalectomized frog in the terminal asthenic stage; number of animals in each group given in brackets.

These symptoms of adrenal insufficiency in the frog are in many ways similar to those occurring in mammals. Data are scarce on the extent of the disturbance of carbohydrate metabolism following adrenalectomy. Maes (1937) found that there was little change in either blood sugar level or liver and muscle glycogen content until asthenia and other marked terminal symptoms supervened (table 18). We have found that hypophysectomy of *R. temporaria* in February and March does not result in a change in blood sugar level. Injections of ACTH will, however, produce hyperglycaemia in the hypophysectomized winter frog and will restore the oedematous, asthenic, hypophysectomized summer frog to an outwardly normal appearance (Smith and Chester Jones, unpublished). Marenzi (1936) showed that the asthenia of the hypophysectomized toad disappeared and liver and muscle glycogen values were restored to normal 5 to 7 days after the injection or grafting of the anterior lobe of the pituitary. Investigation of carbohydrate metabolism in the frog is complicated by its seasonal variation (Smith, 1950). Liver glycogen content drops to low levels during the breeding season in *R. temporaria* and, interestingly enough, it remains low in the post-spawning period despite active feeding. Glycogen begins to be deposited during September, and the liver of the winter frog contains high amounts. These seasonal changes may account for Fowler's failure to obtain glycogen deposition in the adrenalectomized post-spawning frog with injection of hydrocortisone.

The extent of the disturbances in water and salt-electrolyte metabolism following adrenalectomy depend on the season of the year during which the operation is done (Fowler and Chester Jones, 1955). Immediately after adrenalectomy in the summer frog, or at the onset of marked symptoms of adrenal insufficiency in the winter type, there is a progressive increase in body weight due to the accumulation of water (table 19). Angerer and Angerer (1942, 1949) found a 28% increase in body weight 7 days after adrenalectomy, and a 33% increase 12 days after. We found a 22·3% increase in body weight 24–72 hr. after adrenalectomy of the summer frog (table 19). Although adrenalectomized frogs tend to retain water they cannot withstand the stress of dehydration as well as normal frogs (Angerer and Angerer, *loc. cit.*). The winter frog, although outwardly unaffected two weeks or more after adrenalectomy, has a significant loss of muscle sodium (table 19). The values of potassium likewise decline, compared to normal, when considered in terms of wet weight of muscle. This is a reflexion of abnormal water accumulation in the inter-cellular spaces and within the cells themselves. Expressed in terms of dry weight of muscle, the potassium content of the winter frog fourteen days after adrenalectomy is unchanged compared with normal; the fall in sodium content is still evident. Similar considerations apply to the adrenalectomized summer frog. The marked oedema immediately following the operation is accompanied by a real decline in the sodium content of muscle. The fall in the potassium content in terms of muscle wet weight is an apparent one only; muscle dry weight figures reveal normal or higher than normal potassium values. Marenzi and Fustinoni (1938) found, in the adrenalectomized toad, a fall in the

TABLE 19. *The effects of adrenalectomy, total hypophysectomy, adenohypophysectomy, injection of ACTH and of DCA on the body weight and water, sodium and potassium content of gastrocnemius muscle of winter and summer frogs (Fowler, unpublished)*

| | No. of animals | Body weight | | Muscle | | | | |
| | | Start of experiment | End of experiment | Water (%) | Wet weight mEq./kg. | | Dry weight | |
					Na	K	Na	K
Summer frogs (1 to 3 days after operation)								
Normal	10	19·45±0·83	—	80·65±0·31	25·81±1·36	83·66±1·75	130·0	430·0
Totally hypophysectomized	10	22·02±2·27	25·47±1·37	88·10±0·61	16·20±1·40	48·00±2·60	130·0	400·0
Adenohypophysectomized	9	24·59±1·45	29·25±1·96	86·50±1·10	12·34±3·10	59·90±2·00	92·0	440·0
Adrenalectomized	8	27·87±1·87	30·87±2·15	86·50±1·49	9·30±0·34	62·50±2·00	69·0	460·0
Sham-operated, no oedema	8	29·72±1·46	28·30±1·47	81·60±1·66	21·20±2·00	86·30±2·74	110·0	470·0
Sham-adrenalectomized, with oedema	5	26·15±1·38	28·45±2·00	86·96±1·11	7·96±1·70	54·60±1·33	61·0	420·0
Sham-hypophysectomized, with oedema	5	27·65±1·61	30·80±2·39	85·35±0·31	8·30±0·37	62·72±1·90	57·0	420·0
Winter frogs								
Normal	12	24·49±1·79	23·70±2·00	77·95±0·41	25·00±0·54	80·54±1·55	110·0	360·0
Adrenalectomized 14 days	8	25·51±1·79	24·65±1·52	79·49±2·04	14·50±2·00	73·91±2·18	71·0	370·0
Adenohypophysectomized 14 days	12	27·92±2·37	27·63±2·35	80·54±1·55	17·20±0·95	76·41±0·82	88·0	390·0
ACTH and DCA injected								
Control	9	23·37±1·02	22·29±1·33	78·76±0·65	25·32±1·24	77·79±0·41	120·0	360·0
Injected with 1 i.u. ACTH/day for 10 days	7	20·30±0·98	19·53±1·04	79·87±0·77	30·40±1·08	67·90±1·04	150·0	330·0
Injected with 2 mg. DCA/day for 10 days	10	20·14±0·77	19·20±0·68	79·81±0·46	24·76±1·00	76·37±2·23	120·0	370·0

Amphibia

potassium content of heart and liver but not in muscle (table 20). The values were given in terms of organ wet weight so that variation in water content was not considered. Muscle water content can, in fact, increase by about 6 %, as we show in table 19. The summer frog reacts unfavourably to any sort of surgical interference. Sham-operation is sometimes, therefore, accompanied by water and salt-electrolyte disturbances similar to those following adrenalectomy (table 19). It is presumed in these cases that the adrenal cortex is exhausted during the response to the stress of the sham-operation. Both adrenalectomized and sham-operated summer animals can be maintained by placing in isotonic saline. On subsequent removal, several days later, to tap-water the former die while the latter do not. Saline can thus maintain both the adrenalectomized frog and the adrenalectomized rat.

TABLE 20. *The sodium and potassium content of serum and muscle, heart and liver of the adrenalectomized toad 48–72 hr. after operation. (The data of Marenzi and Fustinoni (1938) recalculated)*

		Serum (mEq/l.)		Muscle	Liver (mEq/kg. wet wt.)	Heart
		Na	K	K	K	K
Normal	3	130·9	4·64	108·0	97·6	103·9
Sham-operated	6	125·6	4·70	104·8	90·0	101·0
Adrenalectomized	7	95·0	6·28	104·3	56·5	72·4

It would be expected that plasma electrolytes would show changes after adrenalectomy in the frog as they do after this operation in mammals. Only the data of Marenzi and Fustinoni (1938) are available. They found that after adrenalectomy of the toad, the level of serum potassium rose by about 33·9 % to 35·3 % and the sodium level fell by 24·4 % to 27·4 % (table 20). (It will be noted that these authors give 4·64 mEq/l. for normal serum potassium. It is curious to find that Urano (1908) has 4·6, Fahr (1907/8/9) 3·34, Chistovich (1931) 3·21, Zwarenstein (1933) 5·8, Fenn (1936) 2·5, Gerschman (1943) 3·63, Cicardo (1947) 2·57 to 5·90, Fowler (unpublished) 4·73.)

Replacement therapy with exogenous adrenocortical hormones in adrenalectomized anurans has not been thoroughly investigated as yet. Fustinoni (1938*b*) failed to increase the survival time of toads after adrenalectomy by the injection of adrenocortical extract. Neither was injection of deoxycortico-sterone acetate very successful, as Clark, Brackney and Miliner (1944) found some alleviation of adrenal insufficiency in only four out of twelve adrenalectomized frogs injected with this steroid. Large doses of DCA in the intact summer frog have no effect on the terminal values of water, sodium and potassium contents of the muscles compared to normal. We would expect, however, that aldosterone and large doses of Kendall's com-pounds B, F and E would have positive action on the adrenal-ectomized frog. This supposition has some support in the results of experiments with isolated tissues. Cicardo (1947) found that potassium was lost more quickly from muscles of adrenalecto-mized *B. arenarum* than of normal animals when perfused with potassium-free frog Ringer's solution. This loss of cellular potas-sium was reduced in muscles of toads pre-treated with DCA. Likewise, the rate of oedema formation in perfused frog hind limb preparations was reduced by Kendall's compound A, water-soluble deoxycorticosterone, and Kendall's compound E, in this order of efficiency (Hyman and Chambers, 1943).

The winter frog after hypophysectomy, although it survives the operation equally as well as adrenalectomy, shows a de-crease of the muscle content of sodium (table 19). The muscle potassium content remains the same or slightly higher than normal values. The increased water content of the muscle allows an apparent decrease of potassium content of muscle in terms of wet weight as was the case in the adrenalectomized animals. The fall in the amount of muscle sodium after anterior lobec-tomy was not as great as that after adrenalectomy. It may be that the presence of some undegenerated cells in animals after the former operation allowed small amounts of adrenocortical hormones to be secreted so that signs of adrenal insufficiency develop only slowly. This probably accounts also for the long survival after hypophysectomy which has been recorded (Atwell, 1935; van Oordt, Sluiter and van Oordt, 1951).

Removal of the anterior lobe of the pituitary in the summer frog is followed by symptoms of marked adrenal insufficiency in 24–48 hr. (table 19). The consequences of this operation are

Amphibia

very similar to those of adrenalectomy. Similarly, sham-operation brings about these symptoms. It seems more satis-factory to suppose that they are due to adrenal exhaustion, as was thought to be the case following sham adrenalectomy, rather than induced brain lesions (Jarkowska, 1932).

Big changes in the salt-electrolyte content of the blood in hypophysectomized Amphibia would not be expected unless symptoms of adrenal insufficiency had supervened. Gerschman (1943) found little difference between the serum potassium level of normal toads (at 3·64 mEq/l.) and adenohypophys-ectomized animals (at 3·4 mEq/l.). It is surprising to find, therefore, that Zwarenstein (1933) obtained a *fall* of 22 % in plasma potassium level after hypophysectomy of *Xenopus laevis*. As, however, the *Xenopus* were kept in water this may reflect not so much a loss of blood potassium as a haemodilution due to an increased rate of entry of water after operation (see below).

Total hypophysectomy, when both anterior and posterior lobes of the pituitary are removed, is not necessarily followed by a marked decline in the sodium content of the muscle, as is the case when the anterior lobe alone is removed. This seems to indicate that both the posterior lobe of the pituitary and the adrenal cortex have some influence on water and salt-electro-lyte metabolism. The tissues upon which these two sets of hor-mones play are (i) the skin, and (ii) the kidney. The skin is relatively impermeable and forms a barrier so that the water of the environment does not rush in to dilute the body fluids. As the frog does not drink, except when placed in hypertonic saline, all the water (apart from that in food) enters through the skin. Mammalian neurohypophysial extracts and the amphibian 'water balance principle' itself act on the skin to increase its permeability to water (see Heller, 1956; Sawyer, 1956; Capraro and Garampi, 1956). Adrenocortical hormones enter into this mechanism apparently in two ways. In the first place, the normal water balance response does not take place in the absence of the adenohypophysis. Jones and Steggerda (1935) reported a complete loss of the response in frogs hypo-physectomized more than one week. Chen, Oldham and Geiling (1943) found that the response declined to 25 % of nor-mal ten days after removal of the anterior lobe of the pituitary.

The Adrenal Cortex

The water balance response can be restored in hypophysectomized frogs to nearly normal, however, by the injection of ACTH; it can be completely restored by the injection of thyroxine together with ACTH (Levinsky and Sawyer, 1952). This means that the skin requires the presence of adrenocortical hormones and of thyroid hormone for normal metabolism. In the second place, increased permeability of the skin to water occurs, whether posterior lobe hormones are present or not, in the adrenalectomized animal. The converse, that injection of high doses of adrenocortical hormones renders the skin more impermeable to water and counterbalances neurohypophysial effects, has not been demonstrated. The possibility is indicated by Dow and Zuckerman (1939). They found that adrenocortical extract and DCA have a variable effect on the body weight of axolotls but that they generally produce a reduction. On the other hand, Hsieh (1950) could not demonstrate an antagonistic action between pitressin and DCA.

The frog skin acts as a barrier not only to water but also to sodium. The skin has a very low permeability to sodium, so that normally little or none is lost from the body by this route. It contains, however, an active transport mechanism and sodium is taken into the body against an adverse osmotic gradient (Ussing, 1949, 1952; Linderholm, 1952). The transport mechanism is an intrinsic property of the skin itself and from Jørgensen's (1947) experiments it would be supposed that its activity is unaltered after adenohypophysectomy. Later, however, Johnsen and Ussing (1949) found that injections of ox ACTH induced a pronounced increase in total and net uptake of sodium through the skin of the axolotl. Although this experiment was not conclusive, further investigation may well show that the enzyme systems involved in active transport of sodium through the skin can be influenced by the adrenocortical hormones. It is not clear from the available evidence whether or not the amphibian neurohypophysis *normally* exerts a control on sodium transport. Commercial preparations of the mammalian neurohypophysis induced increased total and net uptake of sodium chloride through the skin of the axolotl (Jørgensen, Levi and Ussing, 1946). This was confirmed by Fuhrman and Ussing (1951) who also found that these extracts,

which increase skin permeability, increased the potential difference across the skin. On the other hand, Adolph (1925)
observed that the relative impermeability of the frog skin was
associated with a high P.D. Moreover, Angerer (1950) noted
that adrenalectomy is associated with a fall in P.D. which
accompanied the increased skin permeability in adrenal insufficiency. We have to decide if the injection of mammalian
posterior lobe extracts gives us physiological, as opposed to
pharmacological, information about the frog. The doses used
are generally large. Boyd and Young (1940) make the telling
and cautionary remark that the amounts of mammalian neurohypophysial extract used in amphibian experiments range
about the equivalent order of the injection of one pint of
pituitrin into man.

The second tissue which hormones may influence is the
kidney. The mesonephros lacks a loop of Henle and this has
been associated with the absence of tubular water reabsorption. If this is so, and as the urine of Amphibia is hypotonic,
there must be reabsorption of osmotically active substances,
mainly sodium chloride; it is suggested that this takes place in
the distal tubule (see H. Smith, 1951). It is to be noted that
the renal tubules of the mesonephros also have a double blood
supply, firstly from the efferent glomerular arterioles and
secondly the peritubular capillaries from the renal portal
system. The water and sodium taken in through the skin of the
frog leave, therefore, by way of the kidney. The neurohypophysis, which in the mammals is thought to effect water-
reabsorption at the level of the distal tubule, is considered in
the Amphibia to act only by constriction of the efferent glomerular arterioles (Sawyer, 1951 b). This would reduce the
glomerular filtration rate and depress urine flow. At the same
time the neurohypophysial secretion would increase skin permeability, allowing more water to enter. Protection against dehydration would then be obtained by posterior lobe action
in both amphibia and mammals, but its actions would be on
differing sites (Heller, 1945, 1950, 1956). It is possible, however,
that some water reabsorption takes place in the distal tubule.
Pasqualini (1938) observed an antidiuretic effect in the frog,
despite unconstricted efferent glomerular arterioles. No direct

observations record the influence of adrenocortical hormones on glomerular filtration rate. These hormones must, however, affect the glomerular filtration rate if only by controlling blood pressure, haemodilution and muscle tone. It seems clear that the cortical hormones control, in some measure, the reabsorption of sodium from the distal tubule. The adrenalectomized frog loses sodium, more rapidly than normally, through the kidney. Moreover, adrenalectomized frogs given a load of sodium chloride excrete much more sodium over a given time than do normal frogs under similar circumstances (Fowler, unpublished). In addition to the sodium loss through the kidney after adrenalectomy, there is also some loss through the skin. This is due to the increased permeability of the skin to sodium occurring in the absence of the adrenocortical hormones (Jørgensen, 1947).

The sequence of events after adrenalectomy may be as follows. Absence of adrenocortical hormones results in increased skin permeability, and a greater amount of water enters. In the body the increased amount of water is not distributed normally but accumulates in the intercellular spaces and in the cells themselves. The active transport of sodium through the skin continues though at a reduced rate. At the same time the skin is more permeable to sodium and there is sodium loss through the skin. The major part of the sodium loss is via the kidneys, where sodium reabsorption in the distal tubule does not go on so efficiently. Muscle sodium content declines mainly because the level of sodium in the intercellular fluid falls; there may also be a loss of cell sodium. The level of blood potassium rises. Perhaps the excretion of potassium through the kidney is reduced.

It is possible to speculate that when the anterior lobe alone is removed, the continued secretion of the posterior lobe results in continued or enhanced water reabsorption, but also increased sodium loss at the kidney. When both anterior and posterior lobes are removed, the additional absence of neurohypophysial secretion may mean reduced water reabsorption and reduced rate of sodium loss. In both cases, however, the greatly reduced amount of cortical secretion would be the overriding consideration. The relationship between the neurohypophysis and the

PLATE V

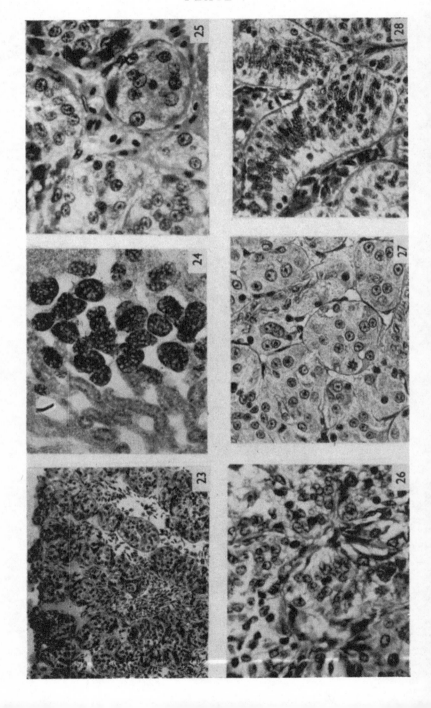

Amphibia

adrenal cortex probably varies within the Amphibia itself. We may suppose that the rate of secretion of the neurohypophysis is mediated through osmoreceptors as in mammals (Verney, 1947). Amphibians which live entirely in water may only rarely call upon the water balance principle. On the other hand, those that live away from ponds may require a constant and fairly high secretion from the neurohypophysis. It is significant in this respect that after hypophysectomy the toad has polyuria but not the frog. The success of the adaptation of the amphibian to drier environments would rest, therefore, on the extent the neurohypophysial secretion was able to alter skin and kidney control of water and salt-electrolyte movement, the fundamental plan of which is controlled by the adrenocortical hormones.

EXPLANATION OF PLATE

PLATE V

Fig. 23. Part of the adrenal cortex of the frog (*R. temporaria*) in transverse section (× 100; Bouin, H. and E.). Most of the cords of cells are cut transversely and show as lobules, surrounded by wide sinuses; no capsule; no zonation. The kidney, not shown, would lie at the bottom of the photograph.

Fig. 24. As fig. 23 (× 100; formalin, sudan black). Kidney to left of photograph, sudanophilic lobules of cortical cells.

Fig. 25. As fig. 23 (× 300). Lobules of cortical cells surrounded by blood sinuses in which lie corpuscles. Chromaffin cells can be seen, characterized by their basophilic cytoplasm.

Fig. 26. Part of the adrenal cortex of the summer frog hypophysectomized for 48 hr. (× 300; Bouin, H. and E.). Collapse of lobules, pycnosis of many nuclei, vacuolation of cytoplasm.

Fig. 27. Part of the adrenal cortex of the summer frog injected with mammalian ACTH (1 i.u. per day for 10 days) (× 300; Bouin, H. and E.). Increased size of lobules, mitotic divisions.

Fig. 28. Part of the adrenal cortex of a summer frog injected with deoxycorticosterone acetate (2 mg. per day for 10 days) (× 300; Bouin, H. and E.). Section somewhat oblique showing cord-like formation. Some collapse of cells, pycnosis of nuclei, vacuolation of cytoplasm.

CHAPTER IV

REPTILIA

THE earliest reference to the adrenals of reptiles seems to have been made by Perrault (1671). Morgagni (1763) found the 'Renum succenturiatorum excretorios ductus' (p. 64), the 'corpora heterogenea' of Swammerdam, present in the viper: 'In vipera autem, et fluviali testudine eos esse inter eosdem renes, et testes nexus membraneos, ut verisimile sit, per hos similia decurrere vascula ab illis renibus exorta.' Later Cuvier (1805) described the position of the adrenals in the snake, Bojanus (1819) and Nagel (1836) in the tortoise and Retzius (1830) in snakes and lizards. Ecker (1846), Frey (1849) and Leydig (1853) published comparative accounts of the adrenal. The histology of the gland gradually emerged with the work of Milne-Edwards (1862), Eberth (1872), Braun (1879), Pettit (1895) among others. Vincent in 1896 (see also 1924) brought together the findings on the reptile adrenal up to that date and added to them by investigation of the lizard *Uromastix*, the adrenal of which he took to be characteristic of the gland in reptiles as a whole. Kohn (1902) recognized the homology between the medulla of the eutherian adrenal and the reptilian chromaffin tissue and used this latter descriptive term. Although Bojanus used the term 'suprarenal', Poll (1904a, b) seems to have been the first to employ the term 'interrenal' in reference to the adrenal cortex of reptiles and 'phaeochrome' for chromaffin tissue. As I have suggested in previous chapters, it is preferable to call the gland the adrenal and its component parts the adrenal cortex and chromaffin tissue.

Wright (unpublished) has recently investigated the West African lizard, *Agama agama* L., the European green lizard, *Lacerta viridis* (Lacertilia), and the grass snake, *Natrix natrix* (Ophidia). The adrenal glands of the male *Agama* are separate elongate bodies, the right being generally a little longer than the left. The dorsal, lateral and ventro-lateral surfaces of the adrenal are heavily pigmented while the remainder is yellow-

168

Reptilia

white in colour. The glands lie near the anastomosis of the left and right posterior venae cavae and are situated, well forward of the kidney, immediately behind the gonads. In the male each adrenal is closely apposed to the vasa deferentia and lies enclosed by the mesorchium. On its ventro-lateral surface the right adrenal is closely applied to the right posterior vena cava and is, indeed, incorporated in the wall of the vein. A number of small ports connect the adrenal to the vein and seem to be confined to the posterior part of the gland (cf. text-fig. 24). The ports which can be seen anteriorly do not appear to open into the vein but into the ventral adrenal sinus. The left adrenal is applied to an anterior diverticulum of the vena cava, draining into it by ports similar to those of the right side. In some animals minute rests of adrenal tissue extend caudally along the posterior venae cavae.

The arterial blood supply of the right adrenal comprises an artery branching from the dorsal aorta at the level of the posterior end of the testis (cf. text-fig. 24). After dichotomizing, one twig of the artery runs to the anterior end of the adrenal, the other enters the gland dorsally just posterior to the mid part. The posterior end of the adrenal receives a branch arising dorsally from the aorta. The left adrenal is supplied by three small branches arising from the aorta on the side opposite to the posterior mesenteric artery. The most posterior branch of the three runs to the anterior part of the gonad and there dichotomizes; each twig then breaks up into smaller arteries. The anterior branch of the three also serves the fore part of the gland, while the middle branch runs to the mid area. The posterior part of the left adrenal is supplied by a small artery, originating from the aorta at the same level, which divides into five or six twigs as the gland is reached.

The female adrenal, enclosed by the mesovarium, has a venous system similar to that of the male; while the arterial supply of the former is somewhat simpler (cf. text-fig. 24). On the left an artery diverges from the aorta at the level of the posterior part of the adrenal. After dividing at the gland, the anterior twig supplies the anterior part; the posterior twig enters the dorsal part of the adrenal near the body wall. The right adrenal is supplied by two branches, one arising at the level of the anterior part of the gland, the other at the posterior. The latter branch divides dorsad of the adrenal, one twig entering the gland, the other passing to the ovaries.

In the lizard (*L. viridis*) the anterior part of the adrenal is supplied by an artery, originating at the same level, from the dorsal aorta, which divides on entering the adrenal dorsally. At the level of the posterior part of the gland, an artery arises from the dorsal aorta and, running anteriorly, is joined at the mid part by another aortic branch. Two fine arteries enter between

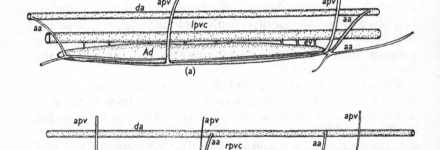

Text-fig. 24. Dorso-lateral view of the blood supply of the left and right adrenal glands of the female grass snake (*N. natrix*). (*a*) Left adrenal; (*b*) right adrenal: *da*, dorsal aorta; *apv*, adrenal portal veins; *lpvc*, left posterior vena cava; *aa*, adrenal arteries; *Ad*, adrenal gland; *ev*, adrenal efferent veins; *rpvc*. right posterior vena cava.

the posterior and mid part of the adrenal. The venous supply of the male lizard is essentially similar to that of the agamid.

While in the green lizard and in the agamid the adrenal venous return is by ports to the venae cavae, in the snake (*N. natrix*), on the other hand, several (about ten) short veins leave the adrenal and enter directly into the venae cavae (text-fig. 24). In addition a vein from the body wall runs to the posterior end of the adrenal and thence anteriorly to be joined midway in its course by a second vein from the body wall. These veins break up into venules and enter the adrenal along its length. This constitutes the so-called 'adrenal portal system' first noted by Gratiolet (1853) and found to be widespread by

Reptilia

Pettit (1895) and Vincent (1896, 1924). More recently Spanner (1929) described the system in Chelonia and in Squamata. In general adrenal portal veins are not markedly obvious in the Lacertilia but they do exist (Wright, unpublished). The adrenals in the grass snake, *N. natrix*, are well cephalad of the kidneys, the right adrenal being more so than the left. The posterior end of the former is nearly at the level of the anterior end of the latter. The arterial supply to the adrenal in the grass snake comprises an artery which originates from the dorsal aorta and divides into three twigs at the anterior end of the gland. One twig, on each side, runs to the gonad, one enters the adrenal without divergence, and the third runs posteriorly and enters the adrenal in the mid part. The posterior part of the adrenal is supplied by two arteries branching from the dorsal aorta. They join posteriorly and the confluence enters the gland (text-fig. 24).

In general the adrenal blood supplies of the lizards and the snake described above set the pattern for those of the whole class of reptiles. Thus in *Gerrhonotus* (Retzlaff, 1949) both adrenal glands are located near the cephalic poles of the gonads. The blood supply to the adrenal comes from arterial branches arising chiefly from the aorta, renal, and mesenteric arteries. The venous drainage of each adrenal passes by way of a large vein lying in contact with the ventral portion of the caudal half of the gland. Both these vesssels drain into the posterior vena cava on the right side, the left crossing the midline of the body in order to do so. The blood system of the adrenals of the Yucca night lizard, *Xantusia vigilis*, described by Miller (1952) is very similar to that of the agamid and green lizards.

In the Order Crocodilia, the alligator, *Alligator mississippiensis*, has yellowish, elongate adrenals between and anterior to the kidneys; they are closely associated with the anterior two-thirds of the gonads and with the aorta and the posterior vena cava (Reese, 1931). Lawton (1937) found that a small branch of the segmental artery from the dorsal aorta arises just posterior to the first lumbar vertebra and enters the anterior end of the adrenal gland together with three additional branches. Two small arteries arise medially from the dorsal aorta and enter the right adrenal gland, while one from the same region passes to the left. Ports drain the venous blood into the vena cava as described for the lizards.

Hartman and Brownell (1949) illustrate the variations of the same pattern for the American chameleon (*Anolis carolinensis*), the black iguana (*Ctenosaura multispinis*), American crocodile (*Crocodylus acutus*) and the Gila monster (*Heloderma suspectum*). The adrenals of the Chelonia seem to be not such discrete glands as in the other orders of the reptiles but lie on the ventral surface of each kidney or, in the marine tortoise *Thalassochelys caretta* according to Holmberg and Soler (1942), are, for the most part, merged together into a single body. The adrenal arterial supply in those chelonids which have been investigated (Pettit, 1896; Spanner, 1929; Thomson, 1932; Hartman and Brownell, 1949; Bachmann, 1954) comprises branches from the dorsal aorta passing directly to the glands. The venous return is by several veins to the posterior vena cava.

Adrenal weights are known for some species of reptiles and most of the known data are given in tables 21 and 23 (from Wright, unpublished; Holmberg and Soler, 1942; Naccarati, 1922; Valle and Souza, 1942; Quiring, 1941). It should be remembered that these data are not obtained from homogeneous populations even within the same species. It is rare to know the age of a reptile, and many other variants such as season, state of reproduction, whether or not actively feeding, are usually not known. As these factors may cause body weight to vary, it is better to consider the variations in adrenal weight expressed in terms of milligrams per 100 g. body weight. Such data as Wright has so far accumulated indicate that the adrenal of the snake *N. natrix* in winter weighs less than that of the snake in summer. It would be reasonable to suppose that this variation represents the inactivity of the adrenal in winter and its activity in summer and is correlated with quasi-hibernation in the former season, and the more active feeding period of the latter. In one of the smallest species of lizard, *Xantusia vigilis*, where normal adults weight 1·0 to 1·1 g., the adrenals vary from 0·7 mm. to 1·5 mm. in length and 0·2 to 0·3 mm. in diameter, the combined weight being 0·75 mg. Bimmer (1950) obtained the relative volumes of cortical and chromaffin tissue in three species of adult lizard (table 22) using transverse sections

TABLE 21. *Gravimetric data for Reptiles*

Animal (and season where known)	n	Body weight (g.)	Adrenal weight (mg.)	Adrenal weight (mg./100 g. body wt.)
(a) Naccarati (1922)				
Chelonia				
Emys	—	250	—	10
Testudo	—	320	—	10
Lacertilia				
L. viridis	—	50	—	40
Ophidia				
N. natrix	—	70	—	50
Zamenis	—	220	—	40
(b) do Valle and Souza (1942)				
Male				
Dryophylax				
Dec. and Jan.	6	81·3 ± 8·7	18·85± 3·75	24·15± 5·20
Feb.	11	76·1 ± 10·1	29·73± 6·7	43·4 ± 7·5
Tomodon				
Feb.	4	46·0 ±11·5	13·0 ± 0·90	34·20± 9·00
Non-pregnant females				
Dryophylax				
Dec.	10	110·9 ±11·2	31·9 ± 3·5	30·1 ± 3·0
Jan.	6	100·17±23·00	25·5 ± 2·3	29·15± 3·5
Feb.	10	46·7 ± 9·35	18·10± 0·45	36·33± 5·00
Tomodon				
Dec. to Feb.	4	45·00± 3·1	15·75± 5·7	33·9 ± 8·5
Pregnant females				
Dryophylax				
Feb.	4	85·5 ±21·5	49·0 ±13·25	55·6 ±10·7
Oct. and Nov.	3	152·3 ±17·2	41·7 ± 9·7	27·37± 5·70
Tomodon				
Jan. and Feb.	3	76·3 ±19·4	10·7 ± 0·78	15·43± 3·3
May and July	3	68·7 ±23·4	14·0 ± 4·05	23·83± 6·8
(c) Holmberg and Soler (1942)				
Testudo chilensis	—	675	—	9·5
Tuinambis teguixin	—	1,955	—	33·7
Varanus salvator	—	16,700	—	14·0
Caiman latirrostris	—	29,000	—	19·0
Caretta caretta	—	73,000	—	33·0
(d) Quiring (1941)				
Turtles	66	—	—	23·77
Gila monster	1	—	—	24·96
Lizards	71	—	—	36·39
Alligators and crocodiles	12	—	—	36·76
Horned toads	3	—	—	53·58
Snakes	14	—	—	81·68

measured by planimeter at 600 magnifications. She found that the volume of both types of tissue varied seasonally. For example, in *Lacerta agilis* the volume of cortical tissue changed from 64,000 such units in June to 77,000 in July to 132,000 by the end of August; at the same time the chromaffin tissue increased from 18,000 in June to 40,000 in July to 111,000 in August. She comments that, during the quiescent period of winter, there was a disproportionate reduction in the amount of both tissues, so that in the spring the volume of chromaffin tissue was much less relatively than that of the cortex (cortex 64,000, chromaffin tissue 18,000). Her seasonal data for *L. vivipara* and *L. serpa* were not complete enough to be sure of similar seasonal variations.

The micro-anatomy of the adrenal gland is very similar throughout the Lacertilia and, as far as we know at present, is

TABLE 22. *Relative figures for cortical and chromaffin tissue volume in three species of lizard. The figures in columns 5 and 6 can be converted into actual volume in cm.^{3} by division by 600^3 (from Bimmer, 1950)*

Sex	Length of captivity	Month killed	Body length and condition of animal	Volume of Cortical tissue	Volume of Chromaffin tissue	Ratio of cortical to chromaffin tissue
			Lacerta agilis			
Male	—	July	2·5 cm. mature	7,956	4,846	1·65 : 1
Male	4 months	August	7·3 cm. mature	132,421	111,589	1·19 : 1
Female	83 days	July	5·9 cm. 2? years	77,815	40,066	1·8 : 1
Female	41 days	June	6·5 cm. pregnant	64,164	18,089	3·5 : 1
Female	9 days	August	8·2 cm. laying eggs	127,031	95,114	1·3 : 1
			Lacerta vivipara			
Male	—	July	2·05 cm.?	2,579	2,415	1·06 : 1
Female	Freshly caught	June	3·4 cm. 1 year	4,432	2,760	1·2 : 1
Female	11 days	May	5·5 cm. ripe eggs	33,562	7,893	4·25 : 1
Female	5 days	June	6·7 cm. pregnant	26,399	9,776	2·7 : 1
Female	11 days	June	6·2 cm. pregnant	29,925	13,067	2·2 : 1
Female	61 days	July	5·3 cm. pregnant	18,809	10,167	1·8 : 1
			Lacerta serpa			
Male	About one month	March	9·0 cm.	123,245	12,681	9·7 : 1
Male	About one month	Oct.	7·6 cm.	164,528	51,123	3·2 : 1
Female	About one month	August	7·3 cm. laying eggs	84,700	36,761	2·3 : 1

Reptilia

of the same basic pattern in all reptilian orders (Pl. VI, fig. 29). In *L. viridis* and *A. agama* frontal and sagittal sections of the gland show an elongate club-shaped form and transversely a roughly oval shape. The gland is surrounded by a connective tissue capsule which is particularly well developed peripherally in support of blood vessels. The main body of the adrenal is made up of cortical tissue comprising a loosely connected, continuous network of cords of cells. These irregularly anastomosing cords run in all three planes with a general orientation of their long axis lying more or less parallel to the anterior/posterior axis of the gland. Thus a short cord cut absolutely sagittally reveals itself as being composed of a double row of cells, while in transverse section the appearance is that of a round solid tube, two cells in diameter, made up of more or less radially arranged cells. The criss-cross anastomosing of cords means that in transverse section round cords predominate, in sagittal section oval to long groups of cells are the majority. A layer of chromaffin tissue extends along the entire dorsal surface of the gland, enclosing the poles and extending thinly laterally over the gland. Numerous islets of chromaffin tissue are scattered throughout the cortical tissue, and, in addition, 'tongues' penetrate down from the dorsal layer (text-fig. 34). It is interesting to note that the dorsal chromaffin tissue produces noradrenaline while the tongues and the islets secrete mainly adrenaline (Wright and Chester Jones, 1955). Blood sinuses are well developed in the gland and cover three areas:

(i) a sinus, which varies in development at different points, surrounds the cortical tissue (lying therefore dorsally between the cortical and the chromaffin tissue) and has an outlet ventrally;

(ii) small lenticular sinuses abound throughout the cortical tissue;

(iii) one, two or three central sinuses occupy a relatively large volume of the cortical mass.

The extent of their development varies in different animals and under different conditions. In some, seen in sagittal section, the central sinuses may occupy as much as a third of the gland, in others they may be much reduced.

The connective tissue capsule which surrounds the gland is

175

made up principally of collagenous fibres with some reticular and elastic tissue. Strands of collagen penetrate the dorsal chromaffin layer which is itself marked off from the cortex by a thickened layer of this type of connective tissue. The cortical cords are outlined by thin collagen fibres which do not penetrate between the cells or cell groups. Reticular fibres run irregularly through the gland forming a lattice-work encompassing groups of cortical cells. These fibres vary from thick to filamentous; some commence and end as thick fibres while others terminate in single fibres or filamentous arborizations; they penetrate not only the chromaffin tissue but also course between the cortical cells. At the posterior end of the adrenal, against the connective tissue and slightly ventro-lateral in position, there is a large round ganglion from which nerve fibres enter the gland.

The normal cortical cell is columnar. The round or oval basiphilic nucleus lies at the peripheral border of the cell in the round cord near the basement membrane which is, in turn, against a venous sinus. The nucleus is of the order of 8μ in diameter and contains one to three nucleoli. The cytoplasm is usually slightly acidophilic, giving a faint red colour after eosin for example, and, after routine methods containing fat solvents, it is very vacuolated. After methods which preserve fat, the cytoplasm is seen to contain lipid droplets or globules which take up sudan and osmic acid. The lipid includes cholesterol, the Schultz modification of the Liebermann-Burchardt reaction giving a blue-green colour more or less co-extensive with that obtained by sudan black. Birefringent cholesterol digitonides can be precipitated from the fat droplets, the intensity of the birefringence varying from gland to gland, and within any one gland itself. The Ashbel-Seligman reaction gives a very fine purple granulation distributed evenly throughout the main part of the cortical tissue. Silver-nitrate-reducing substance is present in the cortical cells; this may be ascorbic acid and if so is displayed in particularly fine granules.

Scattered among the normal cells of the adrenal cortex of the captive green and West African lizards are cells with intensely basophilic nuclei and shrunken cytoplasm and these can be regarded as atrophic. In the active lizard it may be

Reptilia

that no such atrophic cells would be seen (and some experimental evidence for this is given below); certainly Miller (1952) did not find this type of cell in the adrenal of *Xantusia*. Furthermore, in Wright's material, there is a narrow area of atrophic cells lying on the periphery of the cortex, particularly obvious against the dorsal chromaffin tissue and at the poles, but nevertheless extending right round the main cortical mass. This area will be referred to as the 'peripheral zone'. Such a zone has been briefly noted in *Gerrhonotus* (Retzlaff, 1949) and in *Philodryas* (Uchoa Junqueira, 1944). Miller, on the other hand, did not find an atrophic region in the adrenal of *Xantusia*, either from freshly caught or starved material; whether this is due to the minute size of the xantusid adrenal is not known. In the peripheral zone the cytoplasm is more heavily sudanophilic, the fat droplets larger than those found in the major portion of the cortical tissue, and the Ashbel-Seligman reaction is negative. Recent descriptions of lizard adrenals do not differ greatly from that given above. Miller (1952) finds in *Xantusia* that the interrenal cords are composed of radially arranged tall columnar cells 20 to 25 μ in height and 5 to 7 μ in width. The nuclei are basal and abut upon the confining tubular membrane adjacent to the venous sinus. The nuclei are 4 to 5 μ in diameter. The cytoplasm has a slightly eosinophilic reticulated appearance, with osmiophilic fat globules varying from 2 to 4 μ in diameter. In *Gerrhonotus*, Retzlaff (1949) found that the shape and the location of cell nuclei varied. Some nuclei are spherical and placed in the centre of the cell; others are elongated. The number of nuclei varied from one to seven per cell. Mitochondria were in the form of short rods and granules, the latter sometimes appearing to be joined by fine filaments, and either concentrated close to the nucleus or widely scattered among the fat droplets. Golgi bodies (after da Fano's method) were either a heavy network near the nucleus or in a dispersed granular form.

Among the Ophidia, *N. natrix* has long, tapering, cylindrical adrenals, 1 to 3 cm. in length, $\frac{1}{2}$ to 1 mm. in width. The basic pattern of the adrenocortical cell is similar to that already described for the lizards. The cords of cells although arched in various planes have a general longitudinal orientation. The chromaffin cells form smaller groups than in the lizard adrenal but the snake adrenal also possesses a dorsal aggregation of this tissue which, in contrast to the islets, seems to produce noradrenaline. As far as can be judged from the older literature and from da Costa (1913), and Lawton (1937), the micro-anatomical form of the adrenal cortex in the Crocodilia is similar to that of the Squamata and this holds good, also, for the Chelonia. I have had an opportunity of examining some adrenals from *Sphenodon* (Rhynchocephalia) through the courtesy of Professor H. Waring, and here too the adrenal conforms to the general reptilian pattern.

Clear-cut data on cortical function are scarce. The reptile adrenal cortex depends on secretions of the anterior lobe of the pituitary for normal histological appearance and presumably normal functioning. Wright (unpublished) removed the anterior lobe of male agamids, together with a variable amount of intermediate and posterior lobes. The lizards lived well after the operation and fed on house-flies and meal-worms. A month after adenohypophysectomy the chromaffin tissue was unchanged but the adrenal cortex was markedly shrunken (cf. Pl. VI, fig. 30), the weight declining by 58%. Histologically these reduced adrenals have some spindle-shaped cells, with reduced cytoplasm and pycnotic nuclei, others with normal nuclei but with the cytoplasm heavily vacuolated, and various intermediate stages between these two types can be seen. The shrunken collapsed cell seemingly represents the end stage of degeneration but, in the material so far examined, however degenerated the adrenal, there were still some cells of normal appearance scattered haphazardly throughout the cortex. The lipid is distributed in large globules of varying size, the peripheral zone consisting entirely of globules of the largest size—about 10μ in diameter. Birefringent digitonides and cholesterol were still present in, or formed by the adrenal fat of the hypophysectomized agamid. The adrenal gland of the green lizard presents the same picture after hypophysectomy as does that of *Agama*.

In the lizard, *Xantusia*, Miller (1952) found that, 7 days after hypophysectomy, the adrenocortical tissue showed some loss of lipid, decrease in cell size and pycnosis in some areas. There

Reptilia

was a tendency for some cells to remain well filled with fat, while adjacent cells were completely evacuated. The cortex of the snake shows a similar dependence on the presence of the pituitary for maintenance of its micro-anatomy (Pl. VI, fig. 30). Schaefer (1933) found that the adrenals of *Thamnophis sirtalis* and *T. radix* were reduced to about two-thirds of the normal size after hypophysectomy. The grass snake, *N. natrix*, hypophysectomized for a month, has an adrenal weight loss of about 32 % (table 23). Data are not yet available to say whether there is a seasonal difference in response, nor has any sex difference become apparent. Histologically the adrenal cortex of the hypophysectomized snake shows degeneration of the cells some of which have increased accumulation of lipid in large globules, others shrunken cytoplasm and in both various stages of pycnosis of the nuclei; in addition there are, as in the adrenal of the hypophysectomized lizard, a few cells of normal appearance occurring here and there in the cortex.

Implantation of hypophyses into hypophysectomized snakes went some way in maintaining a normal appearance of the adrenal (Schaefer, 1933) and Wright has recently found that small doses of mammalian ACTH will restore the shrunken cortex of the hypophysectomized *Natrix* to normal. Significantly, also, unilateral adrenalectomy in this animal allows the other adrenal to become hypertrophic (Pl. VI, fig. 31). Other experiments using adrenocorticotrophin and cortical steroids have so far only been done on intact animals. The reptile adrenal will react violently to mammalian ACTH and can become exhausted and degenerate. Miller (1952) found that moderate doses of ACTH (Armour) caused lipid depletion in 2 days but after 4 days' treatment the cortical cells had a normal appearance; larger doses (two to five injections of 1/20 mg.) produced degenerative changes with pycnosis of the nuclei, lipid depletion and general cellular degeneration. Wright, too, found that using a large dose of ACTH (2·5 Armour units per day for 20 days), the adrenal lost about 18 % of its weight (table 23). Histologically there were patches of degeneration among large areas of healthy active tissue. The cells of the degenerative areas are syncytial and the nuclei pycnotic; they

TABLE 23

(a) The effects of hypophysectomy (30–33 days after operation) and of the injection of ACTH and cortisone into normal animals on the adrenal weights and on the sodium and potassium content of plasma and muscle and on the water content of muscle in the grass snake, *N. natrix* (from Wright, unpublished).

		Body weight		Adrenal weight		Plasma (mEq./l.)		Water (%)	Muscle mEq./kg. Wet weight	
		Start	End	(mg.)	(mg. %)	Na	K		Na	K
Controls	8	—	49·6 ±5·05	—	31·39±0·42	167·06±2·21	7·16±0·94	81·3 ±0·47	58·92±3·2	83·4±2·2
Hypophysecto-mized	11	—	48·9 ±3·56	—	21·40±2·5	163·5 ±4·4	6·31±0·46	80·49±0·44	60·01±3·17	82·33±3·04
Normal plus ACTH	8	50·75±3·29	42·65±2·75	10·90±0·91	25·58±2·83	186·1 ±8·7	5·71±0·40	76·43±0·60	52·13±5·41	87·17±3·40
Normal plus cortisone	6	41·83±1·72	43·25±2·29	9·48±1·96	21·26±3·37	188·10±13·00	7·70±0·60	79·00±1·23	57·97±2·86	88·67±5·8

(b) The effects of hypophysectomy (32 days after operation) on adrenal weights, sodium and potassium values of plasma and muscle and on the water content of muscle, in the male West African lizard, *A. agama.*

				Adrenal weight		Plasma (mEq./l.)		Water (%)	Muscle mEq./kg. Wet weight	
Controls	7	—	47·2 ±2·3	6·99±0·47	15·09±1·35	179·2 ±6·85	5·19±0·30	78·4 ±0·34	18·98±0·51	106·1±3·30
Hypophysecto-mized	8	—	47·31±2·80	2·91±0·29	6·14±0·55	160·2 ±5·09	5·09±0·33	81·28±0·71	22·51±1·63	109·1±1·8

180

Reptilia

contain large globules of lipid. In the active cells the lipid is in small droplets congregated particularly towards the distal pole of the cell (Pl. VI, fig. 32). The cords of cells vary in the degree of degeneration: some are completely atrophic, others display a few degenerate cells in otherwise normal tissue. It seems, therefore, in Reptilia as in Amphibia, that mammalian ACTH may cause a hypertrophy of the adrenal followed by an over-stimulation effect leading to degeneration. However, most available mammalian ACTH preparations are contaminated with posterior lobe hormones. We do not know, therefore, how far these latter mask or alter the ACTH effects in lower vertebrates.

The injection of steroids much used in mammalian work, cortisone and DCA, produces a degenerative reaction in the adrenal cortex of reptiles. Injections of cortisone, 1/20 mg. for 2 days or 1/40 mg. for 4 days, caused a loss of lipid, shrinkage of cells and an over-all diminution of cortical volume together with increased vascularity in *Xantusia* (Miller, 1952). In the same species DCA, 1 mg. for 2 days, caused lipid loss from about half the cortical cells, while 1/30 mg. for 4 days gave almost complete lipid loss but little evidence of cellular degeneration. In the grass snake, 5 mg. cortisone for 8 days effected a decline in adrenal weight of about 33 % with considerable degeneration (Pl. VI, figs. 33 and 34). Whole areas of the gland may be atrophic or the degeneration confined to a few cords. All the nuclei appear slightly crenated even when other signs of atrophy are not present. Lipid droplets are large and very sudanophilic.

Although we have so far little evidence, it seems reasonable to suppose that the reptile adrenals depend on hypophysial secretions resembling mammalian ACTH and that one of the hormones secreted by the cortex may be corticosterone (Phillips, unpublished).

Physiological data helping us to interpret the function of the adrenal cortex are scarcer than those obtained from histology. There are a few reports on the consequences of adrenalectomy. In 1929, Adams and Harland found that bilaterally adrenalectomized lizards (*Anolis*) lived for varying periods from 8 hr. to 6 days, and that they developed anorexia, inanition and paralysis. Using the same genus, Kleinholz (1938) had survival time of only 10 to 24 hr. after adrenalectomy. Holmberg and Soler (1942) adrenalectomized three large *Iguana*. Two died after 11 and 24 hr. and the third after 5 days, with a loss in weight of about 3 %. Valle (1945) noted that of eighteen bilaterally adrenalectomized snakes (*Philodryas*) sixteen lived for 7 days. Wright has found, in this laboratory, that the adrenal venous return by ports to the adjacent venae cavae renders the operation of adrenalectomy in the green lizard difficult and the predominant symptoms after operation are those arising from operative trauma and haemorrhage. On the other hand, in the grass snake, the small adrenal efferent veins can be tied off and adrenalectomy is a successful and neat operation in this species. Adrenalectomized non-feeding captive snakes lived for about 10 days. Up to the time of death the activity and appearance of the adrenalectomized animals differed little from the controls. These latter, however, were non-feeding captive snakes during a cold summer at a temperature of about 50° F. and were not very active. The adrenalectomized snakes were not seen to have any progressive symptoms of adrenal insufficiency but presumably underwent a rapid crisis ending in death.

There is a suggestion in the literature that reptile adrenocortical secretions may be associated with carbohydrate metabolism. Hernandez and Coulson (1952) found that alligators have a pronounced hypoglycaemia and anorexia in winter, and that injection of whole sheep pituitary powder (Armour) produced a hyperglycaemia. More factors than the adrenal cortex are involved but this gland may well be implicated. In 1953, Coulson and Hernandez investigated the effect of ACTH and of cortisone on the level of blood glucose in alligators (*A. mississippiensis*). Five alligators received 5 mg. ACTH/kg. per day for 45 days, five animals 10 mg. cortisone/kg. per day, and five animals were controls. The daily variation in the level of blood glucose was considerable but, cautiously interpreted, their data indicate that cortisone induced a moderate hyperglycaemia and that ACTH was ineffective in this respect (text-fig. 25). The experiment was somewhat prolonged, though the alligator apparently requires about ten times as much time as the average mammal to show metabolic changes and it may be,

Reptilia

also, that the doses were inadequate as these animals weigh up to 8 kg. (Coulson and Hernandez, 1953).

The other facet of adrenocortical activity we have been accustomed to look for, from analogy with mammalian work, is water and salt-electrolyte metabolism. The normal figures in the agamid lizard for the sodium and potassium content of plasma and muscle and the water content of muscle are very similar to those found in most vertebrates (table 23); and in the alligator, Coulson *et al.* (1950) found the plasma sodium

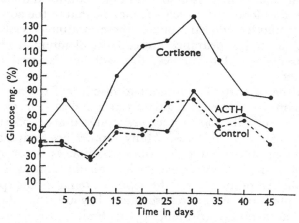

Text-fig. 25. Effect of daily injections of cortisone (10 mg./kg.) and ACTH (5 mg./kg.) on the level of blood sugar in the alligator, *A. mississippiensis*; five animals were used in each group (Coulson and Hernandez, 1953).

concentration to be 154 (146–162) mEq./l. and the potassium 4·2 (2·6–6·4) mEq./l. Cicardo (1947) gives a range of 5·1–7·7 mEq./l. K for reptiles. The grass snake has similar values except that the sodium content of muscle is surprisingly high, about 58 mEq./kg. wet weight Na, whilst a figure of about 24 mEq./kg. wet weight can be regarded as a typical figure for vertebrate muscle. The reason for this difference is not known; certainly 10 days after adrenalectomy the sodium content of snake muscle drops to 43 mEq./kg. wet weight, the potassium content of the muscle of these non-feeding adrenalectomized animals remaining at the normal value of 83 mEq./kg. wet weight; the water content was also normal at 81 % (table 23).

The Adrenal Cortex

Further investigation of the metabolic effects of adrenalectomy in the reptile is clearly required. Hypophysectomy of both the lizard and the snake effects no significant change in the water and salt-electrolyte picture of plasma and muscle except that the water content of the hypophysectomized agamid is somewhat higher than that of the normal (table 23). These approximately normal values for plasma and muscle are maintained in the face of adrenal atrophy and this calls to mind the capacity of the unstressed hypophysectomized eutherian to maintain reasonably normal sodium and potassium values for blood and muscle. It may be that in the adrenal of the hypophysectomized reptile those scattered cells of apparently normal appearance are capable of minimal secretion of adrenocortical hormones; this will be discussed in a later chapter.

It would be of particular interest to discover if the adrenal cortex acts on the kidney to regulate water and salt-electrolyte metabolism and if some neurohypophysial-adrenocortical relationship exists in this class. Many reptiles are adapted to arid conditions and even those whose habitat is in or near water have no movement of water through the skin as is the case with the Amphibia, nor is the reptilian renal tubule specifically adapted for water reabsorption beyond the plasma osmotic concentration (Burian, 1910; Smith, 1932, 1951; Heller, 1950, 1956). The alligator is very responsive to the antidiuretic effect of neurohypophysial extracts and this lies in the pitressin rather than the pitocin fraction as in frogs (Burgess et al. 1933; Sawyer and Sawyer, 1952). Antidiuresis is effected, at least partly, by lowering the glomerular filtration rate, pitressin influencing the smooth muscle of the afferent glomerular arteriole. Moreover, metabolic waste products, and especially the most important one, uric acid, are eliminated not only by glomerular filtration but probably to a much greater extent by tubular secretion (Marshall, 1932; Heller, 1950). It is attractive to speculate that adrenocortical secretions have a role to play at the renal tubular level in reptiles and that the excretion and conservation of sodium, among other electrolytes, is one of their functions. It is hoped that this brief mention of an untilled field will stimulate active investigation.

PLATE VI

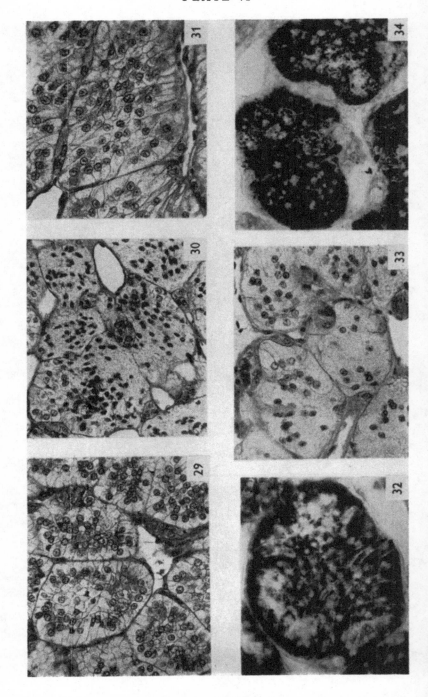

Reptilia

EXPLANATION OF PLATE

PLATE VI

Fig. 29. Part of the adrenal gland of a normal snake (*N. natrix*) (× 300; Bouin, Heidenhain's iron haematoxylin).

Fig. 30. Part of the adrenal gland of a snake hypophysectomized for 30 days (as fig. 29). Collapse of the lobules, pycnotic nuclei.

Fig. 31. Part of left adrenal gland of a snake from which the right adrenal had been removed for 11 days (as fig. 29). Compensatory hypertrophy.

Fig. 32. Part of the adrenal gland of a snake injected with mammalian ACTH (2·5 i.u. per day for 20 days) (× 300; formalin, sudan black). Distribution of sudanophilia does not differ a great deal from normal except that there is heavier staining in the peripheral part of the lobules.

Fig. 33. Part of the adrenal gland of a snake injected with cortisone for 8 days (as fig. 29). Some collapse of lobules and some nuclear degeneration.

Fig. 34. As fig. 33 (formalin, sudan black). Dense sudanophilia.

CHAPTER V

AVES

MANY of the older workers whom we noted as having described the adrenal glands of animals in the other classes of vertebrates also included the bird in their surveys. Thus Perrault (1671) a least saw the adrenals in the bird and Haller (1766) described them correctly. Later Meckel (1806), Nagel (1836)—who discusses Meckel's findings in some detail—Stannius (1846), Pettit (1896) and Vincent (1898b, 1924) added to our knowledge of the anatomical position and histology of the adrenals in this class.

In most birds the adrenals are distinct paired organs lying, often wholly or partly covered by the gonads, at the anterior end of the kidneys just behind the lungs. The glands measure, in the domestic hen for example, about 13 mm. in length, 8 mm. in width and 4–5 mm. in thickness (Müller, 1929). In some species the adrenals lie very close together and in a few there is actual fusion, the stork (*Euxenua manguari*), rhea (*R. americana*), and the gannet (*Sula variegata*) being noted by Holmberg and Soler (1942), and the bald eagle (*Haliaetus leucocephalus*) and the common loon (*Gavia immer*) by Hartman and Brownell (1949). These latter authors add that variation can be considerable within one species so that in the herring gull (*Larus argentatus smithsonianus*) and the hairy woodpecker (*Dryobates villosus*), for example, both fused and separate adrenals were found in different individuals of the same species. The shape of the adrenals differs greatly and is not necessarily constant for one species; they may be oblong, oval or pyramidal but are usually flattened and somewhat irregular. Müller (1929) likened the shape of the right adrenal of the domestic hen and dove to a blunt-edged three-sided pryamid, pointed caudally, while he found the shape of the left gland to be less constant. Either the one gland or the other may be advanced anteriorly a few millimetres relatively to each other. The colour has been described by several authors (see Müller, 1929; Hart-

186

man and Brownell, 1949) as being yellow, orange, pink, greyish yellow to dark reddish brown; Müller (1929) found that the adrenals of the domestic fowl and dove were a grey-yellow to red-yellow. This variation in colour must in part be a reflexion of the activity of the gland and in part on other factors. Thus Findlay (1920) considered that variation in colour of the gland in hens depended primarily on the amount of adrenal lipochrome which in turn was determined by the lipochrome content of the food.

The adrenal veins and arteries are short, as the adrenals lie near the midline by the kidneys. Thus in the domestic fowl the

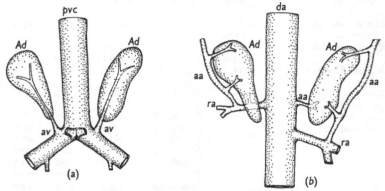

Text-fig. 26. (*a*) The venous system; (*b*) the arterial system of the adrenal glands of the domestic fowl (from Hays, 1914). *Ad*, adrenal gland; *pvc*, posterior vena cava; *av*, adrenal vein; *da*, dorsal aorta; *ra*, renal artery; *aa*, adrenal artery.

glands lie just anterior to the bifurcation of the posterior vena cava, the blood draining to this point by one short adrenal vein on each side (text-fig. 26) (Hays, 1914). The arterial blood may enter the left gland directly from the aorta or through the anterior or posterior twigs of a branch of the renal artery which runs along the lateral border of the gland. The supply of the right gland differs from that of the left in that there is no direct connexion with the aorta but arises solely from a branch of the right renal artery (text-fig. 26) (Hays, 1914). This general pattern of vascular supply is true for most birds studied (Müller, 1929; Hartman and Brownell, 1949; Hartman and Albertin, 1951).

TABLE 24. *The adrenal weights of various species of birds taken from the more extensive lists of Hartman (1946) and of Crile and Quiring (1940). The latter's data on the Leghorn fowl are included to show the change in adrenal and body weight with advancing age*

	After Hartman		After Crile and Quiring			
					Adrenal weight	
		Adrenal weight (mg./100 g. body weight)		Body weight	Absolute	(mg./100 g. body weight)
	n		n			
Pelican (*P. occidentalis*)	6	40·5 ±5·6	2	3290	993	31·0
Herring gull (*L. argentatus*)	8	15·3 ±6·2	2	535	133	24·9
Phoebe (*S. phoebe*)	10	14·6 ±4·4	1	17·5	5·5	31·4
			1	18·1	5·0	27·6
Robin (*T. migratorius*)	14	11·7 ±4·6	2	69·3	21·3	30·7
Sparrow (*P. domesticus*)	28	9·7 ±2·7	11	23·6	6·9	29·3
Leghorn fowl (*G. bankvia*)	—	—	Female 7	43·7	9·5	21·7
			Male 13	46·4	14·7	31·7
			Female 6	68·2	16·3	23·6
			Male 14	80·1	18·5	22·0
			Female 15	119·7	19·8	16·5
			Male 15	119·7	26·3	22·0
			Female 10	359·6	86·4	24·0
			Male 10	353·2	113·5	32·1

The weight of the adrenals in various species has had some attention, although, apart from domestic breeds, the population samples were not homogeneous enough for valid comparisons to be made (table 24). Moreover, many data were presented before standard statistical treatment was usual. Before hatching, the adrenal of the embryonic chick shows no sex difference as regards weight and the gland does not increase in weight as fast as the body. In the White Leghorn chick, before hatching, Sun (1932) found that the paired adrenals weighed 0·3 mg. at 9 days, and 6·2 mg. at 20¼ days; and Venzke (1943) gave the change of body and endocrine gland weights for chick embryos of the same breed from the age of 10 days to 21 days at daily intervals (table 25). The growth and develop-

Aves

ment of the single-comb White Leghorn and Rhode Island Red chicks from hatching until 30 days of age have been studied by Breneman (1941). Both body weight and adrenal weight increase rapidly after hatching, the former at a relatively more rapid rate so that the adrenal weight in mg. per cent at 30 days of age is the lowest of the series (table 26). No real sex differences are apparent. For adult birds, data for the common

TABLE 25. *Body and adrenal weights of male and female chick embryos at 10 and 21 days of age.* (*Extracted from Table 1 of Venzke, 1943*)

	n	Body weight	Adrenal weight	
			Absolute (approx.)	Relative (mg./100 g. body weight)
		Males		
Aged 10 days	10	2·26±0·05	0·79	3·5±0·9
Aged 21 days	10	33·70±1·34	7·75	2·3±0·3
		Females		
Aged 10 days	10	2·01±0·07	0·80	4·0±0·7
Aged 21 days	10	33·75±1·19	8·1	2·4±0·4

TABLE 26. *Body weight and adrenal weight of White Leghorn cockerels and pullets at ages up to 30 days after hatching* (*Breneman, 1941*)

Age in days	n	Body weight	Adrenal weight	
			Absolute (approx.)	mg./100 g. body weight
		Males		
5	28	49·1	8·3	17·1
10	32	76·8	13·1	17·1
15	18	106·0	15·8	14·9
20	36	175·2	27·1	15·4
25	29	190·7	31·0	16·2
30	30	272·3	34·6	12·7
		Females		
5	14	47·8	8·9	18·7
10	14	57·8	14·0	24·2
15	19	95·4	14·6	15·3
20	12	135·2	22·4	16·6
25	12	212·6	26·4	12·4
30	14	266·6	32·8	12·3

domestic types are available in Elliott and Tuckett (1906) for Black Minorcas, in Latimer (1924, 1925) for the White Leghorn (text-fig. 27), in Juhn and Mitchell (1929) for the Brown Leghorn, in Leonard and Righter (1936) for the Bantam, in Landauer and Aberle (1935) for the single-comb White Leghorn and the Frizzle fowl. In addition weights are given in the various protocols of the papers cited below in the discussion of experimental work. Crile and Quiring (1940), Hartman (1946), Hartman and Brownell (1949) range over a

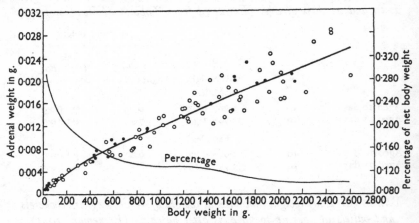

Text-fig. 27. Relative and absolute weights of the adrenal glands of single-comb White Leghorn chickens, plotted against gross body weight. There is no apparent sex difference. (From fig. 24, Latimer, 1924. Latimer noted that the gland in the female is lodged at the base of the ovarian ligament and when the ovary is fully developed the ligament is so strong and tough that removal of the left adrenal without injury is difficult.) ●, males; ○, females.

vast number of species and a small extraction from their data is given in table 24. There is a great deal of variation and it underlines one point at least that, as in Mammalia, both the absolute and relative endocrine weights should be considered and this preferably in animals of known condition, age and metabolic state. The data available show that there is no clear sex difference in the weight of the adrenal gland as there is so markedly in the Eutheria. It seems that young birds have heavier adrenals, relatively to body weight, than do adults (text-fig. 27) (Elliott and Tuckett, 1906; Latimer, 1924).

Aves

Both cortical and chromaffin tissue are present in the adrenal gland of the bird, but they do not bear a constant relationship to each other. The amount of chromaffin tissue varies greatly and lies in varying degrees intermingled with the cortical tissue.

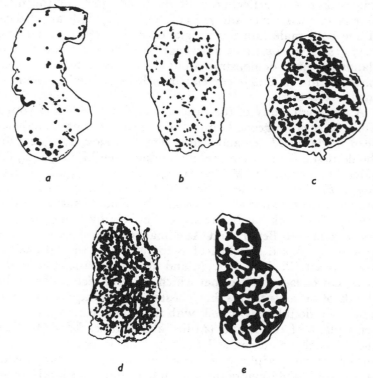

Text-fig. 28. Examples of the varying amount of chromaffin tissue in the adrenal glands of different species of birds (from Hartman *et al.* 1947). Black, chromaffin tissue; white, cortical tissue.

(*a*) *Dendroica aestiva*; (*b*) brown pelican (*P. occidentalis carolinensis*); (*c*) eastern cowbird (*Molothrus ater ater*); (*d*) Cooper's hawk (*Accipiter cooperi*); (*e*) black-throated green warbler (*D. virens*).

Hartman, Knouff, McNutt and Carver (1947) have drawn examples of the patterns of the intermingling of the two kinds of tissue in different species and text-fig. 28 gives some of these. Elliott and Tuckett (1906) gave a proportion of 0·8 : 1 for cortical : chromaffin tissue in the fowl adrenal but Latimer and

The Adrenal Cortex

Landwer (1925) found considerable variation. Miller and Riddle (1942) estimated that in the young white Carneau pigeon 65 % of the gland is cortex, 32 % chromaffin tissue and 3 % blood, lymph, etc., no sex difference being apparent. In single-comb White Leghorns, 86 days old, 40 % of the male adrenal gland is cortex and 71 % of the female, while at 156 days of age both male and female glands have 43 % cortex (Kar, 1947 a, b). No consistent difference between male and female glands has been demonstrated but there may well be transient variation in the female correlated with reproduction (Riddle, 1923; see below).

A thorough study of the histology of the adrenals of young white Carneau pigeons (1·9 to 2·4 months from hatching) has been made by Miller and Riddle (1938, 1939 a, b, 1942); for the domestic fowl, among recent workers (see Müller, 1929 for older literature), by Müller (1929), Uotila (1939 a, b), Kar 1947 a, b); for the domestic dove by Müller (1929); and for the brown pelican (*Pelecanus occidentalis*) by Knouff and Hartman (1951). The duck has been used extensively by Benoit and his co-workers (see Benoit and Assenmacher, 1953) and he very generously placed histological preparations of the adrenal at my disposal (Pl. VII, figs. 35 and 36). The basic unit of the avian cortex is similar to that which we have seen so far in the glands of animals from other classes, comprising a cord composed of a double row of cells with their long axes in the transverse plane of the strand (Miller and Riddle, *loc. cit.*). One end of each columnar cell faces the blood channel. In the pigeon, at the periphery of the gland the cells are large, contain considerable cytoplasm and full round, or slightly oval, nuclei; the chromatin network is thin and diffuse and there are usually two nucleoli. Towards the middle of the gland the cells elongate and eventually become smaller.

Lipid droplets occur in the cells throughout the cortex. The lipid in the adrenal of all birds so far studied is sudanophilic and osmophilic; it gives a positive phenylhydrazine reaction, contains cholesterol and can form birefringent crystals. These reactions are usually found with adrenal lipid where it occurs in the different vertebrate groups and, as we have noted before, are difficult of interpretation in terms of adrenocortical steroid hormone production but can sometimes, with caution, be used as a reflexion of the functional state of the gland. In the bird adrenal there are all degrees of gradation among the cells, showing much or little lipid. Miller and Riddle considered that as the amount of lipid increased the number of mitochondria decreased, and Knouff and Hartman agree for the pelican, but Kar (1947 a, b) did not find such a reciprocal relationship in the White Leghorn adrenocortical cells. Certainly small granular mitochondria abound in avian cortical cells but estimation of their variation in number as between cells of differing size with little lipid and those with large droplets must prove difficult without elaborate counting and statistical techniques. The Golgi apparatus has been seen in the pigeon cortical cell as either a short strand, a small network or a U-shaped filament; it is sometimes larger with one or more strands extending into the cell cytoplasm and may even become an intricate network of very fine strands lying between the lipid droplets.

Uotila (1939 a, b) was particularly concerned with the appearance of the fowl adrenal cortex after using stains containing acid fuchsin (Masson's, or 0·5 % acid fuchsin or Mallory's; see Kar, 1947 b). Fuchsinophilia of cells cannot yet be interpreted in the light of function although its use with this hope, arising from an erroneous generalization from Broster and Vines' (1933) idea of the fuchsinophil cell producing male sex hormone in man, lingers on (cf. Blackman, 1946). Uotila found that cortical cells are of two types, the pale cell and fuchsinophil cell. The pale cell has cytoplasm with a large number of lipid droplets, occasional scattered fuchsinophil granules, a lightly staining nucleus and two or three nucleoli, with few granular or vacuolated mitochondria. Fuchsinophil cells are similar except that they also contain numerous fuchsinophil granules, many of which are vacuolated. There are numerous granular or vacuolated mitochondria which are most abundant at the capillary pole of the cell. A variant of the fuchsinophil cell is one with diffuse fuchsinophilia and increased pigmentation and pycnosis of the nucleus; these cells are often stellate in shape and found at the periphery of the cortical cords.

The bird adrenal is made up of cortical cords which loop against the outer connective tissue capsule and here the cells frequently show differences from those lying more centrally. In the duck (Pl. VII, fig. 35) the lipid droplets are larger in the cells of the loop than elsewhere with the remaining cytoplasm particularly acidophilic. In the fowl, on the other hand, Kar (1947b) noted that the cortical cells of the peripheral region of the gland were larger with a considerable amount of cytoplasm, numerous mitochondria, and rounded nuclei having diffuse chromatin material and two to three nucleoli. In contrast, the cortical cells in the central region of the gland were smaller with the appearance of fewer mitochondria but, apart from some cortical cells just below the capsule which showed little or no lipid, distribution of fat droplets was similar peripherally and centrally. There were more cells with pycnotic nuclei in the latter part of the gland.

The possibility of distinguishing, histologically, two or more zones in the bird adrenal brings in its train conjectures as to differentiation of function and consideration of the theories put forward to account for the zoning of the eutherian gland. These points will be considered in a later chapter, but it is relevant and helpful to give in detail here Knouff and Hartman's careful work on the adrenal of the brown pelican which they consider shows, uniquely among the birds so far studied, clear-cut cortical zonation. They report that the pelican adrenal is embedded in and connected to the surrounding parts by a membranous sheath of loose connective tissue, in which branches of the main vessels and nerves ramify, penetrating the true capsule at various points. The capsule itself comprises reticular fibres with few collagenous and elastic fibres present in the walls of the larger capsular vessels. The reticular fibres pass inwards to form a supporting network to the adrenal. The main adrenal arteries branch freely in the pericapsular sheath. A few large branches pass directly by way of the septa to the central region of the gland where their terminals connect with central venous sinuses. Smaller arterial branches from the pericapsular sheath enter the capsule to form an arteriolar plexus from which twigs run to terminate quickly in peripheral venous sinuses. The central sinuses drain into several large venous channels which pass outwards either in septa or directly through the parenchyma to the periphery where, after receiving large capsular tributaries, they empty into larger veins in the pericapsular sheath. Peripheral sinuses connect directly with capsular veins at many points over the entire glandular surface. The usual intermingling of chromaffin and cortical tissue is apparent in the pelican adrenal, but the cortex in this case takes up roughly 85–90 % of the gland. The masses of chromaffin tissue are small and widely scattered in the cortex but this tissue may not be less, in actual amount, than in other species; it is more likely that, as the pelican adrenal is proportionately large (table 24) the cortical tissue is relatively more abundant. It is this which makes the pelican adrenal attractive for the histology of the cortex though this bird may not be so readily available for laboratory work as is the domestic fowl.

Knouff and Hartman find that the pelican cortical tissue is most clearly organized into three zones which correspond in position and general appearance but not in relative proportions to the zona glomerulosa, zona fasciculata and zona reticularis of the eutherian adrenal cortex. The zona reticularis appears in section to be composed chiefly of convoluted strands of cortical cells which branch and anastomose freely to form a network in the meshes of which lie the vascular channels and the widely scattered clusters of chromaffin cells. At widely scattered places oval or elliptical groupings resembling those in the zona glomerulosa of the mammal are encountered. The zona fasciculata in sections vertical to the glandular surface presents a radial pattern of parallel columns as in mammals; however, it is relatively much narrower in birds and is sometimes absent. Anastomoses and branchings of the columns are infrequent peripherally but more numerous centrally where the zona fasciculata passes without a sharp line of demarcation into the zona reticularis. Peripherally a zona glomerulosa is usually well defined. It is composed of circular, ovoid or arch-shaped masses of variable size; the larger ones appear to be peripherally expanded ends, or the bent reflected portions, of the fasciculate columns; the smaller appear to be knob-like projections of the larger ones. These formations when sectioned transversely show a radial cellular arrangement resembling the tubules of an exocrine gland.

The cytology of the pelican cortical tissue is very similar to that of other birds. In most cells, in the glomerulosa and outer fasciculata the cell body between the nucleus and the free border is packed with droplets of uniform size, but there is less lipid in the zona reticularis cells. The deeply staining fuchsinophil cells of Uotila (1939a), with pycnotic nuclei, few lipid droplets and clumped mitochondria, are to be found in the zonae glomerulosa and fasciculata but are most frequent in the zona reticularis. The basophilia is removed by ribonuclease and, on these grounds, may be regarded as nucleo-protein. These cells are considered to be degenerating both by Uotila and Knouff and Hartman, and this seems a reasonable assumption.

TABLE 27. *Carneau pigeons (1·9 to 2·4 months old). Nestlings 0·9 month old. (Some results extracted from Miller and Riddle, 1942)*

Type	n	Preparation injected	For no. of days	Dose mg./day	Weight-paired adrenals (mg.)
Nestlings	11	—	—	—	39·7
Normal	88	—	—	—	30·2
Hypophysectomized	51	—	—	—	24·2
Normal	4	Corticotrophin 730	16	10	46·4
Hypophysectomized 22 days	4	Corticotrophin 799	10	5	41·1
Normal	3	DCA	10	3–pellet	21·3
Hypophysectomized	5	DCA	10	1–pellet	16·1
Hypophysectomized	4	Cortical extract	10	(0·1 c.c.)	21·4

Although several workers have given us some idea of the histological appearance of the bird adrenal, data on the function which the cortical cells perform are few. It is clear, however, that the avian adrenal cortex depends on the secretions of the anterior lobe of the pituitary for normal functioning. Hypophysectomy (total or adenohypophysectomy) is not difficult in many species. Miller and Riddle (1942) for the pigeon, and Benoit and Assenmacher (1953) for the duck, have given descriptions of the consequences of this operation and their findings may well apply to the Aves as a whole. Miller and Riddle found that the adrenals of the hypophysectomized white Carneau pigeon lost about 25 % of their weight some 10 days after operation, the cortical tissue then constituting about 53 % of the gland, and occupying a less proportion *pari passu* with time (table 27). The cords of the cortex become reduced after operation. The cells of the central portion of the gland generally are smaller, have smaller nuclei, a more compact chromatin network, atrophic Golgi apparatus and less cytoplasm, and stain more heavily with fat stains than those in the normal adrenal. The cords at the periphery are often less affected though sometimes they too change in the same way as the central ones. In a second and more frequent type, the peripheral cells remain of normal size, and there is a Golgi apparatus of the usual appearance, but with numerous mitochondria and less lipid. The atrophic changes in the cortex of the hypophysectomized duck (Rouen and Pekin races) are similar to those of the operated pigeon (Pl. VII, fig. 36). For example, normal Rouen ducks which weigh about 1½–2 kg. have a paired adrenal weight of the order of 115 mg./100 g. body weight, but 53 days after hypophysectomy adrenal weights of 67 mg. % are characteristic. Histologically the adrenal cortex of the hypophysectomized duck, 24 to 53 days after operation, has a shrunken appearance. Stained in a Masson's modification (Benoit and Staubli, 1951) the cells of the loops of cords to the periphery of the gland are more fuchsinophilic than those placed centrally and are also more so than those of the normal animal. This fuchsinophilia marks off a peripheral zone from the more central region. Cortical cells have vacuolated cytoplasm after routine treatment (lipid droplets) and pycnosis of the nuclei is widespread though with a tendency to be more obvious centrally. Not all cells show these signs of degeneration; there remain some with a normal appearance here and there in the cords.

Both pigeons (Schooley, 1939) and ducks survive well after hypophysectomy while, on the other hand, the fowl is not so resistant (Mitchell, 1929; Hill and Parkes, 1934). No research has been done into the metabolic state of the hypophysectomized bird. In the cases where the survival is good, we do not know if the salt-electrolyte metabolism is near normal, as it frequently is in many operated eutherians, and whether some cells of the atrophic cortex continue to secrete despite lack of an adrenotrophin. Certainly there is some histological evidence that the peripheral zone is less affected by hypophysectomy than is the central region of the gland, and it is attractive to regard the outer zone as akin to the zona glomerulosa of the mammalian adrenal, where the zone can be persistent after hypophysectomy (see ch. VIII).

The technique of hypophysectomy and the use of crude preparations of beef adrenotrophin (Bates, Riddle and Miller, 1940) have been employed to demonstrate the existence of a relationship between the anterior lobe of the pituitary and the adrenal cortex (Miller and Riddle, 1942). By the injection of suitable doses of this particular mammalian adenohypophysial extract, the adrenal weight of normal birds can be increased by 30 %, the weight of gland maintained or increased in hypophysectomized animals and post-hypophysectomy atrophic adrenals can be restored to normal weights (table 27). Increase in weight is accompanied by cell enlargement, some increase in the number of mitoses and vascularity of the gland. Stimulated cortical cells are characterized by a large increase in the number of mitochondria, in the amount of basophilic cytoplasm and enlargement of the Golgi apparatus. They resemble, in this, the peripheral cells of the adrenal of the hypophysectomized pigeon. This stimulation by an adrenotrophin of foreign origin indicates that, probably, a similar factor could be obtained by fractionation of bird adenohypophyses. Certainly unilateral adrenalectomy allows the increase in size of the contra-lateral adrenal as is the case of those animals known to be dependent on an ACTH for normal adrenal functioning. Moreover, disease in birds is often accompanied by adrenal hypertrophy as it is in mammals (Riddle, 1923). Young chick adrenals are not so responsive to noxious substances, nor will they show depletion of ascorbic acid on stimulation with fairly pure preparations of ACTH or with adrenaline (Jailer and Boas, 1950). This is probably to be referred to the age of the animals rather than to be taken as indicative of basically different mechanisms of pituitary-adrenal relationships in the bird compared with those of mammals. The biological action and biochemical nature of avian ACTH is unknown, however, and the chance that the biochemists will present us with varying preparations of differing biological action, as they have in the mammal, foreshadows an intriguing complication of the future.

No full study of the metabolic consequences of adrenal-

Aves

ectomy has yet been made in the bird. There is no doubt that the gland is necessary for the continued survival of the bird and death after operation is generally rapid. Adrenalectomy is, however, a difficult operation in the bird, the glands being placed, as we have seen, close to main vessels and to the kidneys and lungs. Gourfein (1896) removed the adrenals of the pigeon and found death supervened from 5 to 24 hr. later. Parkins (1931) briefly noted that the survival period of 18 fowls adrenalectomized in two stages, with an interval of from 6 to 8 days between operations, ranged from 38 to 146 hr. Following Parkes and Selye's (1936) observation that, in their hands, adrenalectomy of the duck and the fowl is rapidly fatal in 6 to 20 hr., Bülbring (1937a, b, 1940) decided to use the operated drake as a test for adrenocortical extract potency. She found that the survival time varied from 5·25 to 15·2 hr. This survival time, interestingly enough, depended on the season of the year, being slightly longer in July than in November, January or March (text-fig. 29). Herrick and Torstveit (1938) found that adrenalectomy, performed in a two-stage operation using male single-comb White Leghorns, mostly five to six months old, caused death in 6 to 15 hr.

This agreement about the short survival time of the untreated adrenalectomized bird does not accord with Leroy and Benoit's (1954) recent investigation of the effects of adrenalectomy in drakes (Pekin, Khaki Campbell and Rouen breeds). The Rouen birds received sustaining therapy. Eight Pekin birds adrenalectomized at fourteen months of age were killed three weeks after operation when six had cortical remnants weighing 15 to 38 mg., while in two no cortical tissue was found. Seven adrenalectomized Khaki Campbells were exposed to light for a month and then killed when it was found that five had cortical remnants weighing 15 to 25 mg., and two were completely operated. Thus these authors had four birds which survived three weeks to a month after adrenalectomy. Bülbring pointed out that when an untreated adrenalectomized bird lasted as long as 20 to 30 hr., a remnant of cortical tissue was found and she notes (1940): 'The removal of every trace of gland is not always easy and a piece no longer than a pinhead can prolong the survival time considerably.'

The Adrenal Cortex

It appears that both surgical problems and possible cortical rests combine to make the study of the adrenalectomized bird difficult. Many of the same stages were gone through in establishing the results of adrenalectomy in other laboratory animals, so that more work and more workers will establish a common basis for discussion of the endocrinological problems of the bird as they have for those of the mammal. With present data we are able to say that the untreated adrenalectomized bird dies a short time after operation, the animal appearing relatively normal until the terminal symptoms of rapid respiratory rate, muscular weakness, and convulsions appear.

The adrenalectomized bird is similar to the adrenalectomized eutherian, particularly the rat, in that it is possible to maintain it for some time by giving saline. Though Bülbring (1937 a, b) did not find this true for drakes, Herrick and Torstveit (1938) found that adrenalectomized Leghorns maintained on saline (with cortical extract injections discontinued 2 to 3 days after operation) lived from a few days to as long as 82 days. Riddle, Smith and Miller (1944) fed adrenalectomized Carneau pigeons with gelatin capsules containing 0·75 g. NaCl and 0·125 g. NaHCO$_3$ twice daily, the average survival time of nearly 150 adrenalectomized birds with the treatment being 9 days. The pigeons lost weight slowly at first and rapidly during a day or two before death. Daily food intake in the ten-day period after operation was reduced to about one-fifth that of normal unoperated pigeons. Water intake was apparently unaffected by adrenalectomy but it was increased to about twice normal value when sodium salts were given (Miller and Riddle, 1943; Riddle et al. 1944), a similar pattern to that of the adrenalectomized rat (Chester Jones and Spalding, 1954). Sodium chloride goes some way in helping the adrenalectomized bird, perhaps by tending to preserve a more normal distribution of salt-electrolytes.

It seems that corticosterone is present in fowl adrenal venous blood (Phillips, unpublished) and experiments using mammalian adrenocortical extracts and DCA indicate that the secretions of the avian adrenal cortex are similar to those of the mammal. Miller and Riddle (1942) have shown that the injection of DCA into both normal and hypophysectomized pigeons produces adrenal atrophy, in the latter case the adrenals being smaller than in the hypophysectomized controls (table 27). Injection of adrenocortical extract has a similar effect on the adrenal of the hypophysectomized bird. The atrophic adrenal produced by these two types of injection is characterized by cortical cells with few mitochondria, atrophic Golgi apparatus and dense staining with fat colorants. Adrenalectomized birds can be maintained with some success on steroids of mammalian origin or DCA. Adrenalectomized pigeons can be restored to a normal growth rate and food intake by the injection of 2 mg. DCA daily, regression occurring on stopping the injections (Miller and Riddle, 1943). Bülbring (1940) kept adrenalectomized drakes alive with 5·0 mg. of DCA a day for as long as the injections were continued. Injection of adrenocortical extract can also be fairly successful in the same way and no doubt the more potent preparations available today would be even more effective. Only Leroy and Benoit (1954), however, have used cortisone with adrenalectomized birds, supporting operated Rouen drakes for three weeks with 50 mg. per day; unfortunately no metabolic data were given. The less potent adrenocortical extract available to Bülbring (1937 *b*, 1940) only prolonged the life of the adrenalectomized drake to 23·5 hr., but she brought to light an interesting seasonal variation in the amount of extract required to prolong the life of the operated bird to 16 hr. (chosen for convenience as a constant period). This variation is shown in text-fig. 29 and indicates that there is a correlation between testes size and sensitivity to adrenalectomy. The differences are not due to testosterone as such, but probably large testes mean increased overall metabolism and consequently greater sensitivity to the absence of adrenocortical secretions, particularly in April and May, and this is the same type of phenomenon which we saw in the frog. Adrenalectomized, castrated drakes, too, require 2·33 times less maintenance dose of cortical extract than do adrenalectomized drakes in April, treated in the same way.

The whole question of the inter-relationship of the secretions of the gonads and of the adrenal cortex is of great interest. In the mammals, where numerous experiments have been done to elucidate this relationship, the position is still obscure (Chester Jones, 1955) and this is equally true of birds. Many of the effects are due to overall metabolic changes rather than precise interaction between the two sets of secretions. Thus, salt-maintained adrenalectomized fowls lose comb colour and size so that the birds have the appearance of typical capons and at the same time the testes become smaller (Herrick and Torst-

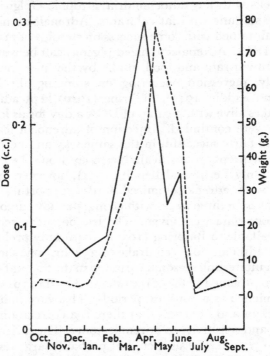

Text-fig. 29. Parallelism between the change in the weight of testes (dotted line) and the hourly dose of cortical extract (continuous line) required to keep adrenalectomized drakes alive for an average period of 16 hr. at different times of the year (from Bülbring, 1940).

veit, 1938), often a 50 % loss in weight occurring (Hewitt, 1947). The number of degenerating basophils in the anterior lobe of the pituitary increased in this type of adrenalectomized fowl, possibly representing an enhanced turnover of cells correlated with a higher rate of adrenotrophin secretion. Castration cells did not appear, however, despite testicular atrophy (Herrick and Finerty, 1940). It is clear that, in the absence of

Aves

adrenocortical steroids, normal metabolism of the bird, or any animal, is not achieved. Hewitt (1947) shows this with his experiments on White Leghorn hens. In these, removal of the one fully developed ovary allows the ovotestis to develop but the double operation of ovariectomy and bilateral adrenalectomy prevents the hypertrophy of this rudiment. Presumably the ovotestis in the presence of adrenocortical hormones responds to pituitary gonadotrophins, while in their absence this 'permissive' action is lacking despite the continued existence of the relationship between gonadotrophins and ovotestis. Naturally, such 'permissive' action is one facet of an overall metabolic effect of the adrenal steroids, and this is shown by the results of Leroy and Benoit's (1954) experiments where testis size was variably affected in totally and sub-totally adrenalectomized drakes.

TABLE 28. *The weights of the adrenal glands of male, female and castrated pure-bred Brown Leghorns. The males and females were all hatched on the same day and killed, in March to June, 8 to 11 months later; the capons were all castrated between 6 and 12 weeks of age and killed after a similar period. (Data re-calculated from Juhn and Mitchell, 1929)*

| | n | Body weight | Adrenal weight | |
			mg.	mg./100 g. body weight
Females	13	841·46±26·31	153·7±19·0	17·80±1·64
Males	16	1146·6 ±12·5	203·7±12·1	18·22±1·42
Castrates	9	1024·0 ±66·83	158·5±14·0	15·7 ±1·63

Removal of the testes themselves generally produces changes in the bird adrenal cortex. Long-term castration of male Brown Leghorns (that is, for 8 to 11 months after the operation, performed between 6 and 12 weeks of age) caused a fall in adrenal weight (Juhn and Mitchell, 1929) (table 28). On the other hand, young cockerels of the single-comb White Leghorn strain, castrated at 44 days of age and killed when 86 days old, showed a clear increase in adrenal weight (Kar, 1947b). Histological examination of the adrenals showed that there was cortical hypertrophy, this tissue now occupying 70 % of the entire gland, a proportion characteristic of coeval females, while in male controls there was about 40 % cortex in the gland. The cortex of the castrate adrenal, moreover, showed enlargement of the cells, increase in fuchsinophilia, in the number of mitochondria and in the amount of lipid. The indications are that the castrate adrenal is an active one, though with the reduced capon comb size (table 29) there is no suggestion of adrenal androgens being secreted. Breneman (1941) castrated male chicks at 4 days of age and found, at 30 days of age, a marked hypertrophy of the adrenal (17·7 mg./ 100 g. body weight) compared with coeval males (12·7 mg. %).

TABLE 29. *The effects of castration and of the injection of testosterone propionate or of diethylstilboestrol on male single-comb White Leghorns castrated at 44 days of age, injections started 8 days thereafter and continued for 33 days. All birds killed at 86 days of age (from Kar, 1947 b)*

	n	Body weight (g.)		Adrenal weight (mg.)		Comb size (l+h)
		Initial	Autopsy	Absolute	(mg. %)	
Normal	13	403·7± 8·8	1038·07±12·3	86·1±0·1	8·2	14·1±5·6
Castrates	11	387·5±11·5	1010·0 ±13·7	114·8±0·4	11·2	7·0±0·04
Castrates and testosterone propionate	10	436·6± 5·8	1050·4 ±11·0	71·3±0·2	6·7	13·2±1·4
Castrates and stilboestrol	12	423·0±12·1	980·5 ±14·3	83·04±1·1	8·4	6·7±0·07

l+h : length+height in cm.

I can find no reports on the influence of ovariectomy on adrenal weight or histological appearance. There is, however, work on the influence of the injection of gonadal steroids on the avian adrenal. In old birds this influence appears to be slight. Kar (1947 a), using Brown Leghorn hens, 22 months old, found no change in adrenal weight with injections of diethylstilboestrol or of testosterone propionate. Histologically there was an impression of large lipid droplets in cortical cells of the oestrogen-treated birds, and minute ones in the androgen-treated ones. Neither did Munro and Kosin (1940), who injected six-day-old male and female chicks with oestrone, oestradiol, oestradiol benzoate and oestradiol propionate for 10 days, obtain variations in adrenal weights which could be differentiated statistically. On the other hand, Miller and

PLATE VII

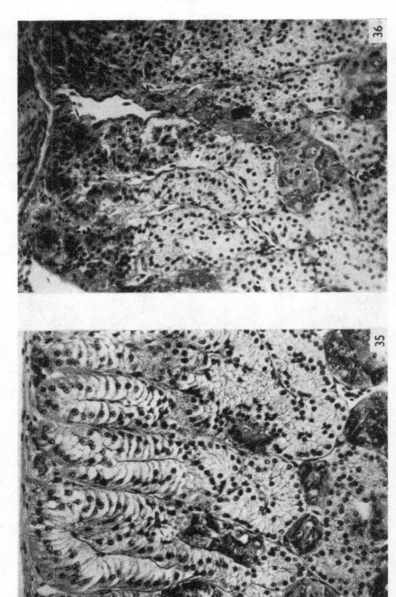

Aves

Riddle (1939 *a*, *b*, 1941, 1942) found that large doses of oestrogen administered to normal pigeons increased the weight of the adrenals above that of the controls. Breneman's data (1941) seem to show (with standard errors not given) that oestrogen and diethylstilboestrol cause an increase of adrenal weight of 30-day-old cockerels when injected from day 3 to day 29; the effect was most clearly shown with 0·200 mg. of stilboestrol daily. It is clear from the data of Kar (1947 *b*), moreover, that the adrenal hypertrophy of young cockerels after castration can be prevented both by stilboestrol and by testosterone propionate (table 29).

It seems that some birds follow the pattern of most mammals, in that castration is followed by adrenocortical enlargement which can be reduced by androgens. Oestrogens, too, can have a depressant effect on the bird cortex, as they can on the mammalian gland. Possibly oestrogen-induced adrenal hypertrophy may be seen in birds as it is sometimes in mammals, as Riddle (1923) found that increased adrenal weight was coincident with ovulation in doves and pigeons. Reinvestigation of these results to establish whether cortical hypertrophy is involved and to redefine the required conditions would be very well worth while.

EXPLANATION OF PLATE

PLATE VII

Fig. 35. Part of the adrenal gland of a normal duck (× 300; Bouin, Masson's). The outer connective tissue capsule is to the top of the photograph. The cords of cells loop peripherally. To the centre of the gland the cords are more compact. Islets of chromaffin tissue, distinguished by their basophilic cytoplasm, are scattered among the cortical cords.

Fig. 36. As fig. 35. Part of the adrenal gland of a duck hypophysectomized for 27 days. General collapse of the cortical cords, the peripheral loops being particularly affected.

CHAPTER VI

PROTOTHERIA AND METATHERIA

(I) PROTOTHERIA

Our knowledge of the adrenal glands of the Monotremata is confined to some observations on their histology. In the Australian echidna, *Tachyglossus*, the adrenals lie on the medial aspect of the cranial part of each kidney (McKenzie and Owen, 1919), and these authors give measurements for the gland in the adult of 0·75 cm. in length by 0·5 cm. in width, at the widest parts. In the platypus, *Ornithorhynchus*, the adrenals lie by the kidneys (Owen, 1847). The left adrenal, in the two specimens examined by Pettit (1894, 1896), lay obliquely to the long axis of the body and with the cranial end curved towards the kidney, and the right adrenal formed a sort of bonnet over the inner anterior portion of the kidney (Bourne, 1949). Pettit found the adrenals of one specimen to measure (the left gland) 1·0 cm. in length by 0·45 cm. in width, (the right) 0·7 cm. by 0·6 cm.; and of the other specimen, 2 cm. by 0·9 cm. and 1·5 cm. by 0·9 cm. The arterial blood supply came from a branch of the renal artery on the left side, while on the right the gland was served by a branch of the aorta in the region of the coeliac trunk. The venous return in both glands was by two veins joining the renal veins (Pettit, *loc. cit.*).

Elliott and Tuckett (1906) were the first authors to describe the micro-anatomy of both the echidna and platypus adrenals, and their descriptions, although from old alcohol-preserved material, are particularly good. Kolmer (1918) briefly described the gland in echidna and Kohno (1925) the adrenals of both echidna and proechidna (*Zaglossus* of New Guinea) which were females over ten years old. Basir (1932) described formalin-fixed echidna adrenals which had been preserved for some time in alcohol, while Bourne (1949) examined the adrenals in twenty echidna, about half the glands being freshly fixed in formalin, and the glands of thirteen platypus, ten of which were freshly preserved.

Prototheria

I have had the advantage, by Professor Waring's kind co-operation, of examining histologically the adrenal glands of echidna and platypus freshly fixed in Bouin's fluid and in formalin (Chester Jones and Wright, unpublished). In platypus, median sagittal sections were about 1·8 cm. in length by 0·8 cm. in depth; transversely the gland presented the appearance of an isosceles triangle with a basal concavity which is against the kidney. The whole gland is surrounded by a thick outer connective tissue capsule. The parenchyma is drained by numerous small sinuses which lead to the venous drainage through the chromaffin tissue. The latter may either occupy a small part of the venous or lower pole, or take up most of the gland when it thrusts through the cortex, tapering out about the mid-portion (text-fig. 30). The chromaffin tissue is composed of short cords of cells, surrounded by thin collagen fibres; the granular cell cytoplasm is varyingly chromophilic; the nucleus is large and centrally placed in the cell. The chromaffin tissue interdigitates freely with the cortex but is separated from it by a marked development of connective tissue (Pl. VIII, fig. 37).

The cortical tissue is clearly divided into three main groups of cells. Group I follows closely the contours of the chromaffin tissue, varying in extent of development in a parallel way. The cells have a faintly acidophilic cytoplasm with yellow-brown granulation and prominent basophilic nuclei which, when oval in shape, are about 9μ by 5μ, but also appear kidney-shaped or constricted in the middle (Pl. VIII, fig. 37). Group I tissue as it projects a short way under the outer connective tissue capsule is seen to be composed of cords, one to two cells in width, surrounded by a few collagen fibres. In the majority of the group, however, as it runs with the chromaffin tissue, the cord-like arrangement is rarely apparent. The major part of the cortex is occupied by two other groups of cells and appears, in sections after routine methods, as alternate light and dark areas. Group II, the more lightly staining, is composed of narrow cords of cells arranged in areas each set about a small blood sinus. Group II cells also run directly under the outer capsule except for that small portion taken up by group I. Group II cells have scanty granular cytoplasm and this, together with the basophilic nuclei, gives a general closely packed

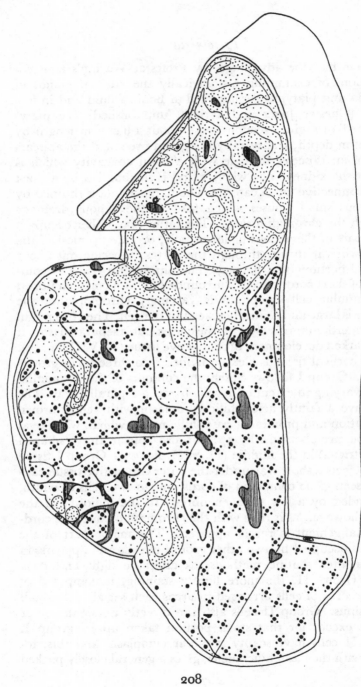

Text-fig. 30. A reconstruction from serial sections of the adrenal gland of the platypus. ✕ chromaffin tissue; ⣿⣿ group I tissue; ▦▦ group II tissue; ∴∴∴ group III tissue; veins and sinuses cross-hatched.

appearance. The nuclei vary in shape very much as do those of group I cells, and are of about the same size (Pl. VIII, fig. 38). Where group II meets group I peripherally, the composing cords of cells run at right angles to each other and, wherever they meet, the boundary is marked by a slightly stronger development of collagen fibres and the presence of much smaller cells with very sparse cytoplasm and deeply basophilic nuclei (fig. 37). Group III forms the deeply staining areas. The cells comprising it are directly continuous with group II and there is a gradation from the one to the other (fig. 38). The cells themselves are larger than those of the other two types, with abundant, acidophilic cytoplasm which may be granular, finely reticulate or vacuolated. Occasionally dense basophilic granulation is seen.

The appearance and size of the nucleus vary considerably (Pl. IX, figs. 42 and 43). Most of the nuclei have a distinctive form in that the nucleoli are clumped together so that after routine methods there is a clear space in the nucleoplasm (Pls. VIII and IX, figs. 38 and 42). The nuclei are sometimes small (c. 6 μ), round or oval bodies, and this type merges to those which are larger and then to a third type which may be very large indeed (up to c. 15 μ), the outline of which shows a varying amount of crenation. Vacuolated cytoplasm most frequently accompanies this third type of nucleus. In addition, some nuclei possess a distinct globule which occasionally occupies almost the whole of the nucleus (Pl. IX, fig. 43). Groups I and II tissues show a faint background sudanophilia; group III possesses cells which are markedly sudanophilic (Pl. VIII, fig. 39) and these are positive for the Schultz cholesterol method.

The adrenal in echidna has a bulbous lower pole and tapers anteriorly; the undersurface of the gland is ridged (text-fig. 31). The gland is surrounded by a well-developed outer connective tissue capsule. The chromaffin tissue occupying a third to a half of the whole gland is confined to the lower, bulbous pole, and is separated from the cortex by a marked development of connective tissue, as in the platypus. A large mass of nervous tissue lies peripherally posterior to the chromaffin tissue to which run nerve fibres. Blood drains from the whole gland by an interconnected system of sinuses which lead

The Adrenal Cortex

into two large veins, one of which passes from the capsule on the dorso-lateral part of the gland through the cortex to join the other smaller one; the large vein so formed passes through the chromaffin mass to leave the gland at the lower pole. The chromaffin tissue is similar to that of platypus and to that of the vertebrates as a whole.

The cortical tissue in echidna is not differentiated into groups as in platypus, but has a generally homogeneous appearance;

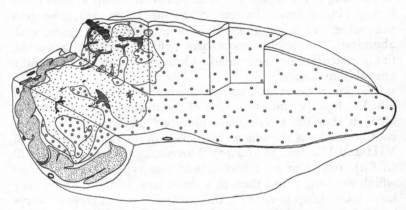

Text-fig. 31. A reconstruction from serial sections of the adrenal gland of *Echidna*. ∴ chromaffin tissue; o o o o cortical tissue; nervous tissue; veins and sinuses, cross-hatched.

indeed, looking rather 'reptilian', apart from the absence of chromaffin islets. The cortex is made up of closely packed, irregularly placed cords often separated by blood sinuses. Two types of tissue can be distinguished. Firstly, a relatively wide peripheral area consisting of irregular cords closely packed together but separated by blood sinuses. The cell cytoplasm is faintly acidophilic, granular or finely reticulate. The nucleus may be oval (*c*. 10μ by 5μ) or round (*c*. 5μ to 8μ) and contains several nucleoli in a diffuse chromatin background (Pl. VIII, fig. 40). Towards the centre of the gland the cords change gradually to form a second type of tissue. Here the cytoplasm is scanty and the cells packed closely together. The nucleus, which occupies most of the cell, is similar to the peripheral type

Prototheria

(Pl. VIII, fig. 41). Sudanophilic and Schultz-positive lipid is found irregularly distributed throughout the cortex.

It is interesting to find these differences in the histology of the adrenal in the two major species of the Prototheria. Moreover, while the appearance of the echidna cortex is rather reptilian, that of the platypus, and especially the group III cells, recalls the adrenal of some elasmobranchs and amphibians. Dittus (1936, 1941) has shown nuclear stages in *Torpedo* and *Ichthyophis* adrenal cortices, which correspond very closely to those stages seen in the platypus in the variation in nuclear size up to large dimensions and the arrangement of the chromatin, though his stage of 'dehiscence' from the nucleus has not been observed. Furthermore, Fowler (unpublished) has seen the very same stages (except for a 'dehiscent' one) in the adrenal cortex of the summer frog markedly stimulated with ACTH. Also, the occurrence of globules within the nucleus of cortical cells has been seen by Schiller (1944) in several eutherian species, including man, and in *Didelphis*.

These variations in appearance of the nucleus must be bound up in some unknown way with variations in cell secretory activity (Bachmann, 1954). Certainly it would be reasonable to suppose, on the basis of nuclear form and sudanophilic, Schultz-positive lipid, that it is group III cells which actually secrete the adrenocortical hormones. Following this suggestion, then, group II could be regarded as a formative area. Group I, on one hypothesis (ch. VIII), could be regarded as an inactive zone left behind as the group II formative zone thrust out in prongs, associated with blood vessels, to weave the intricate pattern of the platypus adrenal cortex.

(II) METATHERIA

The histology of the adrenal glands in marsupials is well known, largely due to the work of Bourne (1949). Little experimental work has been done in this group, however, so that data are not available for the correlation of changes in adrenal micro-anatomy with age, sex and function.

Bourne found that the relationship of the adrenals to the kidney varies considerably in the different families. In general

The Adrenal Cortex

there are three main types: (i) both glands are situated on the antero-mesial border of the corresponding kidneys to which they are intimately attached, for example, *Setonyx brachyurus*, the short-tailed wallaby, and kangaroos generally; (ii) the right adrenal is adpressed to the vena cava into which opens the venous drainage from the medulla, while the left gland is variably placed in relation to the kidney, for example, species of *Dasyurus*, the 'native' and 'tiger' cats; (iii) the right adrenal is against or actually under the liver capsule while, as in *Trichosurus*, the 'Australian opossum', the left adrenal is firmly attached to the antero-mesial border of the left kidney.

Bourne found that in the nine species of the genus *Phascogale* (*Dasycercus*; Canning's 'little dog') that he examined, all the possible variations of right adrenal position were exhibited. No precise details of the blood supply to the adrenal are available, but it would seem that the adrenal vein, when present, enters the renal vein or the vena cava.

Information on the adrenal weights of marsupials is confined to that of Hartman and Brownell (1949) who found, for three female American opossums, *Didelphis virginiana*, the adrenals were 0·0237, 0·0251 and 0·0171 % of the body weight. These fit in with the range of data for eutherians: in the adult female albino rat, for example, the adrenals are 0·026 % of the body weight.

The most important feature of the marsupial adrenal is that the chromaffin tissue is aggregated centrally to form a medulla of the eutherian type. My own experience of marsupial adrenal histology is based on these glands in the family Macropodidae, sub-family Macropodinae, the kangaroos and wallabies, namely *Setonyx brachyurus*, *Macropus ruficollis*, *M. robustus*, *Protemnodon eugenii derbiana* (sometime *M. eugenii* and *Thylgale eugenii*) and *Lasiorhinus latifrons* (specimens kindly supplied by Professor Waring and Dr Jarrett). In these, a well-marked cortex surrounded the medulla and could easily be differentiated into the three classical layers, the zona glomerulosa, zona fasciculata and the zona reticularis (Pl. IX, fig. 44). In this I disagree with Bourne (1949) who found an indefiniteness of zoning, a 'true characteristic of the adrenal of Macropodinae', and therefore confirm Kolmer (1918) in his statement that the adrenals of the marsupials resemble those of the Eutheria in that the three cortical layers are clearly developed. Bourne, however, did find the eutherian type of cortical zoning in many examples from different families. Thus, to consider but a few, in *Myrmecobius*, the wombat, the adrenal cortex possesses well-defined zonae glomerulosa and fasciculata with the zona reticularis little developed; and this is the case, also, in *Phascolarctos* and *Pseudocheirus*; in *Perameles* and some Phalangeridae all three zones are distinct and in the *Dasyurus* adrenal, although the cortex is narrow, the zones are well marked. On the other hand, Bourne found in the cortex of *Notoryctes* that signs of fasciculation could only be seen with difficulty; in the phalangerids, *Dactylopsila* and *Trichosurus*, while the zona glomerulosa and zona reticularis could hardly be made out, the zona fasciculata was arranged in clearly defined fascicles; in *Phascogales* all three zones are imperfectly differentiated.

In the adrenals I examined, there was certainly variation in the relative proportions of the zones: *Protemnodon* had an obvious zona glomerulosa, a narrow zona fasciculata and a wide zona reticularis. On the other hand, the cortical fasciculation in *Macropus robustus* was most striking, while the adrenal cortex of *Setonyx* was altogether on the eutherian pattern. In formalin fixed material, sudanophilic lipid droplets could be observed in the cortex similar to those in the glands of many Eutheria (Pl. IX, fig. 45). There was variation in the distribution of sudanophilia: in *Protemnodon* the sudanophilia was, for the most part, confined, as coarse lipid droplets, to the zona glomerulosa. In the *Macropus* species and in *Setonyx*, both the two outer zones showed fine sudanophilic droplets in the cells, with sparse larger ones in the zona reticularis. Cholesterol, as shown by the Schultz test, was present in the adrenal cortex of these animals.

It would appear from my data and the very extensive observations of Bourne that in the marsupials we see a similar wide variation of cortical appearance as has been found among eutherian glands. There does not seem to be any real reason why the marsupial adrenal should not be considered as being on the same general plan and with the same overall reactions as have been found in the eutherian gland.

The Adrenal Cortex

It is possible that the cortex of the Australian opossum provides an exception to this generalization, as it shows unique features. Bourne (1934b, 1949) found that the right adrenal is usually much reduced compared with the left. In the male the cortex is narrow and surrounds a big medulla, giving a picture somewhat similar to that of the adult male mouse (cf. Bourne, 1949, Pl. 6, fig. 25; Chester Jones, 1948, Pl. I, fig. 1). In the female, the left gland shows clearly, and the right imperfectly, a complicated growth and subsidence of special zones during pregnancy and lactation. The cortex surrounds the medulla in the usual way and is divided by Bourne into an outer lipid zone, the α zone, and an inner, the β zone which contains little or no lipid. Bourne, therefore, prefers to rely on a lipid differentiation. There are, however, cortical cells arranged in columns (Bourne, 1949) which seem to be co-extensive with the β zone, and there seems no reason why the normal terminology of zona fasciculata should not be used. The zone lying peripherally to the zona fasciculata, Bourne's α zone, does not have the characteristics of a true zona glomerulosa which is therefore declared absent: however, a zona reticularis is present (Bourne, 1934b).

Looking, therefore, at the cortex of the Australian opossum in the light of the cortical morphology of Metatheria and Eutheria in general, the conclusion that it conforms to the same general plan seems to be justified. Where the opossum cortex is unique, in the female, is in the possession of a group of cells present on one side of the gland only and which, at certain times, may encroach on the medulla to occupy, in some adrenals, the major part of the gland. Bourne suggests that this extra mass of cells arises from multiplication of β zone cells (i.e. zona fasciculata) which are small and compact with relatively large, deeply staining nuclei and scanty, rather neutrophilic cytoplasm (Bourne, 1934b). In some adrenals these β cells appear to have metamorphosed into a larger type of cell with an eosinophilic cytoplasm, called δ cells. When the latter constitute the major part of the hypertrophied zone, Bourne refers to it as the δ zone. In the virgin female, the hypertrophied bulge of the cortex is small; in the mature nonpregnant female, it may be larger and some δ cells appear. At the onset of pregnancy the β zone increases slightly and the δ zone greatly in size; at mid-term the latter is reduced. Immediately after parturition, there is a rapid metamorphosis of the β cells into δ cells with a multiplication of the latter leading to an increase of the hypertrophied area. During the early suckling period, there is apparently a rapid destruction of δ cells and replacement by β cells; thence, as suckling continues, δ cells become dominant again, only to be replaced later on by β cells. Finally, at the end of the suckling period, both β and δ cells are present. Bourne points out that the size of this extra mass of cells, when expressed as a percentage of the whole section of the gland, remained constant once maturity was reached, and the phenomenon was confined, in reality, to an interplay between β and δ cells.

We have no direct evidence of the nature of this extra mass, but because it occurs in the female and shows variation with pregnancy and lactation, some sort of cortical reaction to differing titres of circulating gonadotrophins could be suspected. This recalls the variation of the X zone in the mouse. Here the extra zone disappears during first pregnancy, due to the formation of ovarian androgens (Chester Jones, 1952). If we suppose that ovarian androgens are not formed in the ovary of the opossum, then any equivalent of the X zone would be free to respond to pituitary gonadotrophins and to reflect variations in their secretion during the episodes of pregnancy and lactation similar to the adrenal variations in the cat, described by Lobban (1952). Moreover, in the post-pubertally castrated male mouse, the origin of an extra mass of cortical cells—the secondary X zone—can be seen to originate from the zona fasciculata.

It is perhaps on lines like these that further work will reveal the cortex of the female opossum as being but a variation on the general theme of mammalian adrenal pattern. Whether or not the marsupial adrenal has a true X zone cannot be decided on the data available. It is interesting to note, however, that a marked medullary connective tissue capsule is described in the adrenals of many marsupials. Possibly the generalization for the Eutheria, that those adrenals which show a medullary connective tissue capsule had, at one time in their life-history,

The Adrenal Cortex

a primary X zone (Chester Jones, 1949 b), might be extended to include the Metatheria.

Little is known about the physiology of the marsupial, though apparently its carbohydrate metabolism has a similarity with that of the eutherian ruminant. Recently Buttle, Kirk and Waring (1952) have investigated the effect of adrenalectomy on the Rottnest wallaby, *Setonyx brachyurus*. The animals survived only an average of 36 hr. after removal of the second adrenal in a two-stage operation. The symptoms, however, were characteristic of those obtaining in adrenalectomized Eutheria: increasing anorexia, muscular weakness and eventual collapse. In addition, there was a decline in the level of plasma sodium and an increase in that of potassium, for example, their animal number fifteen showed:

	Hours after operation					
	0	5	10	15	20	60
Na (mEq./l.)	137·5	138·1	135·5	124·2	127·2	122·7
K	5·04	—	5·57	5·39	6·03	8·53

It is probable that, as the authors suggest, more experience of the special requirements of marsupials and a longer interval between the removal of the first and second adrenal might well increase the post-operative survival time. On the other hand, the American opossum has distinctive features: principally that the adrenalectomized animal is able to maintain normal levels of serum sodium, and data taken from Britton and Silvette (1937 a) and from Hartman, Smith and Lewis (1943) are given in table 30. It seems as if the opossum is less dependent than other mammals on the adrenal for the control of salt-electrolyte metabolism. The adrenalectomized opossum, however, does show many of the features associated with adrenal insufficiency, for it dies with the characteristic signs (Britton and Silvette, 1937) and with decreased urine flow (Silvette and Britton, 1938 a); it reacts to slightly hypertonic saline and hypotonic saline as does the adrenalectomized rat; both adrenal extract and DCA administration can cause sodium retention (Silvette and Britton, 1938 a; Smith *et al.* 1943) and in long-surviving adrenalectomized animals, there is a tendency for the potassium to rise (table 30).

TABLE 30. *The effects of adrenalectomy, principally on the concentration of serum sodium and potassium in the American opossum,* Didelphys virginiana

	n	Serum (mEq./l.) Na	K	Muscle (mEq./kg.) Na	Water (%)	
Controls	18	146	—	29·6	77·6	⎫ From Britton
Adrenalectomized						
(a) No treatment	11	156·5	—	34·8	72·7	From Britton
(b) No food but water	5	123·5	—	21·7	77·7	⎬ and Silvette
(c) No food or water	3	148·1	—	—	—	(1937a)
(d) Lactating	3	110·5	—	—	—	
(e) With diarrhoea	9	118·3	—	31·7	76·6	⎭
Control values before adrenalectomy	5	141·80± 5·40	4·96± 0·30	—	—	⎫
After adrenalectomy	*	148·03± 3·14	6·47± 0·48	—	—	⎬ Recalculated from Hartman
(* Nineteen readings taken from animals 10 to 288 days after operation)						*et al.* (1943) ⎭

It was from experiments with the opossum as well as the rat that Silvette and Britton came to formulate their idea of adreno-cortical-posterior pituitary antagonism (p. 80 above). It would be possible, therefore, to suppose that the neurohypophysis is not secreting in the adrenalectomized opossum, and a fall in plasma sodium concentration then only occurs with some form of stress (e.g. lactation, see table 30). While the possibility of adrenal accessory tissue has not been finally ruled out, there is no doubt that the American opossum would repay further investigation.

EXPLANATION OF PLATES

PLATE VIII

Fig. 37. Part of the adrenal gland of platypus (× 300; Bouin, H. and E.). Chromaffin tissue at the top of the photograph, then strong connective tissue development against which lies group I tissue. To the bottom of the photograph lies group II tissue.

Fig. 38. As fig. 37 (× 300). To the left of the photograph group II tissue merging into group III tissue on the right. Note large nuclei. One nucleus contains a 'globule' shown in detail in fig. 43.

The Adrenal Cortex

Fig. 39. As fig. 37 (formalin, sudan black). Group II tissue slightly, and group III tissue heavily sudanophilic.

Fig. 40. Part of the adrenal gland of echidna (× 300; Bouin, H. and E.). To the left of the photograph is the outer connective tissue capsule against which lie loops of the outer cortical tissue.

Fig. 41. As fig. 40. Inner part of the gland. To the left, the inner compact cortex, to the right, chromaffin tissue separated by a connective tissue capsule from the cortex.

PLATE IX

Fig. 42. Cells of group III tissue of the adrenal cortex of platypus (× 1200; Bouin, H. and E.). This is one of the large nuclei of this tissue. Compare with Pl. I, figs. 3, 4 and 5.

Fig. 43. As fig. 42. A cell with nucleus containing a 'globule'.

Fig. 44. Part of the adrenal cortex of the female kangaroo (*Macropus*; Euro. type) (× 57; Bouin, H. and E.). Marked outer connective tissue capsule, then zona glomerulosa, zona fasciculata, zona reticularis, medullary connective tissue capsule, medulla.

Fig. 45. Part of the adrenal cortex of *Setonyx brachyurus* (a kangaroo) (× 78; formalin, sudan black). Zona glomerulosa and outer part of the zona fasciculata sudanophilic, remainder of the cortex sparsely stained.

PLATE VIII

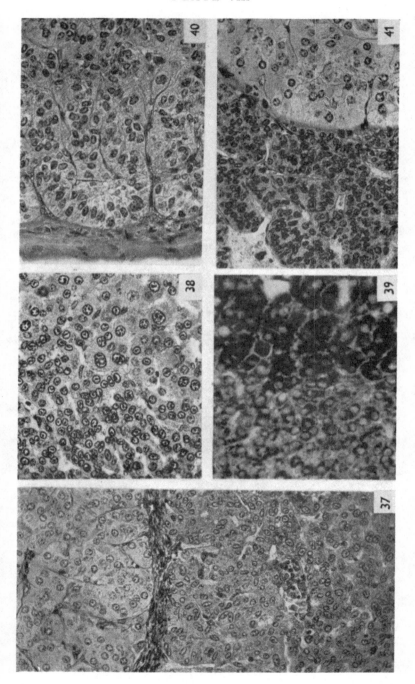

PLATE IX

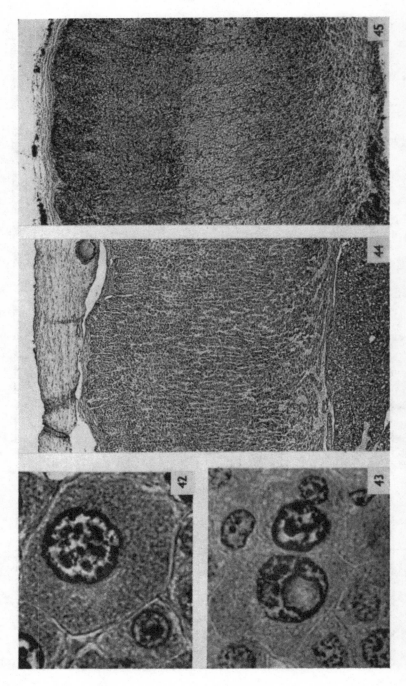

CHAPTER VII

EMBRYOLOGY

ADRENOCORTICAL tissue arises from mesoderm. It is generally considered to come from the columnar epithelial cells which, differentiated from mesoderm, line the coelom. There has been doubt about this in the past, and in recent years Witschi (1951, 1953) has assigned the origin of the cortex to the intermediate mesoderm (see below). Even if the cortical cells do not differentiate from the peritoneal epithelium itself, at least it is clear that they come from the mesenchyme cells in the vicinity of this mesothelium. The development of adrenocortical tissue is intimately bound up with that of the kidney, the gonads and the posterior venous system. We can describe these relationships in topographical terms alone, and only speculate about causality and embryological inductive mechanisms.

In vertebrates as a whole, we find that on each side, the cortical anlagen arise from (or by) the peritoneal epithelium, near the aorta in the region of the angle between the dorsal mesentery and the germinal ridge (text-fig. 32). The most anterior site of origin of the anlagen is in the region of the posterior part of the pronephros, often in the 'Zwischenzone' which lies between the caudal end of the pronephros and the cranial end of the mesonephros. From this point, at the level of the pronephros, it is possible that *potential* corticogenic mesoderm stretches posteriorly, in the mesenterial angle, to the level of the cloaca. The potentiality is never fully realized, but the evolutionary picture is one of varying amounts of development of cortical tissue along this longitudinal line and subsequent amalgamation and contraction of cortical anlagen.

It is attractive to attempt a correlation of the changes in the kidney with those in cortical tissue in the different classes of vertebrates, for both are a story of differential development and ultimate shortening of basic tissue to form compact organs (cf. Harms, 1921). The vertebrate kidney is built on a fundamentally similar plan (Fraser, 1950). The nephrogenic tissue,

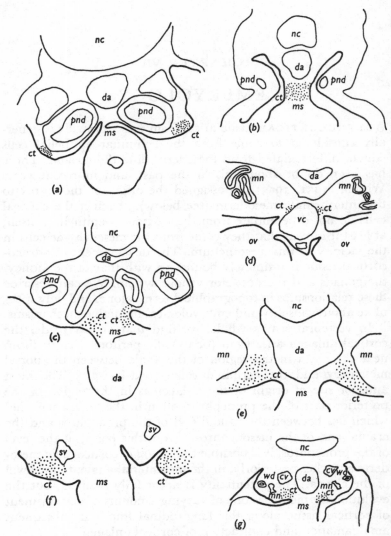

Text-fig. 32. Early stages in the development of cortical tissue in representatives of the different vertebrate classes (redrawn from various authors). *nc*, notochord; *da*, dorsal aorta; *pnd*, pronephric duct; *ms*, dorsal mesentery; *ct*, cortical tissue; *mn*, mesonephros; *wd*, Wolffian duct; *cv*, cardinal vein; *vc*, vena cava; *sv*, subcardinal vein.

(*a*) *Petromyzon* (Giacomini, 1902*b*); (*b*) *Scyllium* (Poll, 1906); (*c*) *Salmo* (Giacomini, 1911*b*); (*d*) *Rana* (Witschi, 1953); (*e*) Turtle (Kuntz, 1912); (*f*) Domestic fowl (Hays, 1914); (*g*) Mouse (Inaba, 1891; Waring, 1935).

Embryology

arising from the mesodermal intermediate cell-mass, extends *potentially* from the level of the second post-otic somite to some distance behind the anus. The actual amount of kidney tissue present in the adult, however, varies considerably. The most anterior portion, the pronephros, persists only in the larvae of Anamnia and in a few adult teleosts (Fraser, 1950). In most vertebrates, the main excretory organ is derived from the nephrogenic blastema forming posteriorly to the pronephros or, if absent as such, the zone of anterior rudimentary tubules. The blastema provides the mesonephros, the functional kidney of the adult Anamnia, and from its hinder end the metanephros of the Amnia. In the latter, the mesonephroi are represented in the adult principally by efferent tubules of the testes. In embryological development, the kidney differentiates out from the mesoderm well before cortical anlagen appear, and it is a possibility that functional nephric tissue is the inductor of the latter. In the cyclostome larvae, the cortical tissue appears at the level of the functional pronephros before there is any sign of the mesonephros. In elasmobranchs, the posterior part of the corticogenic tissue remains as the interrenal lying between the functional part of the mesonephroi which, in this group, are concentrated posteriorly in a dozen segments or less. Likewise, in the Amphibia, the cortex spreads over the mesonephroi following in extent both the elongate kidneys in salamanders and Apoda and the short compact organs of the Anura. Furthermore, Branin (1937) found in *Hemidactylum* that the position of the cortical primordia at different stages of development corresponded very closely to the position of the functional kidney units at different times.

While, in Amniotes, the mesonephros for the most part disappears in the adult, it is functional in embryonic life and the cortical tissue forms from its cephalad end, extending over a few segments posteriorly. These anlagen form small discrete organs over the metanephroi which, by differential growth, have moved forwards to it. The teleosts follow a similar pattern, in that the cortical tissue occurs in the region of the pronephros, but are exceptional in that it does not extend along the functional mesonephros in the adult. Teleosts also differ, in that the kidney does not give up its anterior end for gonadal purposes as it does in most other vertebrates, but much more information about the development of the gonad and the kidney and the distribution of the cortical tissue in all Osteichthyes is required before a clear pattern can be made out. Branin (1937) regards the relationship of the cortical anlagen with the functional kidney as coincidental and she considers that the primary relationship is with the vascular system. It is certainly true that the adrenal cortex is found in close contact with the cardinal veins and their tributaries in fish, or with the inferior vena cava and its tributaries in higher forms. The development of kidney and blood system is, of necessity, a concomitant process and it is therefore on general embryological grounds that I would consider nephrogenic tissue as the more likely evocator of cortical anlagen.

Witschi (1951, 1953) and his co-workers (Witschi, Bruner and Segal, 1953; Segal, 1953) have recently reinvestigated the problem of the origin of the adrenal cortex, especially in the frog. They have found that while the total amount of intermediate mesoderm remains constant, the proportions given over to the formation of mesonephros, of the medullary component of the sex glands, of the efferent tubules of the testes, and of the cortex can be altered experimentally (p. 156). They consider, therefore, that the cells of the mesonephric length of the nephrogenic blastema cord are pluri-potent. This means, for example, that in the early larval stage of *R. temporaria* (Witschi, 1953) the blastema cells still have the potentiality of differentiation into either urinary tubules, adrenal lobules, or efferent and medullary structures of the sex glands. If this hypothesis of Witschi is true, then the commonly accepted view that the cortex comes from peritoneal epithelium is wrong. Furthermore, the cells of the adrenal cortex throughout the vertebrates are arranged into cords and this is consistent with the general plan of the known derivatives of intermediate mesoderm, namely the cords and tubes of the reproductive and excretory systems.

Another interesting facet of the development of the adrenal gland is the variable amount of intermixing of the cortical and chromaffin tissue in the different vertebrate classes. This subject will be returned to in the next chapter. In the elasmo-

branchs, the two tissues remain separate. In most other vertebrates, cells originating in the sympathetic nervous system
become closely applied to the cortical anlagen and later migrate
within them. The origin of the sympathetic cells is in doubt.
Most embryologists consider that they are derived from the
neural crest; others think that they come from a migration of
cells from the neural tube along the anterior root. Fairly early
on in their differentiation, these sympathetic cells, which are to
intermix with the cortex, display the chromaffin reaction.

This section is concluded by a brief review of some of the
salient points in the development of adrenocortical tissue in
representatives of the different vertebrate classes. The classical
work on the comparative embryology of the adrenal is by Poll
(1906). Though his description for teleosts was confined to the
corpuscles of Stannius and we now know more about other
vertebrate classes, his work is unrivalled for its comprehensive
comparative discussion. In cyclostomes, the first signs of cortical tissue occur in 6–7 mm. larvae aged 21 to 24 days. The
diffuse anlagen, six or seven in number, are found at the caudal
end of the pronephros corresponding to spinal ganglia six to
eight. They occur against the coelomic epithelium in the angle
of the pronephric fold. In larvae of 9 mm. in length, the cortical anlagen are separated off from the mesothelium into discrete masses and at this stage the mesonephros begins to form.
By the time *Ammocoetes* is 35 mm. in length, the cortical tissue
has much the same distribution as in the adult, being found as
small lobules in the wall of the cardinal veins and by the ventral wall of the aorta (Giacomini, 1902 *b*).

In elasmobranchs, Poll (1906) found that the first indication
of cortical tissue is in the 7 mm. embryo (*Scyllium stellare*). At
this stage, a group of cells forms from the coelomic epithelium
at each side at the root of the dorsal mesentery. These cells join
to form one mass which commences at the level of the posterior
region of the pronephros and extends, broken up into many
groups though not segmentally, to the cloaca. By the stage
reached in embryos 10 mm. in length, the cortex appears as a
rod of tissue lying below the ventral wall of the aorta and in
the root of the mesentery. The longitudinal extent of the interrenal has, however, diminished, stretching from the seventh

segment behind the end of the pronephros to the edge of the first and second quarters of the cloaca, thus covering twenty segments, as against the twenty-five originally bearing cortical anlagen. The interrenal begins to take up the position in which it is found in the adult. This involves the loss of the anterior cortical cell groups and the growth and amalgamation of the posterior ones, so that in the embryo 24 mm. in length, the gland spreads over only the twelve segments lying anterior to the cloaca. At the same time, the posterior part of the mesonephros has become predominant.

Giacomini (1912, 1920, 1922) investigated the development of adrenocortical tissue in salmonids (*Siphostoma, Nerophis*) and lophobranchs. The sequence of events in *Salmo* species (*lacustris, fario, irideus* and *salar*) is typical of all the types examined. The first sign of cortical tissue is seen in embryos about 4 mm. long, 22 to 23 days after fertilization. The cortical anlagen arise by proliferation of coelomic epithelial cells on each side of the root of the mesentery, ventral to the glomeruli of the pronephros. In embryos about 5 mm. in length, 24 to 25 days after fertilization, the archinephric duct has reached the cloaca and the cortical anlagen have the appearance of round groups of cells. These proliferate and the cortical tissue spreads irregularly in the region of the pronephros and in close association with the mesenteric artery. The pronephros later degenerates and the cortical tissue is found in the adult position of scattered islands in the lymphoid tissue of the vascular head-kidney. It seems that the cortical anlagen are confined to the anterior part of the nephrogenic tissue and abortive anlagen have not been seen more posteriorly in the region of the mesonephros. If we suppose that the basic vertebrate type possesses corticogenic mesoderm lying below all the nephrogenic tissue, then it would be expected that investigation of a wide range of teleost species would bring to light those that showed both anterior and posterior cortical anlagen. Moreover, experiments of the type done by Witschi in Amphibia might reveal intermediate mesodermal potentiality in teleosts even though not shown in their normal development.

Garrett (1942) describes the development of the corpuscles of Stannius in the holostean ganoid (*Amia calva*), in the salmonid (*Salvelinus fontenalis*) and several other teleosts, including the specialized *Platypoecilus maculatus*, and he confirms the earlier work on this subject, especially that of Giacomini. The Stannius corpuscle develops as a bud-like evagination or series of evaginations from the wall of the pronephric duct. While all forms show a fundamental similarity of development, there are differences in detail. The most outstanding one is the reduction of the number of individual evaginations, from forty or fifty in the Holostei to a single pair in nearly all teleosts, except in salmonids where there may be three or four pairs, more or less segmentally arranged. As the corpuscles of Stannius do not occur in the Chondrostei and as they are not considered to represent adrenocortical tissue, their homology is not obvious. The best suggestion so far is that they represent part of the Müllerian duct in Holostei and Teleostei, so that the oviducts in these groups must have a special origin, and there is evidence that this may be so (Garrett, 1942).

A guide to the literature concerning the development of the adrenal cortex in Amphibia is to be found in Dittus (1936), Branin (1937), and Witschi and co-workers (*loc. cit.*). According to the latter authors, the stage immediately preceding adrenocortical differentiation in the frog is the time of the closure of the gill sacs, only one or two days before the larvae begin to feed. The genital ridges are rising, the left and right subcardinal veins have moved together and by fusing have just given origin to the lowest segment of the vena cava. The cortical anlagen arise from the mesodermal tissue bars which lie between the numerous outlets of the mesonephric blood sinuses running into the vena cava. In the four-toed salamander (*Hemidactylium scutatum*), Branin (1937) was not sure whether or not the cortical anlagen arose from coelomic epithelium. She first found them in the 8·8 mm. embryo as paired or unpaired cell groups in the 'Zwischenzone', either in contact with or just beneath the mesothelium, on either side of the dorsal mesentery. Further cortical anlagen arise posteriorly, so that while only two to four are seen at the earliest stage, there are thirty or more at metamorphosis. The increase in number of primordia occurs concomitantly with the development of the vena cava and of the mesonephros. A month after hatching, in larvae of 13–15 mm., the cortical tissue extends over an area bounded by the eighth and fourteenth pairs of spinal ganglia and consists of individual masses of different sizes. At metamorphosis, the cortical tissue, still in islands, is distributed from the level of the ninth to that of the twentieth spinal ganglia. Accompanying and immediately following metamorphosis, there is a progressive coalescence and growth of the individual cortical masses resulting in strands of tissue along the caval system. This leads to the adult form where the cortical islets are more or less interconnected in the region of the mesonephroi, scattered along the inferior vena cava and its tributaries from the cranial end of the pancreas to the caudal end of this vein.

The principal papers on the development of the reptile adrenal cortex are those of Poll (1906), Kuntz (1912) and Bimmer (1950). The latter author brought together the earlier observations and added her own on *Lacerta vivipara* and *L. agilis*. Poll, working with *Emys europaea* (var. *taurica*), found that the cortical tissue arose in the same way as in other vertebrates. The cortical anlagen were found at the earliest stage (10 mm. embryos) stretching from the caudal end of the pronephros over the area of eight to nine spinal ganglia (numbers six to fourteen or fifteen). Poll makes the interesting observation that of the twelve to thirteen somites from the posterior end of the pronephros to the anterior side of the cloaca (spinal ganglia five to seventeen or eighteen), the cortical tissue occupies 66·7 %, while in a corresponding stage in Amphibia (urodeles), it takes up 80 %, and in elasmobranchs 100·%. In the other amniotes, birds and mammals, this percentage is much less, ranging from 25–33 % to 8·3 %. The further development of the adrenal gland in reptiles consists of the joining up of the cortical anlagen and their concentration to a smaller area, so that in the 28 mm. *Emys* embryo, the primordia extend over only two segments. Bimmer (1950) brings out the fusion of cortical anlagen and concomitant concentration very well in *Lacerta*.

In the chick, the best known representative of the birds, the area over which the cortical anlagen form is much less than in the elasmobranchs, amphibians and reptiles. The cortical primordia first appear in somite seventeen, at the caudal end of the pronephroi in somites five to fifteen or sixteen, and extend over to somite twenty-two. The cephalad region of the mesonephroi overlaps the posterior pronephric tubules, originating in somite thirteen or fourteen and reaching to somite thirty. Hays (1914) found that the cortical anlagen are first seen in the ninety-sixth hour of incubation, appearing as a thickening of the peritoneal epithelium, ventral and mesial to the mesonephroi, ventral to the abdominal aorta and dorsal to the hind gut which is open at this time. The developing cells push in dorsally from the epithelium upon which they rest and become larger and more nearly circular in outline than those cells from which they arose. Nine hours later, by the one hundred and fifth hour, the cortical cells have piled up by the peritoneal epithelium, so that a solid body is formed on each side of the base of the mesentery and lies just medial to the ventral side of the mesonephros. A dorsal migration occurs so that cortical cells come to lie slightly dorsal to the ventral level of the aorta, lying then between the aorta and the mesonephros. After 120 hr., all cortical cells have migrated to this point, reaching about the same level on both right and left sides of the aorta. During the next 10 hr. up to 130 hr. of incubation, there is an increase in size of the cortex, so that at 144 hr. it forms a large oval mass on each side of the aorta. By the one hundred and sixty-eighth hour, the cortical cells are arranged in irregular chains and the gland continues to enlarge until hatching. The history of the chromaffin cells is similar to that in all vertebrates except the elasmobranchs. Sympathetic cells migrate singly as large oval cells to reach the cortical mass on each side at 130 hr. incubation. In the bird, the cells enter the cortex in cords which break up so that by 216 hr. incubation the chromaffin tissue is found in small groups as in the adult gland.

In mammals, the lengthwise cortical primordia appear early in development—in the 9–12 mm. guinea-pig embryo, in the 12 day-old mouse embryo, in the 7–8 mm. human embryo (third to fourth week)—characterized by accelerated local proliferation of cells from the splanchnic mesoderm around the notch on either side of the base of the primary dorsal mesentery adjacent to the cephalic pole of the mesonephros. The

evolutionary tendency for the lengthwise restriction of corticogenic tissue reaches its peak in the Eutheria. In the mouse, the
anlagen on each side lie between the sixteenth and seventeenth
somites, the anterior end being at the level of the second tubule
of the mesonephros (Inaba, 1891). Likewise, in the other mammalian species studied, the primordia are short. Towards the
end of the fourteenth day of gestation in the mouse, the adrenal
is a small, compact, almost spherical, organ, flattened against
the anterior end of the kidney with a central blood vessel leading
into the cardinal vein. Patten (1948) notes that, in the pig,
the cortical cells tend to become arranged into cords which are
quite conspicuous by the 15–17 mm. stage (see ch. VIII). Pankratz (1931) found that, in the rat, during the fifteenth and
sixteenth days of gestation, the cortex formed an oval mass
protruding a little into the dorsal coelomic cavity. The clus-
tering and migration of sympathetic cells occur some time after
the formation of the cortical anlagen. Sympatho-chromaffin
cells migrate into the cortex on the fourteenth day in the mouse,
on the sixteenth day in the rat, at the 18 mm. stage in the
guinea-pig, at the 30–35 mm. stage in the pig, and the 16–18
mm. stage in the cat. The presumptive medullary cells then
form nests which gradually take up a central position.

The development of the adrenal in man is specially interesting because of the marked 'foetal' as well as 'permanent' cortical tissue. Both Keene and Hewer (1927) and Uotila (1940)
agree that these two parts of the embryonic adrenal do not
arise at the same time. The first cortical primordia, originating
in the usual sites, grow rapidly and reach a considerable size
in embryos of 8–9 mm. After the complete separation of the
gland from the mesothelium on each side, differentiation of
the cortex, which is the 'foetal' cortex or X zone, continues in
embryos of 10–12 mm. The cells are large with conspicuous
nuclei and with cytoplasm staining more darkly than that of
the neighbouring mesenchymal cells. A capsule is barely distinguishable and remains poorly developed until the permanent
cortex has differentiated. After the separation of the X zone,
the coelomic epithelial cells continue to undergo mitotic division, giving rise to a second proliferation of cells which are
destined to form the 'permanent' cortex. The cells which give

rise to the permanent cortex are smaller and possess less prominent nuclei than those which formed the foetal cortex. The presumptive permanent cortex cells form on the ventral, ventromedial and ventrolateral surface of the mass of X zone and, in the absence of a capsule, attach themselves firmly to the X zone, invade its periphery and spread over it. At the same time, capsular cells form, intermingling with the permanent cortex. By the 18 mm. stage, the major portion of the gland is taken up by the foetal cortex with its cells possessing granular acidophilic cytoplasm surrounded by a narrow layer of permanent cortex, the cells of which are smaller with basophilic cytoplasm and a small, deeply stained nucleus. Cords of sympathetic nerve cells formed by small spindle-shaped, darkly staining cells can be seen on the dorsomedial aspect of the foetal cortex in embryos of 11–12 mm. (six weeks). Invasion commences in embryos of 13–14 mm. (six to seven weeks) and then continues, penetration of the capsule taking place at the medial side. The permanent cortex in embryos of 20 mm. onwards increases in size throughout intra-uterine life, though in the full-term foetus it forms only about 25 % of the gland.

These two separate proliferations have not been seen by other workers in human embryology. It would be interesting to find if they are confined to the Anthropoidea in which the possession of a marked embryonic X zone is also a unique feature. It is impossible to say which cortical proliferation is homologous with the cortical anlagen in other vertebrates. Indeed, surveying the phylogenetic development of the adrenal gland, it is surprising to find in primates such a marked variation of the general pattern. The suggestion that the first proliferation of cortical tissue is evoked by chorionic gonadotrophin, secreted by the placenta in vast amounts early in pregnancy (Chester Jones, 1955), implies that it is the second proliferation which is homologous with the vertebrate cortical primordium. The development of the 'foetal' cortex seems to be equally as mysterious as its undetected function.

CHAPTER VIII

CORRELATION OF STRUCTURE
AND FUNCTION

EXAMINATION of the histological structure of the adrenal cortex has been confined, for the most part, to a few species of the Eutheria. Theories which would undertake the correlation of structure and function in the adrenal cortex have, therefore, been based upon this circumscribed information. The given facts are: (i) the adrenal gland in the Eutheria consists of concentric layers of cells—principally the outer connective tissue capsule (with or without a sub-capsular layer), the zona glomerulosa, the zona fasciculata and the zona reticularis—around a central medulla of chromaffin tissue; (ii) the adrenal cortex secretes steroid hormones concerned in carbohydrate and water and salt-electrolyte metabolism, among other things, and these gluco- and mineralo-corticoid activities overlap and are achieved principally by aldosterone and Kendall's compounds B and F, as far as is known at present; (iii) both the histological appearance of the cortex and the normal secretory pattern depend upon adenohypophysial secretions (principally ACTH) but to a varying extent. The first simple division of the problem can therefore be either (i) that the zones, once formed, are independent of one another, and have different secretory functions; or (ii) that the zones are related to each other in varying degrees and in varying ways as are their secretory capacities.

It is essential to point out a fact which is frequently overlooked in the discussion of this intriguing problem of form and function, namely that the zonation of the eutherian adrenal cortex is based, for better or for worse, on histological criteria—that is, the appearance, shape and grouping of cells after routine methods. The appearance of the adrenal cortex after other methods, such as lipid stains, is not relevant and has been, to a large degree, misleading. While acceptable generalizations may conceivably be advanced from the routine histological appearance of the adrenal cortex throughout the Eutheria, this cannot be based, for example, on variations in sudanophilia, especially in view of the absence of this property in the glands of some species. Furthermore, the danger exists that workers will rely on fat stains alone and take the outermost cortical area to be the zona glomerulosa while, in some conditions, this can be in fact the zona fasciculata reaching right to the outer connective tissue capsule—a decision to be based on the histological appearance of radial cords reaching to the periphery.

It was Gottschau in 1883 who, examining the zonation of the adrenal as laid down by Arnold (1866), considered that there was an inward migration of cells from the periphery of the cortex with degeneration in the 'zona consumptiva', using this latter term to include inner cortex and medulla. With the distinction established between medulla and cortex, the 'cell migration' or 'escalator' theory conceived of cortical cells degenerating in the zona reticularis at the medullary border and constantly being renewed by multiplication at the periphery together with migration inwards, with cytomorphosis, of the cells so formed. This theory found many adherents over the years and Graham (1916) was the first to attempt experimental confirmation, followed by the studies of Hoerr (1931), Bennett (1940) and Williams (1947) among others. Bennett, in accepting the theory, designated the outer zona fasciculata as the actively secreting part, so that the cortex could be divided into pre-secretory, secretory, post-secretory and senescent zones, roughly, but not completely identical with, the zonae glomerulosa, outer fasciculata, inner fasciculata and reticularis.

Authors who accepted the idea of centripetal cell migration in principle have modified it in regard to the place of origin of new cells. The simplest idea was that the new cells formed in the outermost layer, the zona glomerulosa, and thence passed inwards. It was noticed, however, that the majority of mitotic divisions occurred at the inner part of the zona glomerulosa and the outer part of the zona fasciculata (text-fig. 33). Thus, the origin of new cells was assigned to this area and the zona glomerulosa considered a reserve layer of cells (Mulon, 1903;

The Adrenal Cortex

Bernard and Bigart, 1902; da Costa, 1913; Bouin, 1932; Giroud and Leblond, 1934; see Bachmann, 1954). The growth of the adrenal cortex, therefore, could be directed both outwards and inwards from the zona fasciculata (Stoerk and Haberer, 1908 a, b). Later on, workers inferred an additional or sometimes an alternative source of new cells, namely the

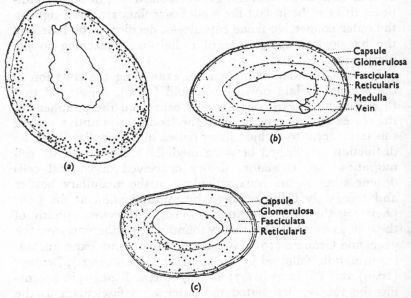

Text-fig. 33. The distribution of mitotic divisions in the adrenal cortex in the rat. Colchicine causes arrest of mitotic divisions and makes their numeration easier. The injection may itself be a stress and may cause enhancement of mitotic activity before arrest.

(a) From Mitchell (1948): 19 day-old rat; colchicine injected; camera lucida drawing of the distribution of mitoses in one entire 6μ section. (b), (c) From Baxter (1946): camera lucida drawings of sections through the middle of the left adrenal of two adult rats treated with colchicine 16 hr. before death.

outer connective tissue capsule itself (Zwemer, Wotton and Norkus, 1938; Baker and Bailif, 1939; Salmon and Zwemer, 1941; Wotton and Zwemer, 1943; Gruenwald and Konikov, 1944; Elias, 1948; Baker, 1952). It was considered that there was transformation into glomerulosa-type adrenocortical cells from the capsule fibroblasts as such, or as an additional or alternative suggestion, from the sub-capsular elements in those species in which a differentiation between outer elongated fibroblasts and inner cells with more rounded nuclei can be made. Direct experimental proof of the movement of adreno-cortical cells is lacking; Salmon and Zwemer's attempt to mark cells with trypan blue for this purpose did not succeed (Calma and Foster, 1943; Baxter, 1946).

The establishment of the cell-migration theory depends, essentially, on finding (i) cell division occurring more plentifully in the peripheral part of the adrenal cortex than in other parts; (ii) evidence of cell death in the zona reticularis, particularly against the medullary border. As time went on, more and more doubts about the validity of the cell-migration theory crept into the literature; some animals had little or no zona reticularis (e.g. the adult male mouse), some a prominent one (e.g. the guinea-pig); some animals had an obviously demarcated zona glomerulosa (e.g. the horse and dog), others a poorly developed one (e.g. the bat). Furthermore, it began to be felt that the zona reticularis might not contain numerous dead cells.

The discernment of a dead or dying cell by histological methods is not always easy. Acid azo dyes such as trypan blue, injected *intra vitam* will enter into and stain diffusely the cytoplasm of degenerating cells (Ludford, 1933) or will colour the nucleus alone of such cells (Darlington, 1937). By this test Bennett (1940) did find that the zona reticularis of the cat contained dead and dying cells, and others, for example Baxter (1946) on the rat, have confirmed him. Alternatively, it might be argued that zona reticularis cells instead of being about to degenerate (as Hoerr (1931) supposed of the light and dark cells) might equally well represent different actively functional stages. The increased pigmentation and siderophilia frequently seen in the zona reticularis, although often thought to be an accompaniment of senescence, might merely indicate a different phase of activity. On the other hand, it is not denied that unquestionably dead cells do occur in the zona reticularis, and these are characterized by advanced pycnosis of the nuclei and either cytolysis, chondriolysis, caryolysis and loss of pigment and lipid or by shrinking and homogenization of the cytoplasm, chondriomegaly and retention or increase of lipid and pigment content (Hoerr, 1931). It is felt, however, that the relative number of unquestioned dead cells in the reticularis is not large enough to reflect a process of centripetally moving cortical cells, dying as they reach the medulla. Moreover, dead cells have been seen in the zona fasciculata by Bennett and others, though generally not in the zona glomerulosa. However, when dead cells do occur in the adrenal cortex, the majority will be found in the zona reticularis.

Another doubt which assailed investigators was occasioned by the finding that although it was indeed true that the majority of cell divisions do occur in the region of the zona glomerulosa and outer zona fasciculata (e.g. Hoerr, 1931; Whitehead, 1933; Bennett, 1940; Blumenthal, 1940; Gruenwald and Konikov, 1944; Baxter, 1946; Chester Jones, 1948; Mitchell, 1948), mitoses are also found in the deeper cortical regions including the zona reticularis (text-fig. 33; Baxter, 1946; Mitchell, 1948). It was felt, therefore, that, even if cell degeneration did take place in the zona reticularis, the area was not solely devoted to cell death but also must include, to some degree, cell replenishment.

Against the background of these difficulties in substantiation of the cell-migration theory, the hypothesis that the three main zones of the adrenal cortex were independent secretory layers gained in favour. The stimulus to this was the finding that although adrenalectomy was followed by a well-known electrolytic imbalance this did not occur after hypophysectomy when the zona glomerulosa persisted undegenerated at least for an appreciable time. Swann (1940) briefly suggested, therefore, that there may be a division of secretory function between the outer and inner layers of the cortex and that these were not equally dependent on the anterior lobe of the pituitary. Sarason (1943), from his experimental work on the rat, supported this view, and Deane and Greep developed it (Deane and Greep, 1946; Greep and Deane, 1947; Deane, Shaw and Greep, 1948; Greep and Deane, 1949*b*; Greep and Chester Jones, 1950*a*). The zonal theory, as it was named by Chester Jones (1948), is widely supported and the cell-migration theory regarded as somewhat old-fashioned (Yoffey, 1953). On the zonal theory, the zona glomerulosa is considered to secrete

Correlation of Structure and Function

deoxycorticosterone-like mineralocorticoids, largely independently of ACTH control, and the zona fasciculata, controlled
by ACTH, is thought to produce the glucocorticoids. With the
subsequent identification of aldosterone, the secretion of this
compound would be assigned, on the zonal theory, to the zona
glomerulosa and its secretory rate would be predicted to vary
independently of the amount of circulating ACTH; from the
zona fasciculata would arise substances typified by Kendall's
compounds B and F. Although Greep and Deane did not
stress the zona reticularis as a separate entity, it must be so on
the zonal theory, and it is thought to produce sex hormones,
principally androgens of which androsterone could be regarded
as the most characteristic (Albright, 1943; Vaccarezza, 1945,
1946; Blackman, 1946).

A compromise between the two theories seems to have been
arrived at by Tonutti (1941, 1942 a, b, c, 1953; Tonutti, Bahner
and Muschke, 1954) in his hypothesis of the 'transformation
fields' of the adrenal cortex. Tonutti considers that variations
in the gland take place by change in two cortical areas, namely
the outer region comprising the capsule and the zona glomerulosa, and an inner region, comprising the zona reticularis and
to a varying extent the zona fasciculata. These two regions he
calls the outer and inner transformation fields. By progressive
transformation, the outer and inner fields can be changed into
actively secreting zona fasciculata—the functional part of the
gland—under the influence of adrenotrophin. By regressive
transformation when the demand for adrenocortical hormones
is relatively small, the zonae glomerulosa and reticularis reappear with consequent decrease in size of the zona fasciculata.
The type of regressive transformation depends on the species;
it can be characterized by cell shrinkage and the appearance of
connective tissue, or by storage of substances in the cells of both
the inner and outer field. Essentially, then, the zona glomerulosa and the zona reticularis can be differentiated from the
zona fasciculata, acting in a reserve capacity when adrenocortical secretion is not high and these two zones reform into
zona fasciculata, under the influence of ACTH, when there is
demand for active secretion. Tonutti's attitude to the variations
of the zona glomerulosa in relation to the zona fasciculata is

235

unexceptionable, but his inclusion of the zona reticularis as another reserve zone may not meet with immediate approval (Dempster, 1955). In contradistinction to Tonutti, for example, Yoffey considers that the zona reticularis is an actively functioning zone, which expands, not diminishes, with cortical hypertrophy (Yoffey and Baxter, 1949; Yoffey, 1953, 1955).

In the formulation of any theory concerned in the correlation of cortical structure and function, cognizance must be taken of the regenerative capacity of the adrenal cortex. The gland, particularly of the rat and mouse, can be squeezed *in situ* so that all its contents are removed, leaving only the outer connective tissue capsule and such zona glomerulosa cells as remain adhering thereto. This is the operation of 'enucleation' and the enucleated adrenal remaining *in situ* after the operation will regrow into an adrenal cortex, without a medulla, but with zones of normal appearance. It is undisputed that in this case all the zones have been derived from the periphery, although there is a division of opinion whether the new cortex arose solely from the remaining zona glomerulosa cells at enucleation (Greep and Deane, 1949 a, b; Chester Jones and Spalding, 1954; Chester Jones and Wright, 1954a), or solely (or in addition to glomerulosa origin) from the capsular cells (Ingle and Higgins, 1938; Turner, 1939; Elias, 1948; Hartman and Brownell, 1949; Baker, 1952). While all are agreed that the zonae fasciculata and reticularis arise peripherally after enucleation, whether it be from capsule or zona glomerulosa, divergence of opinion occurs between those who consider this merely a special case and those who feel it is related to the normal manifestations of the gland. Those who do not work with the rat and the mouse are more prone to take the former attitude, since the enucleation procedure in the dog, for example, has not so far been followed by successful cortical regeneration (Dempster, 1955). It may well be, however, that in the dog the surgical interference with the copious arterial supply necessitated by the enucleation procedure accounts for the failure rather than any inherent lack of regenerative capacity of dog adrenal cortex when well vascularized.

A further fact which must be borne in mind is that the zona glomerulosa is not independent of adrenotrophin nor unresponsive to it (Chester Jones, 1949*b*; Wexler, Rinfret, Griffin and Richardson, 1955). In the intact animal rising titres of ACTH are accompanied by a diminution of the zona glomerulosa (Selye and Stone, 1950; Baker, 1952; Chester Jones and Wright, 1954*b*; Tonutti, Bahner and Muschke, 1954; Krohn, 1955). I have found that in the adrenals of the rat, the zona glomerulosa can be made to disappear with adequate ACTH stimulation, and this is true of similar material of Burton Baker and of Ganong and Hume for the dog which these workers kindly allowed me to examine. In a recent series of experiments I found that a 50 % or more increase in adrenal weight achieved by the injection of long-acting ACTH was invariably accompanied by an extinction of the zona glomerulosa (Chester Jones, unpublished); the criterion of this must necessarily be the appearance of the radial columns of the zona fasciculata reaching right up to the outer connective tissue capsule. Furthermore, it has been often testified that the adrenal cortex of the hypophysectomized animal, consisting as it does mainly of persistent zona glomerulosa, will respond to exogenous ACTH with the re-formation of normal zones. In the mouse, for example, ninety days after hypophysectomy, the adrenal cortex still gave a gland of normal appearance with ACTH (Chester Jones and Roby, 1954), leaving no doubt that in this special case the zonae fasciculata and reticularis were formed from peripheral tissue (the atrophic remnants of the original zonae fasciculata and reticularis having disappeared by this time).

Consideration of the changing form of the adrenal cortex in the different vertebrate classes throws light on the particular appearance of the eutherian gland. Two major processes are at work: (i) the tendency to change from scattered chromaffin islets to aggregation into groups, and finally into a central mass (text-fig. 34); (ii) the formation of a discrete adrenal cortex and its gaining a more compact shape with encapsulation and thickening of the outer connective tissue capsule. These processes operate upon the basic unit of the vertebrate cortex which is, essentially, a cord of cells (text-fig. 35). Generally, cell division goes on from one part of the cord, but stimulation by adrenotrophin extends this. The cords may appear as

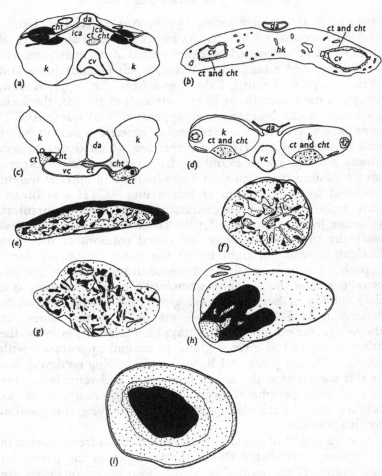

Text-fig. 34. The intermingling and centralization of chromaffin tissue within cortical tissue throughout the vertebrates. (a) The dogfish (from Young, 1950); (b) perch (from Baecker, 1928); (c) Amphibia, *Ichthyophis* (from Dittus, 1941); (d) frog; (e) lizard; (f) Crocodilia; (g) pigeon; (h) echidna (from Bourne, 1949); (i) rat.

Stippled, cortical tissue; black, chromaffin tissue; *k*, kidney; *hk*, head-kidney; *cv*, cardinal vein; *vc*, vena cava; *cht*, chromaffin tissue; *da*, dorsal aorta; *ica*, intercostal artery; *ct*, cortical tissue.

such, with anastomosing strands, very much coiled but largely unconfined by restricting connective tissue and not orientated in reference to any plane, as in teleosts and many elasmobranchs.

In some elasmobranchs, such as the rays, the adrenal cortex may be gathered together as a compact organ distinct from chromaffin tissue and in these there may be slight evidence of cortical zonation. In the Amphibia, reptiles and birds, several cortical cords may be gathered together to form cylinders so that in transverse section across them the cells are radially arranged. In Amphibia, the cords run in a longitudinal direction with anastomoses in all three planes; hence longitudinal and sagittal sections demonstrate the cord-like make-up and show also transverse and oblique sections especially of the anastomoses, while transverse sections demonstrate the radial cellular arrangement of many cords together, of course, with sections in other planes. There is no bounding connective tissue capsule and copious blood sinuses allow change of cortical size without restriction.

While the reptilian adrenal is clearly built on the same plan as the amphibian with the over-all direction of the cords being longitudinal, it shows some distinctive features. In addition to the scattered islets of chromaffin tissue intermingling with that of the cortical, as it does in some teleost and all amphibian glands, there is a peripheral mass of chromaffin tissue. Against this barrier there is some change of cortical form which we have interpreted as degeneration. The bulk of the reptilian cortical cords have little restriction, however, as the gland possesses abundant blood sinuses and a thin outer connective tissue capsule.

The adrenal gland of birds is a much more compact organ than that of the lower groups, with a relatively firm outer connective tissue capsule with scattered chromaffin islets but not gathered together in any large masses. The cortical cords, which in the fish cortex looped, branched and anastomosed haphazardly, and which in Amphibia and reptiles were orientated in a general longitudinal direction, in birds loop against the outer connective tissue capsule (text-fig. 35). The looped cords grow centripetally and towards the centre of the gland the anastomosing cords are compressed and give the appearance of a zona reticularis. The loops of the cords as they lie against the outer connective tissue capsule are out of the main stream of growth and become resting cortical cells to be stimulated into activity with high amounts of circulating ACTH. It is suggested that this is an essential accompaniment of 'looping' in a more compact gland.

In the monotremes the chromaffin tissue has aggregated together and is found, of variable extent, at one end of the gland. In echidna the cortical appearance is somewhat avian; in platypus it seems as if groups of cords have formed prongs against the medulla and interlaced to form a complicated pattern.

In all these groups the cortical cords have a direct relationship to an adrenotrophic secretion of the pituitary, but not all the cells of a cord are equally responsive. It would seem as if there is a place of cell growth which buds off cells to form the basic cortical unit and at any one time there are cells in a cord of different ages after formation. In the loose unrestricted adrenals of the lower groups, the cells are difficult, if not impossible, to differentiate on histological grounds. That such a differentiation does exist is witnessed by the effects of hypophysectomy where some cortical cells are completely atrophic and others are of more normal appearance. It is only, however, when the cortex becomes more compact that some orientation is imposed on the cords. The peripheral looping of the cords in the bird adrenal carries with it, it seems, the formation of inactive cells in the loop and then leading off through the active region to an inactive region which, due to the compactness of the gland, is of necessity, pressed together centrally.

Text-fig. 35. The organization of the basic cord of adrenocortical cells throughout the vertebrates. (*a*) The basic unit: a cord of cells capable of growth in each direction; (*b*) transverse sections of (*a*) showing that the cord can be either a single line of cells or grouped to give a radial pattern; (*c*) simple looping of the basic cord; (*d*) more complex looping (as in teleosts); (*e*) illustrating the capacity to form loops, all parts of which are not equally capable of growth: it is easier for growth to take place in the direction away from the loop; (*f*) organization of cords of cells in longitudinal runs (as in Amphibians and reptiles); (*g*) looping of cords peripherally against a well-formed outer connective tissue capsule (*c*) (as in birds) when the consideration given in (*e*) applies; (*h*) looping of cords peripherally against a well-formed capsule (*c*) (when (*e*) again applies) and also the termination of the cords against a central medulla (*md*) (with necessary alteration of growth pattern)—as in Metatheria and Eutheria. (*zg*, zona glomerulosa; *zf*, zona fasciculata; *zr*, zona reticularis.)

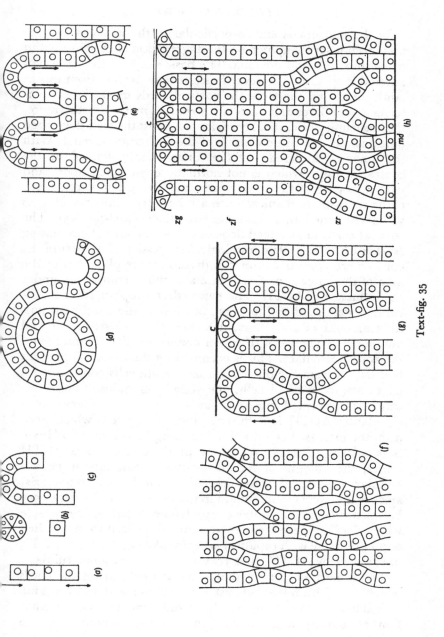

Text-fig. 35

The Adrenal Cortex

In the marsupials and in particular in the Eutheria, we see these trends superimposed on the aggregation of the chromaffin tissue into a central medulla. The basic shape of the eutherian adrenal is spheroid, however much it may deviate from this in adult form, and therefore the cortical cords must place themselves circumferentially around a central mass. Furthermore, the basic unit, the cortical cord, loops against the well-developed outer connective tissue capsule of the eutherian adrenal. With the inner circumference of the cortex rigidly defined by the central medulla which is not normally encroached upon, and with a looping of the cords necessitated by a firm capsule, the cords must of mechanical necessity be forced into radial lines when contained in a spheroid-like shape (text-fig. 35). The cortical cords so arranged depend on the amount of circulating ACTH for their precise histological expression. The part of the cord where normally most cell division takes place lies in the area of the outer zona fasciculata and this is completely under the influence of ACTH. The basic adrenotrophin/adrenocortical-hormone relationship does not support the maximum possible amount of zona fasciculata and there exists a potential annular space between the outer connective tissue capsule and the zona fasciculata. This is occupied by the loops of the cords budded off from the outer zona fasciculata cells constituting a backwater of cortical cells, the zona glomerulosa. This zone can be transformed into zona fasciculata when the amount of circulating ACTH is sufficient, and further cells which arise from the increased cell-division following increased ACTH will be similarly changed. When the production of extra ACTH ceases, the amount of ACTH available maintains a smaller annular volume of zona fasciculata (and zona reticularis) against the unchanging inner circumference, and the difference between this and the outer circumference (that is, the capsule which does not necessarily re-accommodate itself to diminished cortical volume with rapidity) is again taken up by cortical cells, unchanged by ACTH, thus reforming the zona glomerulosa.

The innermost part of the cortex is made up of the cords which are compressed against the medulla and form a zona reticularis which comprises cells in different stages of declination of activity and therefore of varying responsiveness to

Correlation of Structure and Function

ACTH. That the zona reticularis is made up of the continuation of the fasciculate cords, and that the cords are of the type common throughout the vertebrates, is best seen in the reforming enucleated glands where the cords, initially less compressed, grow in a way reminiscent of less hampered cortical cords in lower vertebrate glands. Increased amounts of ACTH can stimulate the cells of the zona reticularis to various degrees, depending on the extent of their slow declination of activity. Moreover, it is suggested that in some special circumstances gonadotrophins may be circulating in such considerable amounts that the inner zona fasciculata cells as they decline in responsiveness to ACTH may react to them—as in the postpubertally castrated male mouse. Of course, such inner fasciculate cells would respond to greater amounts of ACTH also, and post-castration enhancement of gonadotrophin secretion may stimulate synergistically. The zona intermedia, on this suggested theory, would not then be a blood stop-cock mechanism of capillary compression (Lever, Cater and Stack-Dunne, 1953; Cater and Lever, 1954; Lever, 1954, 1955) but would represent the products of cell division, particularly those of the outer zona fasciculata. The number of zona intermedia cells at any one time would depend on the amount of circulating ACTH. Normally some zona intermedia cells would remain unstimulated as zona glomerulosa, or be transformed by ACTH into fasciculate cells. Increasing ACTH would increase this latter transformation, with concomitant disappearance of the zona glomerulosa. On this theory both the glomerulosa and fasciculata would secrete aldosterone, the former minimal and the latter zone, responding to ACTH, copious amounts of other corticosteroids.

The evolutionary changes in cortical morphology are bound up closely with the distribution of chromaffin tissue. Its aggregation into a central medulla in higher forms may be connected with the dependence of the methylation of noradrenaline to adrenaline on cortical secretion. In the dogfish, where cortical and chromaffin tissues are entirely separate, the latter contains wholly noradrenaline (Coupland, 1953). In amphibians, where the two tissues are intermingled, noradrenaline forms 50 to 69 % of the total catechol content in the frog (Coupland, 1953)

The Adrenal Cortex

and 35 to 58 % in the toad (Houssay, Gerschman and Rapela, 1950). In lizards the peripheral layer of chromaffin tissue seems to secrete only noradrenaline while the islets and 'tongues', intermixing with cortical tissue, form adrenaline (Wright and Chester Jones, 1955). In the Eutheria, where the medulla receives the venous drainage of the cortex, mostly adrenaline is produced (Hillarp and Hökfelt, 1954; Coupland, 1953; Eränkö, 1955). The fowl adrenal is exceptional in that, despite close intermingling of cortical and chromaffin tissues, it contains 70 to 80 % noradrenaline (Shepherd and West, 1951). The physiological significance of varying proportions of adrenaline to noradrenaline is not known. If one is revealed it may help to resolve the problem of the differing association of cortical with chromaffin tissue shown in the vertebrate series.

Consideration of the possibility of evolutionary change in cortical function requires the answers to two questions: firstly, are similar adrenocortical hormones secreted throughout the vertebrates? and secondly, are their activities essentially similar? We do not know the answers to either of these questions. We may speculate, however, that, as the body fluids and tissues have a similar ionic constitution in all vertebrates (with the exception of the Myxinoidea; Robertson, 1954), a fine control of the inorganic ion content would be required. This, it is suggested, is provided by the adrenocortical hormones, having as their primary site of action the epithelium of the kidney tubule.

The bony fish in fresh water, faced with the problem of an external medium hypotonic to its body fluids, has not only to excrete water but also to conserve salts which tend to diffuse outwards. A hypotonic urine is formed, so that there is selective resorption of salts but not water in the kidney tubule. It is this selective resorption which we consider is controlled by the cortical hormones, constituting a pre-requisite for the establishment of the early fish in fresh water.

Bony fish in sea water and elasmobranchs have other extrarenal excretory methods 'superimposed' on this basic mechanism. Marine teleosts, in a medium hypertonic to their body fluids, must conserve water and excrete salts. Nevertheless they form a hypotonic urine and this implies that the suggested action of the cortical hormones on the kidney tubule is still taking place. We do not know if the excretion of salts extra-renally is under hormonal control (Smith, 1932). It may be that the tendency to take in sea water, hypertonic to body fluids, might trigger off neurohypophysial secretion and in-fluence gill epithelium to form 'chloride-secreting' cells. Per-haps with such an excretory mechanism at work there is less demand on the kidney, and this may account for the apparent quiescence of the interrenal observed in some teleosts.

The elasmobranchs, though they are exceptional in accumu-lating urea giving body fluids hyperosmotic to sea water, also produce hypotonic urine, and the salt content of the blood tends to be regulated to narrower limits than either the urea or the total osmotic pressure (Smith, 1932). Again we may suppose that an extra-renal mechanism has been added to the basic pattern of excretion by the kidney. In the cyclostomes, while the Petromyzontidae have values of body fluid ions charac-teristic of those in vertebrates generally, the Myxinoidea are isosmotic or hyperosmotic with sea water. The mechanism of ionic regulation is not known, but Robertson considers that resorption and secretion of ions may well occur in the kidney tubules so that, in both groups of cyclostomes, adrenocortical hormones may have their role to play. In vertebrates from amphibians onwards the cortical hormones act at the kidney level but also enter more and more into relationship with neurohypophysial secretions. By this means a more varied and finer control of water and salt-electrolyte metabolism is achieved.

The hypothesis that the prime necessity for the evolution of vertebrates was ionic regulation at the kidney level by cortical hormones seems to indicate that their control of carbohydrate metabolism is only a secondary function. It is probable, how-ever, that as electrolyte regulation is essential for the processes concerned in energy production, the influence of the cortical hormones on carbohydrate metabolism goes hand in hand with that on salt-electrolyte metabolism. The picture here will be clearer when we have more facts about the lower vertebrates. It is the hope that this book will stimulate workers to find them.

REFERENCES

ABELOUS, J. E. and LANGLOIS, P. (1891 *a*). La mort des grenouilles après la destruction des deux capsules surrénales. *C.R. Soc. Biol., Paris,* **43**, 855.

ABELOUS, J. E. and LANGLOIS, P. (1891 *b*). Note sur les fonctions des capsules surrénales chez la grenouille. *C.R. Soc. Biol., Paris,* **43**, 792.

ABELOUS, J. E. and LANGLOIS, P. (1892). Action toxique du sang après destruction des capsules surrénales. *C.R. Soc. Biol., Paris,* **44**, 165.

ABOIM, A. N. (1939). La graisse de l'organe interrénal des Sélaciens. *Bull. Soc. portug. Sci. nat.* **13**, 61.

ABOIM, A. N. (1944). L'organe interrénal des Sélaciens. Étude cytologique, histochimique et histophysiologique. *Arch. portug. Sci. biol.* **7**, 89.

ABOIM, A. N. (1946). L'organe interrénal des cyclostomes et des poissons. *Portug. acta biol.* **1**, sér. A, no. 4, 353.

ADAMS, A. E. and BOYD, E. M. (1933). Changes in the adrenals of newts following hypophysectomy or thyroidectomy. *Anat. Rec.* **57**, 34 (suppl.).

ADAMS, A. E. and HARLAND, M. (1929). The effects of adrenalectomy in the lizard. *Anat. Rec.* **41**, 42.

ADDISON, T. (1855). *On the constitutional and local effects of disease of the suprarenal capsules.* London: Highley.

ADOLPH, E. F. (1925). The passage of water through the skin of the frog and the relation between diffusion and permeability. *Amer. J. Physiol.* **73**, 85.

ALBANESE, M. (1892 *a*). Recherches sur la fonction des capsules surrénales. *Arch. ital. Biol.* **18**, 49.

ALBANESE, M. (1892 *b*). La fatigue chez les animaux privés de capsules surrénales. *Arch. ital. Biol.* **18**, 17.

ALBERT, S. and LEBLOND, C. P. (1946). The distribution of the Feulgen and 2, 4-dinitrophenylhydrazine reactions in normal, castrated, adrenalectomized and hormonally treated rats. *Endocrinology,* **39**, 386.

ALBERT, S. and LEBLOND, C. P. (1949). Age changes revealed by carbonyl reagents in tissue sections. *J. Anat., Lond.,* **83**, 183.

ALBRIGHT, F. (1943). Cushing's Syndrome. *Harvey Lect.* **38**, 122.

ALDRICH, T. B. (1901). Adrenalin, the active principle of the adrenals. *Amer. J. Phsyiol.* **5**, 457.

ALLERS, W. D. (1935). The influence of diet and mineral metabolism on dogs after suprarenalectomy. *Proc. Mayo Clin.* **10**, 406.

ALLERS, W. D. and KENDALL, E. C. (1937). Maintenance of adrenalectomized dogs without cortin through control of the mineral constituents of the diet. *Amer. J. Physiol.* **118**, 87.

ALPERT, L. K. (1931). The innervation of the suprarenal glands. *Anat. Rec.* **50**, 221.

ALPERT, M. (1950). Observations on the histophysiology of the adrenal gland of the golden hamster. *Endocrinology,* **46**, 166.

References

ALTHAUSEN, T. L., ANDERSON, E. M. and STOCKHOLM, M. (1939). Effect of adrenalectomy and of NaCl on intestinal absorption of dextrose. *Proc. Soc. exp. Biol., N.Y.*, **40**, 342.

AMES, R. G. and VAN DYKE, H. B. (1952). Antidiuretic hormone in the serum or plasma of rats. *Endocrinology*, **50**, 350.

AMOUR, M. C. d' and AMOUR, F. E. d' (1939). Effect of luteinization on the survival of adrenalectomized rats. *Proc. Soc. exp. Biol., N.Y.*, **40**, 417.

ANDERSON, E. M., HERRING, V. V. and JOSEPH, M. (1940). Salt after adrenalectomy. III. Carbohydrate stores in adrenalectomized rats given various levels of sodium chloride. *Proc. Soc. exp. Biol., N.Y.*, **45**, 488.

ANGERER, C. A. (1950). Body weight, survival times, coloration and water content of skeletal muscles of adrenalectomized frogs. *Ohio J. Sci.* **50**, 103.

ANGERER, C. A. and ANGERER, H. H. (1942). The effect of dehydration on the viability of adrenal-insufficient frogs. *Fed. Proc.* **1**, 3.

ANGERER, C. A. and ANGERER, H. H. (1949). The rate and total loss of body water in the survival time of adrenalectomized frogs. *Proc. Soc. exp. Biol., N.Y.*, **71**, 661.

ANGEVINE, D. M. (1938). Pathologic anatomy of hypophysis and adrenals in anencephaly. *Arch. Path. (Lab. Med.)* **26**, 507.

ANSON, B. J., CAULDWELL, E. W., PICK, J. W. and BEATON, L. E. (1947). The blood supply of the kidney, suprarenal gland, and associated structures. *Surg. Gynec. Obstet.* **84**, 313.

ARNOLD, J. (1866). Ein Beitrag zu der feineren Structur und dem Chemismus der Nebennieren. *Virchows Arch.* **35**, 64.

ARTUNDO, A. (1927). La glycémie, le glycogène et l'action de l'insuline chez les rats décapsulés. *C.R. Soc. Biol., Paris*, **97**, 411.

ASLING, C. W., REINHARDT, W. D. and LI, C. H. (1951). Effects of adreno-corticotropic hormone on body growth, visceral proportions, and white blood cell counts of normal and hypophysectomized male rats. *Endocrinology*, **48**, 534.

ASTWOOD, E. B. (1953). Some recent developments in the clinical use of cortical steroids and corticotropin. In *The Suprarenal Cortex. Proc. 5th Symp. Colston Res. Soc.* p. 213. Ed. Yoffey. London: Butterworth.

ASTWOOD, E. B., RABEN, M. S. and PAYNE, R. W. (1952). Chemistry of corticotrophin. *Recent Progr. Hormone Res.* **7**, 1.

ASTWOOD, E. B., RABEN, M. S., PAYNE, R. W. and GRADY, A. B. (1951). Purification of corticotropin with oxycellulose. *J. Amer. chem. Soc.* **73**, 2969.

ATWELL, W. J. (1932). Effects of administration of cortico-adrenal extract to the hypophysectomized anuran. *Proc. Soc. exp. Biol., N.Y.*, **29**, 621.

ATWELL, W. J. (1935). Effects of thyrotropic and adrenotropic principle on hypophysectomized amphibia. *Anat. Rec.* **62**, 361.

ATWELL, W. J. (1937). The effects of administering thyrotropic and adrenotropic extract to thyroidectomized and hypophysectomized tadpoles. *Amer. J. Physiol.* **118**, 452.

The Adrenal Cortex

AXELRAD, B. J., JOHNSON, B. B. and LUETSCHER, J. A., Jr. (1954). Factors regulating the output of sodium-retaining corticoid of human urine. *J. clin. Endocrin.* **14**, 783. (Abst. 43.)

BAAR, H. S. (1954). Foetal cortex of the adrenal glands. *Lancet*, **i**, 670.

BACHMANN, R. (1941). Nebennierenstudien. *Ergbn. Anat. EntwGesch.* **33**, 31.

BACHMANN, R. (1954). *Die Nebenniere. Handb. mikrosk. Anat. Mensch.* **6**, 1. Berlin: Springer-Verlag.

BACILA, M. and BARRON, E. S. G. (1954). The effect of adrenal cortical hormones on the anaerobic glycolysis and hexokinase activity. *Endocrinology*, **54**, 591.

BACSICH, P. and FOLLEY, S. J. (1939). The effect of oestradiol monobenzoate on the gonads, endocrine glands, and mammae of lactating rats. *J. Anat., Lond.*, **73**, 432.

BAECKER, R. (1928). Über die Nebennieren der Teleostier. *Z. mikr.-anat. Forsch.* **15**, 204.

BAKER, B. L. (1950). Modification of body structure by adrenocortical secretions with special reference to the regulation of growth. *Symposium on Pituitary-Adrenal function. Amer. Ass. Adv. Sci.* p. 88. Baltimore: Horn-Shafer.

BAKER, B. L. (1951). The relationship of the adrenal, thyroid and pituitary glands to the growth of hair. *Ann. N.Y. Acad. Sci.* **53**, 690.

BAKER, B. L. (1952). A comparison of the histological changes induced by experimental hyperadrenocorticalism and inanition. *Recent Progr. Hormone Res.* **7**, 331.

BAKER, B. L. (1954). The connective tissue reaction around implanted pellets of steroid hormones. *Anat. Rec.* **119**, 529.

BAKER, B. L. and INGLE, D. J. (1948). Growth inhibition in bone and bone marrow following treatment with adrenocorticotropin. *Endocrinology*, **43**, 422.

BAKER, B. L., INGLE, D. J. and LI, C. H. (1950). Increase in glyceride content of brown fat by treatment with adrenocorticotropin. *Proc. Soc. exp. Biol., N.Y.*, **73**, 337.

BAKER, B. L., INGLE, D. J. and LI, C. H. (1951). The histology of the lymphoid organs of rats treated with adrenocorticotropin. *Amer. J. Anat.* **88**, 313.

BAKER, B. L., INGLE, D. J., LI, C. H. and EVANS, H. M. (1948). Growth inhibition in the skin induced by parenteral administration of adrenocorticotropin. *Anat. Rec.* **102**, 313.

BAKER, B. L. and WHITAKER, W. L. (1948). Growth inhibition in the skin following direct application of adrenal cortical preparations. *Anat. Rec.* **102**, 333.

BAKER, B. L. and WHITAKER, W. L. (1949). Relationship of the adrenal cortex to inhibition of growth of hair by estrogen. *Amer. J. Physiol.* **159**, 118.

BAKER, B. L. and WHITAKER, W. L. (1950). Interference with wound healing by the local action of adrenocortical steroids. *Endocrinology*, **46**, 544.

References

BAKER, D. D. (1937). Studies of the suprarenal glands of dogs. I. Comparison of the weights of suprarenal glands of mature and immature male and female dogs. *Amer. J. Anat.* **60**, 231.

BAKER, D. D. (1938). Comparison of the weights of suprarenals of dogs in oestrus, pregnancy and lactation. *J. Morph.* **62**, 3.

BAKER, D. D. and BAILLIF, R. N. (1939). The role of capsule in suprarenal regeneration studied with the aid of colchicine. *Proc. Soc. exp. Biol.,* *N.Y.,* **40**, 117.

BAKER, J. R. (1946). The histochemical recognition of lipine. *Quart. J. micr. Sci.* **87**, 441.

BALFOUR, F. M. (1878). *A monograph on the development of Elasmobranch fishes.* London: Macmillan and Co.

BALFOUR, F. M. (1882). On the nature of the organ in adult teleosteans and ganoids which is usually regarded as the head kidney or pronephros. *Quart. J. micr. Sci.* **22**, 12.

BALFOUR, W. E. (1953). Changes in the hormone output of the adrenal cortex of the young calf. *J. Physiol.* **122**, 59 P.

BANTING, F. G. and GAIRNS, S. (1926). Adrenal insufficiency in dogs. *Amer. J. Physiol.* **77**, 100.

BARGMANN, W. (1933). Über den Bau der Nebennierenvenen des Menschen und der Säugetiere. *Z. Zellforsch.* **17**, 118.

BARRON, E. S. G. (1951). Thiol groups of biological importance. *Advanc. Enzymol.* **11**, 201.

BARTHOLINUS, C. (1611). *Anatomicae Institutiones Corporis Humani.* Wittenberg.

BARTLETT, G. R., WICK, A. N. and MACKAY, E. M. (1949). Influence of insulin and adrenal cortical compounds on the metabolism of radioactive C^{14} glucose in the isolated rat diaphragm. *J. biol. Chem.* **178**, 1003.

BASIR, M. A. (1932). The histology of the spleen and suprarenals of Echidna. *J. Anat., Lond.,* **66**, 628.

BATES, R. W., RIDDLE, O. and MILLER, R. A. (1940). Preparation of adrenotropic extracts and their assay on two-day chicks. *Endocrinology,* **27**, 781.

BAUMANN, E. J. and KURLAND, S. (1927). Changes in the inorganic constituents of blood in suprarenalectomized cats and rabbits. *J. biol. Chem.* **71**, 281.

BAXTER, J. S. (1946). The growth cycle of the cells of the adrenal cortex in the adult rat. *J. Anat., Lond.,* **80**, 139.

BAYLISS, R. I. S. (1955). Factors influencing adrenocortical activity in health and disease. *Brit. med. J.* **i**, 495.

BAYLISS, W. M. and STARLING, E. H. (1902). The mechanism of pancreatic secretion. *J. Physiol.* **28**, 325.

BEATTY, B. (1940). A comparative account of the adrenal glands in *Rana temporaria* and *R. esculenta. Proc. Leeds phil. lit. Soc.* **3**, 633.

BECKS, H., SIMPSON, M. E., LI, C. H. and EVANS, H. M. (1944). Effects of adrenocorticotropic hormone (ACTH) on the osseous system in normal rats. *Endocrinology,* **34**, 305.

BELL, P. H. (1954). Purification and structure of β-corticotropin. *J. Amer. chem. Soc.* **76**, 5565.

BENEDICT, F. G. (1938). Vital energetics. A study in comparative basal metabolism. *Carnegie Inst. Wash. Publ.* **503**, 3.

BENNETT, H. S. (1940). The life history and secretion of the cells of the adrenal cortex of the cat. *Amer. J. Anat.* **67**, 151.

BENNETT, H. S. and KILHAM, L. (1940). The blood vessels of the adrenal gland of the adult cat. *Anat. Rec.* **77**, 447.

BENNETT, L. L., LIDDLE, G. W. and BENTNICK, R. C. (1953). Does a large intake of potassium modify the metabolic effects of ACTH (corticotropin) in man? *J. clin. Endocrin.* **13**, 392.

BENNETT, L. L. and PERKINS, R. Z. (1945). The maintenance of muscle glycogen in fasted hypophysectomized-adrenalectomized rats. *Endocrinology*, **36**, 24.

BENOIT, J. and ASSENMACHER, I. (1953). Rapport entre la stimulation sexuelle préhypophysaire et la neurosécrétion chez l'oiseau. *Arch. Anat. micr. Morph. exp.* **42**, 334.

BENOIT, J. and STAUBLI, A. (1951). Modification d'une coloration trichrome de Masson pour la mise en évidence supplémentaire du chondriome. *C.R. Ass. Anat.* 38th reunion, 19–21 March 1951.

BENUA, R. S. and HOWARD, E. (1950). A carbonyl reaction differentiating the fetal zona reticularis of the human adrenal cortex from the mouse X zone. *Johns Hopk. Hosp. Bull.* **86**, 200.

BERGNER, G. E. and DEANE, H. W. (1948). Effect of pituitary adrenocorticotrophic hormone on the intact rat, with special reference to cytochemical changes in the adrenal cortex. *Endocrinology*, **43**, 240.

BERLINER, R. W. (1952). Renal secretion of potassium and hydrogen ions. *Fed. Proc.* **11**, 695.

BERNARD, C. (1855). *Leçons de Physiologie expérimentale au Collège de France.* Paris.

BERNARD, C. (1859). *Leçons sur les propriétés physiologiques et les altérations pathologiques des liquides de l'organisme*, **2**, 441.

BERNARD, L. and BIGART, M. (1902). Sur les réactions histologiques générales des surrénales à certaines influences pathogènes expérimentales. *C.R. Soc. Biol., Paris*, **54**, 1219.

BERTHOLD, A. (1849). Transplantation der Hoden. *Arch. Anat. Physiol., Lpz.*, p. 42.

BESSESSEN, A. N., Jr. and CARLSON, H. A. (1923). Postnatal growth in weight of the body and of the various organs in the guinea pig. *Amer. J. Anat.* **31**, 483.

BIBILE, S. W. (1953). The assay of cortical steroids by the mouse eosinophil test. *J. Endocrin.* **9**, 357.

BIEDL, A. (1913). *Innere Sekretion.* 2 Auflage. Teil 2. Berlin und Wien: Urban und Schwarzenberg.

BIERRY, H. and MALLOIZEL, L. (1908). Hypoglycémie après décapsulation, effets de l'injection d'adrénaline sur les animaux décapsulés. *C.R. Soc. Biol., Paris*, **65**, 232.

References

BILLINGHAM, R. E., KROHN, P. L. and MEDAWAR, P. B. (1951 a). Effect of cortisone on survival of skin homografts in rabbits. *Brit. med. J.* i, 1157.

BILLINGHAM, R. E., KROHN, P. L. and MEDAWAR, P. B. (1951 b). Effect of locally applied cortisone acetate on survival of skin homografts in rabbits. *Brit. med. J.* ii, 1049.

BIMMER, E. (1950). Metrische Untersuchungen über die Entwicklung der Nebenniere und der von ihr benachbarten Organe bei Eidechsen. *Anat. Anz.* 97, 276.

BIRMINGHAM, M. K., ELLIOTT, F. H. and VALÈRE, P. H. L. (1953). The need for the presence of calcium for the stimulation *in vitro* of rat adrenal glands by adrenocorticotrophic hormone. *Endocrinology*, 53, 687.

BISHOP, P. M. F. (1954). *Recent advances in endocrinology.* 7th ed. London: J. and A. Churchill, Ltd.

BLACKMAN, S. S. (1946). Concerning the function and origin of the reticular zone of the adrenal cortex. *Johns Hopk. Hosp. Bull.* 78, 180.

BLISS, E. L., SANDBERG, A. A., NELSON, D. H. and EIK-NES, K. (1953). The normal levels of 17-hydroxycorticosteroids in the peripheral blood of man. *J. clin. Invest.* 32, 818.

BLOCH, K. (1951). The biological synthesis of cholesterol. *Recent Progr. Hormone Res.* 6, 111.

BLUMENTHAL, H. T. (1940). The mitotic count in the adrenal cortex of normal guinea pigs. *Endocrinology*, 27, 477.

BLUNT, J. W., Jr., PLOTZ, C. M., LATTES, R., HOWES, E. L., MEYER, K. and RAGAN, C. (1950). Effect of cortisone on experimental fractures in the rabbit. *Proc. Soc. exp. Biol., N.Y.*, 73, 678.

BOBIN, G. (1948). Images histo-cytologiques des corpuscles de Stannius de l'anguille européenne. *Arch. Zool. exp. gén.* 86, 1.

DE BODO, R. C., SINKOFF, M. W., KURTZ, M., LANE, N. and KIANG, S. P. (1953). Significance of adrenocortical atrophy in the carbohydrate metabolism of hypophysectomized dogs. *Amer. J. Physiol.* 173, 11.

BØGGILD, D. H. (1925). The importance of the adrenals in the regulation of blood sugar. *Acta path. microbiol. scand.* 2, 68.

BOJANUS, M. (1819). *Anatomia testudinis.* Wilnae.

BONDY, P. K., INGLE, D. J. and MEEKS, R. C. (1954). Influence of adrenal cortical hormones upon the level of plasma amino acids in eviscerate rats. *Endocrinology*, 55, 354.

BORDEU, T. DE (1775). Recherches sur les maladies chroniques. *Œuvres complètes*, 2, 942.

BOTELLA LLUSIA, J. and CANO MONASTERIO, A. (1950). Investigaciones sobre la zona sexual suprarrenal. III. Acción de la gonadotropina coriónica sobre la vesicula seminal del ratón castrado. *Trab. Inst. nac. Cienc. méd., Madr.*, 13, 81.

BOUIN, P. (1932). *Éléments d'histologie.* 2 ed. Paris.

BOURNE, G. H. (1934 a). A study on the Golgi apparatus of the adrenal gland. *Aust. J. exp. Biol. med. Sci.* 12, 123.

BOURNE, G. H. (1934b). Unique structure in the adrenal of the female opossum. *Nature, Lond.*, **134**, 664.

BOURNE, G. H. (1949). *The mammalian adrenal gland.* Oxford: Clarendon Press.

BOURNE, G. H. (1955). Aspects of the histochemistry of the adrenal cortex. *Ciba Found. Coll. End.* **8**, 1.

BOURNE, G. H. and MALATY, H. A. (1953). The effect of adrenalectomy, cortisone and other steroid hormones on the histochemical reaction for succinic dehydrogenase. *J. Physiol.* **122**, 178.

BOURNE, G. H. and ZUCKERMAN, S. (1940). The influence of the adrenals on cyclical changes in the accessory reproductive organs of female rats. *J. Endocrin.* **2**, 268.

BOYD, E. M. and YOUNG, F. M. (1940). Optical stimuli and water balance in frogs. *Endocrinology*, **27**, 137.

BOYLE, P. J. and CONWAY, E. J. (1941). Potassium accumulation in muscle and associated changes. *J. Physiol.* **100**, 1.

BRADY, R. O. and GURIN, S. (1951). The synthesis of radioactive cholesterol and fatty acids *in vitro. J. biol. Chem.* **189**, 371.

BRADY, R. O., LUKENS, F. D. W. and GURIN, S. (1951). Synthesis of radioactive fatty acids *in vitro* and its hormonal control. *J. biol. Chem.* **193**, 459.

BRANIN, M. L. (1937). The development of the cortical adrenal in the four-toed salamander *Hemidactylium scutatum* (Schlegel). *J. Morph.* **60**, 521.

BRAUN, M. (1879). Ueber Bau und Entwickelung in Nebennieren bei Reptilien. *Zool. Anz.* **2**, 238.

BRENEMAN, W. R. (1941). Growth of the endocrine glands and viscera in the chick. *Endocrinology*, **28**, 946.

BRINK, N. G., BOXER, G. E., JELINEK, V. C., KUEHL, F. A., RICHTER, J. W., and FOLKERS, K. (1953). Pituitary hormones. VII. The nature of corticotropin-B. *J. Amer. chem. Soc.* **75**, 1960.

BRITTON, S. W., KLINE, R. F. and SILVETTE, H. (1938). Blood-chemical and other conditions in normal and adrenalectomized sloths. *Amer. J. Physiol.* **123**, 701.

BRITTON, S. W. and SILVETTE, H. (1931). Some effects of cortico-adrenal extract and other substances on adrenalectomized animals. *Amer. J. Physiol.* **99**, 15.

BRITTON, S. W. and SILVETTE, H. (1932a). The effect of cortico-adrenal extract on carbohydrate metabolism in normal animals. *Amer. J. Physiol.* **100**, 693.

BRITTON, S. W. and SILVETTE, H. (1932b). The apparent prepotent function of the adrenal glands. *Amer. J. Physiol.* **100**, 701.

BRITTON, S. W. and SILVETTE, H. (1937a). Further observations on sodium chloride balance in the adrenalectomized opossum. *Amer. J. Physiol.* **118**, 21.

BRITTON, S. W. and SILVETTE, H. (1937b). A comparison of sodium, chloride, and carbohydrate changes in adrenal insufficiency and other experimental conditions. *Amer. J. Physiol.* **118**, 594.

References

BRITTON, S. W., SILVETTE, H. and KLINE, R. F. (1938). Adrenal insufficiency in American monkeys. *Amer. J. Physiol.* **123**, 705.

BROSTER, L. R. and VINES, H. W. C. (1933). *The Adrenal Cortex.* London: H. K. Lewis.

BROWNE, J. S. L. (1951). ACTH and cortisone in disease. Reported in *Brit. med. J.* **i**, 880.

BROWNE, J. S. L. (1952). *Proc. Conf. on Effects of Cortisone*, p. 27. Merck, N.Y.

BROWNE, J. S. L., BECK, J. C., DYRENFURTH, I., GIROUD, C. J. P., HAWTHORNE, A. B., JOHNSON, L. G., MACKENZIE, K. R. and VENNING, E. H. (1955). Cushing's syndrome. *Ciba Found. Coll. End.* **8**, 505.

BROWN-SÉQUARD, C. E. (1856). Recherches expérimentales sur la physiologie et la pathologie des capsules surrénales. *Arch. gén. Méd.*, série 5, 8, pp. 385 and 572.

BROWN-SÉQUARD, C. E. (1857). Nouvelles recherches sur les capsules surrénales. *C.R. Acad. Sci.* **45**, 1036.

BROWN-SÉQUARD, C. E. (1858). Nouvelles recherches sur l'importance des fonctions des capsules surrénales. *J. Physiol. Path. gén.* **1**, 160.

BROWN-SÉQUARD, C. E. (1889). Expérience démontrant la puissance dynamogénique chez l'homme d'un liquide extrait de testicules d'animaux. *Arch. Physiol. norm. path.* **21**, 651.

BRUNER, J. A. (1951). Distribution of chorionic gonadotropin in mother and fetus at various stages in pregnancy. *J. clin. Endocrin.* **11**, 360.

BUELL, M. V. and TURNER, E. (1941). Cation distribution in the muscles of adrenalectomized rats. *Amer. J. Physiol.* **134**, 225.

BÜLBRING, E. (1937*a*). Relation between size of testes and requirement of cortical extract in adrenalectomized drakes. *J. Physiol.* **91**, 18 P.

BÜLBRING, E. (1937*b*). The standardization of cortical extracts by the use of drakes. *J. Physiol.* **89**, 64.

BÜLBRING, E. (1940). The relation between cortical hormone and the size of the testis in the drake, with some observations on the effect of different oils as solvents and on desoxycorticosterone acetate. *J. Pharmacol.* **69**, 52.

BULLIARD, H., MAILLET, M. and DROZ, B. (1953). Surrénale de *Rana esculenta* et ACTH. *C.R. Ass. Anat.* **39**, 617.

BURGESS, W. W., HARVEY, A. M. and MARSHALL, E. K. (1933). The site of the antidiuretic action of pituitary extract. *J. Pharmacol.* **49**, 237.

BURIAN, R. (1910). Funktion der Nierenglomeruli und Ultrafiltration. *Arch. ges. Physiol.* **136**, 741.

BURRILL, M. W. and GREENE, R. R. (1939). Androgenic function of the adrenals in the immature male castrate rat. *Proc. Soc. exp. Biol., N.Y.*, **40**, 327.

BURROWS, H. (1949). *Biological actions of sex hormones.* 2nd ed. Cambridge University Press.

BUSH, I. E. (1953). Species differences and other factors influencing adrenocortical secretion. *Ciba Found. Coll. End.* **7**, 210.

BUTCHER, E. O. (1939). Effect of adrenalectomy on the growth of mammary glands in underfed albino rats. *Proc. Soc. exp. Biol., N.Y.*, **42**, 571.

BUTTLE, J. M., KIRK, R. L. and WARING, H. (1952). The effect of complete adrenalectomy on the wallaby (*Setonyx brachyurus*). *J. Endocrin.* **8**, 281.

CAIN, A. J. (1949*a*). On the significance of the plasmal reaction. *Quart. J. micr. Sci.* **90**, 75.

CAIN, A. J. (1949*b*). A critique of the plasmal reaction, with remarks on recently proposed techniques. *Quart. J. micr. Sci.* **90**, 411.

CAIN, A. J. (1950). The histochemistry of lipoids in animals. *Biol. Rev.* **25**, 73.

CALLAMAND, O. (1943). L'Anguille européenne (*Anguilla anguilla* L.). Les bases physiologiques de sa migration. *Ann. Inst. océanogr. Monaco*, **21**, 361.

CALLOW, R. K., LLOYD, J. and LONG, D. A. (1954). A chloro-derivative of cortisone with enhanced activity. *Lancet*, **ii**, 20.

CALLOW, R. K. and YOUNG, F. G. (1936). Relations between optical rotatory power and constitution in the steroids. *Proc. roy. Soc.* A, **157**, 194.

CALMA, I. and FOSTER, C. L. (1943). Trypan blue and cell migration in the adrenal cortex of rats. *Nature, Lond.*, **152**, 536.

CANO MONASTERIO, A. (1946). La zona sexual en la corteza suprarrenal. Histofisiología. *Trab. Inst. Cajal Invest. biol.* **38**, 129.

CAPRARO, V. and GARAMPI, M. L. (1956). The hormonal control of water and salt-electrolyte metabolism—studies with the isolated frog skin. *Mem. Soc. Endocrin.* **5**, 60.

CARPENTER, W. B. (1852). Secretion. *R. B. Todd's Cyclopaedia of Anatomy and Physiology, London*, **4**, 439.

CARTER, S. B. (1954). *The influence of oestrogen on the adrenal gland of the female rat*. Ph.D. Thesis, University of Birmingham.

CATER, D. B. and LEVER, J. D. (1954). The zona intermedia of the adrenal cortex. A correlation of possible functional significance with development, morphology, and histochemistry. *J. Anat., Lond.*, **88**, 437.

CATER, D. B. and STACK-DUNNE, M. P. (1953). The histological changes in the adrenal of the hypophysectomized rat after treatment with pituitary preparations. *J. Path. Bact.* **66**, 119.

CHEN, G., OLDHAM, F. K. and GEILING, M. K. (1943). Effect of posterior pituitary extract on water uptake in frogs after hypophysectomy and infundibular lesions. *Proc. Soc. exp. Biol., N.Y.*, **52**, 108.

CHESTER JONES, I. (1948). Variation in the mouse adrenal cortex with special reference to the zona reticularis and to brown degeneration, together with a discussion of the 'cell migration' theory. *Quart. J. micr. Sci.* **89**, 53.

CHESTER JONES, I. (1949*a*). The relationship of the mouse adrenal cortex to the pituitary. *Endocrinology*, **45**, 514.

CHESTER JONES, I. (1949*b*). The action of testosterone on the adrenal cortex of the hypophysectomized, prepuberally castrated male mouse. *Endocrinology*, **44**, 427.

CHESTER JONES, I. (1950). The effect of hypophysectomy on the adrenal cortex of the immature mouse. *Amer. J. Anat.* **86**, 371.

References

CHESTER JONES, I. (1952). The disappearance of the X zone of the mouse adrenal cortex during first pregnancy. *Proc. roy. Soc.* B, **139**, 398.

CHESTER JONES, I. (1953). A reconsideration of the relationship of the histological zones of the rat adrenal cortex to water and salt-electrolyte metabolism. *J. Endocrin.* **9**, xxxviii.

CHESTER JONES, I. (1955). The adrenal cortex in reproduction. *Brit. med. Bull.* **11**, 156.

CHESTER JONES, I. (1956 a). The role of the adrenal in the control of water and salt-electrolyte metabolism in Vertebrates. *Mem. Soc. Endocrin.* **5**, 102.

CHESTER JONES, I. (1956 b). Comparative aspects of adrenocortical-neuro-hypophysial relationships. In: *The Neurohypophysis*. Ed. Heller. London: Butterworth.

CHESTER JONES, I. and ROBY, C. C. (1954). Some aspects of zonation and function of the adrenal cortex. I. The effects of hypophysectomy on the adrenal cortex of the adult male mouse. *J. Endocrin.* **10**, 245.

CHESTER JONES, I. and SPALDING, M. H. (1954). Some aspects of zonation and function of the adrenal cortex. II. The rat adrenal after enuclea-tion. *J. Endocrin.* **10**, 251.

CHESTER JONES, I. and WRIGHT, A. (1954 a). Some aspects of zonation and function of the adrenal cortex. III. Self-selection after adrenal enu-cleation in rats. *J. Endocrin.* **10**, 262.

CHESTER JONES, I. and WRIGHT, A. (1954 b). Some aspects of zonation and function of the adrenal cortex. IV. The histology of the adrenal in rats with diabetes insipidus. *J. Endocrin.* **10**, 266.

CHEVREUL, M. (1815). Recherches chimiques sur plusieurs corps gras, et particulièrement sur leurs combinaisons avec les alcalis. 5th Mémoir *Ann. Chim. (Phys.)* **95**, 5.

CHIFFELLE, T. L. and PUTT, F. A. (1951). Propylene and ethylene glycol as solvents for sudan IV and sudan black B. *Stain Tech.* **26**, 51.

CHISTOVICH, O. F. (1931). Potassium and calcium contents of frog blood serum at different temperatures and resulting from stimulus of the central nervous system. (Strychnine and curare.) *Russ. J. Physiol.* **14**, 320. (See *Chemical Abstracts*, **28**, 6203 (1934).)

CHRISTENSEN, B. G. (1954). The size of the suprarenal medulla in hypo-physectomized rats. *Acta endocr., Copenhagen*, **15**, 247.

CICARDO, V. H. (1947). *Importancia biológica del potasio*. El Ateneo, Buenos Aires.

CLARK, I. (1950). Effect of cortisone on protein metabolism in the rat as studied with isotopic glycine. *Fed. Proc.* **9**, 161.

CLARK, W. G., BRACKNEY, E. L. and MILINER, R. A. (1944). Adrenal-ectomy in frogs and toads. *Proc. Soc. exp. Biol., N.Y.*, **57**, 222.

CLAUSEN, F. W. and FREUDENBERGER, C. B. (1939). A comparison of the effects of male and female sex hormones on immature female rats. *Endocrinology*, **25**, 585.

CLAYTON, B. E. and PRUNTY, F. T. G. (1951). Physiological factors affecting the response of experimental wound healing to ACTH. *J. Endocrin.* **7**, 362.

The Adrenal Cortex

CLEGHORN, R. A. (1932). Observation on extracts of beef adrenal cortex and elasmobranch interrenal body. *J. Physiol.* **75**, 413.

COLE, D. F. (1953a). The action of desoxycorticosterone acetate and of dietary sodium and potassium on skeletal muscle electrolyte in adrenalectomized rats. *Acta endocr., Copenhagen*, **14**, 245.

COLE, D. F. (1953b). Changes of cell sodium in response to administration of hypotonic saline in normal and adrenalectomized rats. *J. Physiol.* **121**, 18P.

COLE, D. F. (1953c). Action of desoxycorticosterone acetate and cortisone on response of adrenalectomized rats to hypotonic load. *Biochem. J.* **55**, xiii.

COLLINGE, W. E. and VINCENT, S. (1896). On the so-called suprarenal bodies in the cyclostomata. *Anat. Anz.* **12**, 232.

COLLINGS, W. D. (1941). The effect of experimentally induced pseudo-pregnancy upon the survival of adrenalectomized rats. *Endocrinology*, **28**, 75.

COLLIP, J. B., SELYE, H. and THOMPSON, D. L. (1933). Gonad-stimulating hormones in hypophysectomized animals. *Nature, Lond.*, **131**, 56.

COLOWICK, S. P., CORI, G. T. and SLEIN, M. W. (1947). The effect of adrenal cortex, anterior pituitary extracts and of insulin on the hexokinase reaction. *J. biol. Chem.* **168**, 583.

COMOLLI, A. (1913). Ricerche istologiche sull'interrenale dei Teleostei. *Arch. ital. Anat. Embriol.* **11**, 377.

CONN, J. W. (1948). Studies upon mechanisms involved in the induction with adrenocorticotropic hormone of temporary diabetes mellitus in man. *Proc. Amer. Diabetes Ass.* **8**, 215.

CONWAY, E. J. (1947). Exchange of K, Na and H ions between the cell and the environment. *Irish J. med. Sci.* Oct.–Nov. p. 593.

CONWAY, E. J. (1949). A redox theory of hydrochloric acid production by the gastric mucosa. *Irish J. med. Sci.* **282**, 801.

CONWAY, E. J. (1951). The biological performance of osmotic work. A redox pump. *Science*, **113**, 270.

CONWAY, E. J. (1952). The biological excretion of sodium and its relation to cellular water. *Ciba Found. Coll. Endocrin.* **4**, 417.

CONWAY, E. J. (1954). Some aspects of ion transport through membranes. *Symp. Soc. exp. Biol.* **8**, 297.

CONWAY, E. J. (1956). Fundamental problems in the hormonal control of water and salt-electrolyte metabolism. *Mem. Soc. Endocrin.* **5**, 3.

CONWAY, E. J. and HINGERTY, D. (1946). The influence of adrenalectomy on muscle constituents. *Biochem. J.* **40**, 561.

CONWAY, E. J. and HINGERTY, D. (1948). Relations between potassium and sodium levels in mammalian muscle and blood plasma. *Biochem. J.* **42**, 372.

CONWAY, E. J. and HINGERTY, D. (1953). The effects of cortisone, desoxycorticosterone and other steroids on the active transport of sodium and potassium ions in yeast. *Biochem. J.* **55**, 455.

References

COREY, E. L. and BRITTON, S. W. (1941). The antagonistic action of desoxy-corticosterone and post-pituitary extract on chloride and water balance. *Amer. J. Physiol.* **133**, 511.

COREY, E. L., SILVETTE, H. and BRITTON, S. W. (1939). Hypophysial and adrenal influence on renal function in the rat. *Amer. J. Physiol.* **125**, 644.

CORI, C. F. (1946). Enzymatic reactions in carbohydrate metabolism. *Harvey Lect.* **41**, 253.

CORI, C. F. and CORI, G. T. (1929). Fate of glucose and other sugars in the eviscerated animal. *Proc. Soc. exp. Biol., N.T.*, **26**, 432.

COSTA, A. C. DA (1913). Recherches sur l'histo-physiologie des glandes surrénales. *Arch. Biol., Paris*, **28**, 111.

COULSON, R. A. and HERNANDEZ, T. (1953). Glucose in Crocodilia. *Endocrinology*, **53**, 311.

COULSON, R. A., HERNANDEZ, T. and BRAZDA, F. G. (1950). Biochemical studies on the alligator. *Proc. Soc. exp. Biol., N.T.*, **73**, 203.

COUPLAND, R. E. (1953). On the morphology and adrenaline- nor-adrenaline content of chromaffin tissue. *J. Endocrin.* **9**, 194.

COURRIER, R., BACLESSE, M. and MAROIS, M. (1953). Rapports de la cortico-surrénale et de la sexualité. *J. Physiol. Path. gén.* **45**, 327.

COWDRY, E. V. (1950). *A textbook of histology.* 4th ed. London: Henry Kimpton.

COWIE, A. T. (1949). The influence of age and sex on the life span of adrenalectomized rats. *J. Endocrin.* **6**, 94.

COWIE, A. T. and FOLLEY, S. J. (1947). The role of the adrenal cortex in mammary development and its relation to the mammogenic action of the anterior pituitary. *Endocrinology*, **40**, 274.

COWIE, A. T. and FOLLEY, S. J. (1948). Adrenalectomy and replacement therapy in lactating rats. 5. The effect of adrenalectomy on lactation studied in pair-fed rats. *J. Endocrin.* **5**, 282.

COWIE, A. T. and STEWART, J. (1949). Adrenalectomy in the goat and its effects on the chemical constituents of the blood. *J. Endocrin.* **6**, 197.

COWIE, A. T. and TINDAL, J. S. (1955). Maintenance of lactation in adrenalectomized rats with aldosterone and 9α-halo derivatives of hydrocortisone. *Endocrinology*, **56**, 612.

CREDITOR, M. C., BEVANS, M., MUNDY, W. L. and RAGAN, C. (1950). Effect of ACTH on wound healing in humans. *Proc. Soc. exp. Biol., N.T.*, **74**, 245.

CRILE, G. and QUIRING, D. P. (1940). A record of the body weight and certain organ and gland weights of 3690 animals. *Ohio J. Sci.* **40**, 219.

CROOKE, A. C. and GILMOUR, J. R. (1938). A description of the effect of hypophysectomy on the growing rat, with the resulting histological changes in the adrenal and thyroid glands and the testicles. *J. Path. Bact.* **47**, 525.

CROOKSHANK, F. G. (1914). Deficiency of endocrine glandular secretion. In Report on Roy. Soc. Med. *Brit. med. J.* **i**, 369.

CSIK, L. and LUDANY, G. v. (1933). Die Zuckungskurve des Muskels nach Exstirpation der Nebennieren. *Pflüg. Arch. ges. Physiol.* **232**, 187.

The Adrenal Cortex

CUTULY, E. (1936). Quantitative study of the adrenals of hypophysecto-mized rats. *Anat. Rec.* **66**, 119.

CUTULY, E., CUTULY, E. C. and McCULLAGH, D. R. (1938). Spermato-genesis in immature hypophysectomized rats injected with androgens. *Proc. Soc. exp. Biol.*, *N.Y.*, **38**, 818.

CUVIER, G. (1805). *Leçons d'anatomie comparée.* Paris.

DALTON, A. J. and SELYE, H. (1939). The blood picture during the alarm reaction. *Folia. haemat. (Lpz.)*, **62**, 397.

DARLINGTON, J. M. (1937). The use of trypan blue in detecting cell death in the perfusion of the mammalian kidney, and the evaluation of some modified Ringer-Locke fluids by this method. *Anat. Rec.* **67**, 253.

DAVIDSON, C. S. and MOON, H. D. (1936). Effect of adrenocorticotropic extract on accessory reproductive organs of castrated rats. *Proc. Soc. exp. Biol.*, *N.Y.*, **35**, 281.

DAVIES, R. E. (1954). Relations between active transport and metabolism in some isolated tissues and mitochondria. *Symp. Soc. exp. Biol.* **8**, 453.

DAVIES, R. E. and GALSTON, A. W. (1951). Rapid rate of turnover of potassium ions in kidney slices. *Nature, Lond.*, **168**, 700.

DAVIES, S. (1937). The development of the adrenal gland of the cat. *Quart. J. micr. Sci.* **80**, 81.

DEANE, H. W. and GREEP, R. O. (1946). A morphological and histo-chemical study of the rat's adrenal cortex after hypophysectomy, with comments on the liver. *Amer. J. Anat.* **79**, 117.

DEANE, H. W. and LYMAN, C. P. (1954). Body temperature, thyroid and adrenal cortex of hamster during cold exposure and hibernation, with comparisons to rats. *Endocrinology*, **55**, 300.

DEANE, H. W. and SELIGMAN, A. M. (1953). Evaluation of procedures for the cytological localisation of ketosteroids. *Vitam. & Horm.* **11**, 173.

DEANE, H. W., SHAW, J. H. and GREEP, R. O. (1948). The effect of altered sodium or potassium intake on the width and cytochemistry of the zona glomerulosa of the rat adrenal cortex. *Endocrinology*, **43**, 133.

DEANESLY, R. (1928). A study of the adrenal cortex in the mouse and its relation to the gonads. *Proc. roy. Soc. B*, **103**, 523.

DEANESLY, R. and PARKES, A. S. (1937). Multiple activities of androgenic compounds. *Quart. J. exp. Physiol.* **26**, 393.

DE GURPIDE, E. H. G. (1953). Influencia de la adrenocorticotrofina y la hipofisectomía sobre la secreción suprarrenal de la 17-hidroxi-cortico-sterona. *Rev. Soc. argent. Biol.* **29**, 15.

DEMING, Q. B. and LUETSCHER, J. A., Jr. (1950). Bioassay of desoxycortico-sterone-like material in urine. *Proc. Soc. exp. Biol.*, *N.Y.*, **73**, 171.

DEMPSTER, W. J. (1955). The transplanted adrenal gland. *Brit. J. Surg.* **42**, 540.

DESAULLES, P., TRIPOD, J. and SCHULER, W. (1953). Wirkung von Elektro-cortin auf die Elektrolyt- und Wasserausschiedung im vergleich zu Desoxycorticosteron. *Schweiz. med. Wschr.* **83**, 1088.

DEUEL, H. J., Jr., HALLMAN, L. F., MURRAY, S. and SAMUELS, L. T. (1937). The sexual variation in carbohydrate metabolism. VIII. The rate of

References

absorption of glucose and of glycogen formation in normal and adrenalectomized rats. *J. biol. Chem.* **119**, 607.

DEVIS, R. (1951). La chimie des stéroïdes cortico-surrénaux. *Acta clin. belg.* **6**, 525.

DIAMARE, V. (1895). I corpuscoli di Stannius ed i corpi del cavo addominale die Teleostei. Notizie anatomiche e morfologiche. *Boll. Soc. Nat. Napoli*, **9**, 10.

DIAMARE, V. (1896). Ricerche intorno all'organo interrenale degli elasmobranchi ed ai corpuscoli di Stannius dei teleostei. *Mem. Mat., Roma*, Ser. III, **10**, 173.

DIAMARE, V. (1899). Sulla morfologia delle capsule surrenali. *Anat. Anz.* **15**, 357.

DIAMARE, V. (1905). Varietà anatomiche dell'interrenale. *Arch. ital. Anat. Embriol.* **4**, 366.

DIAMARE, V. (1933). L'organo interrenale, i corpuscoli di Stannius del mesonefro, i cordoni epiteliali ed il tessuto cromaffine del rene cefalico dei Teleostei—Nuovo contributo alla morfologia degli equivalenti corticali e midollari surrenali. *Atti Accad. Sci. fis. mat. Napoli*, Ser. 2, **19**, no. 6, 1.

DIAMARE, V. (1934). Sull'interrenale vero nel cosidetto 'sistema interrenale'. *Anat. Anz.* **78**, 90.

DIAMARE, V. (1935). Cordoni cellulari pronefrici e corpuscoli de Stannius dei Teleostei. *Atti Accad. Sci. fis. mat. Napoli*, Ser. 2, **20**, no. 6, 1.

DITTUS, P. (1936). Interrenalsystem und chromaffine Zellen im Lebensablauf von *Ichthyophis glutinosus* L. *Z. wiss. Zool.* **147**, 459.

DITTUS. P. (1937). Experimentelle Untersuchungen am Interrenalorgan der Selachier. *Pubbl. Staz. zool. Napoli*, **16**, 402.

DITTUS, P. (1941). Histologie und Cytologie des Interrenalorgans der Selachier unter normalen und experimentellen Bedingungen. Ein Beitrag zur Kenntnis der Wirkungsweise des kortikotropen Hormons und des Verhältnisses von Kern zu Plasma. *Z. wiss. Zool.* **154**, 40.

DIXON, H. B. F., STACK-DUNNE, M. P. and YOUNG, F. G. (1953). The preparation and properties of ACTH. In: *The Suprarenal Cortex. Proc. 5th Symp. Colston Res. Soc.* p. 11. Ed. Yoffey. London: Butterworth.

DIXON, H. B. F., STACK-DUNNE, M. P., YOUNG, F. G. and CATER, D. B. (1951). Influence of adrenotropic hormone fractions on 'adrenal repair' and on adrenal ascorbic acid. *Nature, Lond.*, **168**, 1084.

DOE, R. P., FLINT, E. B. and FLINT, M. G. (1954). Correlation of diurnal variations in eosinophils and 17-hydroxycorticosteroids in plasma and urine. *J. clin. Endocrin.* **14**, 774. Abst. no. 26.

DOLK, H. E. and POSTMA, N. (1927). Über die Haut- und Lungenatmung von *Rana temporaria. Z. vergl. Physiol.* **5**, 417.

DONALDSON, H. H. (1924). The rat. Reference tables and data for the albino rat (*Mus norvegicus albinus*) and the Norway rat (*Mus norvegicus*). *Mem. Wistar Inst. Anat.* no. 6, 2nd ed. Philadelphia.

DONALDSON, H. H. and KING, H. D. (1929). Life processes and size of the body and organs of the gray Norway rat during ten generations in captivity. *Amer. anat. Mem.* no. **14**.

17-2

DONALDSON, J. C. (1919). The relative volumes of the cortex and medulla of the adrenal gland in the albino rat. *Amer. J. Anat.* **25**, 290.

DONALDSON, J. C. (1928*a*). The adrenal glands in pregnancy; cortico-medullary relations in albino rats. *Anat. Rec.* **38**, 239.

DONALDSON, J. C. (1928*b*). Adrenal glands in wild gray and albino rats. Cortico-medullary relations. *Proc. Soc. exp. Biol., N.Y.*, **25**, 300.

DORFMAN, R. I. (1952). Steroids and tissue oxidation. *Vitam. & Horm.* **10**, 331.

DORFMAN, R. I. (1953). The bioassay of adrenocortical hormones. *Recent Progr. Hormone Res.* **8**, 87.

DORFMAN, R. I. (1955). Adrenocortical steroids in humans. Metabolism and generalizations. *Ciba Found. Coll. End.* **8**, 112.

DORFMAN, R. I., HAYANO, M., HAYNES, R. and SAVARD, K. (1953). The *in vitro* synthesis of adrenal cortical steroids. *Ciba Found. Coll. End.* **7**, 191.

DORFMAN, R. I., SHIPLEY, R. A., SCHILLER, S. and HORWITT, B. N. (1946). Studies on the 'cold test' as a method for the assay of adrenal cortical steroids. *Endocrinology*, **38**, 165.

DORFMAN, R. I. and UNGAR, F. (1953). *Metabolism of steroid hormones.* Minneapolis: Burgess.

DOSTOIEWSKY, A. (1886). Ein Beitrag zur mikroskopischen Anatomie der Nebennieren bei Säugetieren. *Arch. mikr. Anat.* **27**, 272.

DOUGHERTY, T. F. (1952). Effect of hormones on lymphatic tissue. *Physiol. Rev.* **32**, 379.

DOUGHERTY, T. F. and WHITE, A. (1944). Influence of hormones on lymphoid tissue structure and function. The role of the pituitary adreno-trophic hormone in the regulation of the lymphocytes and other cellular elements of the blood. *Endocrinology*, **35**, 1.

DOUGHERTY, T. F. and WHITE, A. (1945). Functional alterations in lymphoid tissue induced by adrenal cortex secretion. *Amer. J. Anat.* **77**, 81.

DOUGHERTY, T. F. and WHITE, A. (1947). An evaluation of alterations produced in lymphoid tissue by pituitary-adrenal cortical secretion. *J. Lab. clin. Med.* **32**, 584.

DOW, D. and ZUCKERMAN, S. (1939). The effect of vasopressin, sex hormones and adrenal cortical hormones on body water in axolotls. *J. Endocrin.* **1**, 387.

DRILL, V. A. and BRISTOL, W. R. (1951). Comparison of water and saline hydration on diuretic action of sympathomimetic agents. *Endocrinology*, **48**, 589.

DUCOMMUN, P. and MACH, R. S. (1949). Effet de l'hormone adrenocortico-trope sur la morphologie du cortex surrénalien, son contenu en acide ascorbique et en esters de cholestérol chez le rat normal. *Acta endocr., Copenhagen*, **3**, 17.

EATON, O. N. (1938). Weights and measurements of the parts and organs of mature inbred and crossbred guinea pigs. *Amer. J. Anat.* **63**, 273.

EBERTH, C. J. (1872). *Manual of human and comparative histology.* Ed. S. Stricker. New Sydenham Society, London, **7**, 110.

References

ECKER, A. (1846). *Der feinere Bau der Nebennieren beim Menschen und den vier Wirbelthierclassen*. Braunschweig.

ECKER, A. (1847). Recherches sur la structure intime des corps surrénaux chez l'homme et dans les quatres classes d'animaux vertébrés. *Ann. Sci. nat.*, sér. 3, Zool., **8**, 103.

ELFTMAN, H. (1947). Response of the alkaline phosphatase of the adrenal cortex of the mouse to androgen. *Endocrinology*, **41**, 85.

ELIAS, H. (1948). Growth of the adrenal cortex in domesticated Ungulata. *Amer. J. vet. Res.* **9**, 173.

ELKINTON, J. R., HUNT, A. D., GODFREY, L., McCRORY, W. W., ROGERSON, A. G. and STOKES, J. (1949). Effects of pituitary adrenocorticotropic hormone (ACTH) therapy. *J. Amer. med. Ass.* **141**, 1273.

ELLIOTT, T. R. and TUCKETT, I. (1906). Cortex and medulla in the suprarenal glands. *J. Physiol.* **34**, 332.

ELLISON, E. T. and BURCH, J. C. (1936). The effect of estrogenic substances upon the pituitary, adrenals and ovaries. *Endocrinology*, **20**, 746.

ELMAN, R. and ROTHMAN, P. (1924). Adrenal insufficiency and atrophy of the cortex following venous obstruction of the suprarenal glands. *Johns Hopk. Hosp. Bull.* **35**, 54.

EMERY, F. E. and GRECO, P. A. (1940). Comparative activities of desoxycorticosterone acetate and progesterone in adrenalectomized rats. *Endocrinology*, **27**, 473.

EMERY, F. E. and SCHWABE, E. L. (1936). The role of the corpora lutea in prolonging the life of adrenalectomized rats. *Endocrinology*, **20**, 550.

ENGEL, F. L. (1951). A consideration of the roles of the adrenal cortex and stress in the regulation of protein metabolism. *Recent Progr. Hormone Res.* **6**, 277.

ERÄNKÖ, O. (1955). Distribution of adrenaline and noradrenaline in the adrenal medulla. *Nature, Lond.*, **175**, 88.

ESSELLIER, A. F., JEANNERET, P., KOPP, E. and MORANDI, L. (1954). Evidence against destruction of eosinophils by glucocorticoids as shown by *in vitro* experiments. *Endocrinology*, **54**, 447.

EVANS, G. (1936). The adrenal cortex and endogenous carbohydrate formation. *Amer. J. Physiol.* **114**, 297.

EVANS, G. (1941). Effect of adrenalectomy on carbohydrate metabolism. *Endocrinology*, **29**, 731.

EVANS, H. M., SIMPSON, M. E. and LI, C. H. (1943). Inhibiting effect of adrenocorticotropic hormone on the growth of male rats. *Endocrinology*, **33**, 237.

EVERSE, J. W. R. and DE FREMERY, P. (1932). On a method of measuring fatigue in rats and its application for testing the suprarenal cortical hormone (Cortin). *Acta brev. neerl. Physiol.* **2**, 152.

FAHR, G. (1907/8). Über die Wirkung des Kaliumchlorids auf den Kontraktionsakt des Muskels. *Z. Biol.* **50**, 203.

FAHR, G. (1908/9). Über den Natriumgehalt der Skelettmuskeln des Frösches. *Z. Biol.* **52**, 72.

The Adrenal Cortex

FANCELLO, O. (1937). Interrene, surreni e ciclo sessuale nei Selaci ovipari. *Pubbl. Staz. zool. Napoli*, **16**, 80.

FARQUHARSON, F. R. (1950). *Simmond's Disease. Extreme insufficiency of the adenohypophysis*. American Lecture Series 34. Springfield, Illinois: Charles C. Thomas.

FARRELL, G. L. and LAMUS, B. (1953). Steroids in adrenal venous blood of the dog. *Proc. Soc. exp. Biol., N.Y.*, **84**, 89.

FARRELL, G. L. and RICHARDS, J. B. (1953). Isolation of a potent sodium-retaining substance from adrenal venous blood of the dog. *Proc. Soc. exp. Biol., N.Y.*, **83**, 628.

FARRELL, G. L. and WERLE, J. M. (1954). A sodium-retaining factor in dog adrenal venous blood. *Fed. Proc.* **13**, 42.

FAWCETT, D. W. (1948a). The effect of hypo- and hyperinsulinism on the deposition of glycogen in the adipose tissue of rats. *Anat. Rec.* **100**, 740.

FAWCETT, D. W. (1948b). Histological observations on the relation of insulin to the deposition of glycogen in adipose tissue. *Endocrinology*, **42**, 454.

FAWCETT, D. W. and CHESTER JONES, I. (1949). The effects of hypophysectomy, adrenalectomy and of thiouracil feeding on the cytology of brown adipose tissue. *Endocrinology*, **45**, 609.

FENN, W. O. (1936). Electrolytes in muscle. *Physiol. Rev.* **16**, 450.

FERGUSON, J. S. (1906). The veins of the adrenal. *Amer. J. Anat.* **5**, 63.

FERREBEE, J. W., PARKER, D., CARNES, W. H., GERITY, M. K., ATCHLEY, D. W. and LOEB, R. F. (1941). Certain effects of desoxycorticosterone. The development of 'diabetes insipidus' and the replacement of muscle potassium by sodium in normal dogs. *Amer. J. Physiol.* **135**, 230.

FETZER, S. (1953). Zur Verwendung formolfixierten Materials zur Ausführung der Plasmal-Reaktion. 'Mikroskopie.' *Zent. f. mikr. Forsch. u. Meth.* **81**, 189.

FEULGEN, R. and VOIT, K. (1924). Über einen weitverbreiteten festen Aldehyd; seine Entstehung aus einer Vorstufe, sein mikrochemischer und mikroskopisch-chemischer Nachweis und die Wege zu seiner präparativen Darstellung. *Pflüg. Arch. ges. Physiol.* **206**, 389.

FINDLAY, G. M. (1920). The pigments of the adrenals. *J. Path. Bact.* **23**, 482.

FLEXNER, L. B. and GROLLMAN, A. (1939). The reduction of osmic acid as an indicator of adrenal cortical activity in the rat. *Anat. Rec.* **75**, 207.

FLINT, J. M. (1900). The blood-vessels, angiogenesis, organogenesis, reticulum, and histology, of the adrenal. *Johns Hopk. Hosp. Rep.* **9**, 153.

FLÜCKIGER, E. and VERZAR, F. (1954). Die Wirkung von Aldosteron ('Electrocortin') auf den Natrium-, Kalium- und Glykogen-Stoffwechsel des isolierten Muskels. *Experientia*, **10**, 259.

FOLLEY, S. J. (1952). Lactation. In: Marshall's *Physiology of reproduction*, p. 525. Ed. A. S. Parkes. London: Longmans, Green and Co.

FOLLEY, S. J. (1953). The adrenal cortex and the mammary gland. In: *The Suprarenal Cortex*. Proc. 5th Symp. Colston Res. Soc. p. 85. Ed. Yoffey. London: Butterworth.

FOLLEY, S. J. (1955). Hormones in mammary growth and function. *Brit. med. Bull.* **11**, 145.

References

FONTAINE, M. (1956). The hormonal control of water and salt-electrolyte metabolism in Fish. (i) General. *Mem. Soc. Endocrin.* **5**, 69.

FONTAINE, M., CALLAMAND, O. and OLIVEREAU, M. (1949). Hypophyse et euryhalinité chez l'anguille. *C.R. Acad. Sci., Paris,* **228**, 513.

FONTAINE, M. and HATEY, J. (1953). Recherches sur le controle hypophysaire de l'interrénal antérieur d'un poisson téléostéen, l'anguille (*Anguilla anguilla* L.). 1. Variations pondérales de l'interrénal antérieur. *C.R. Soc. Biol., Paris,* **147**, 217.

FONTAINE, M. and HATEY, J. (1954*a*). Sur la teneur en 17-hydroxycorticostéroïde du plasma de Saumon (*Salmo salar* L.). *C.R. Acad. Sci., Paris,* **239**, 319.

FONTAINE, M. and HATEY, J. (1954*b*). Teneur en acide ascorbique de l'interrénal antérieur des poissons (sélaciens et téléostéens). *Bull. Inst. Oceanogr. Monaco,* **1037**, 7.

FONTAINE, M. and RAFFY, A. (1950). Le facteur hypophysaire de rétention d'eau chez les téléostéens. *C.R. Soc. Biol., Paris,* **144**, 6.

FORSHAM, P. H., THORN, G. W., PRUNTY, F. T. G. and HILLS, A. G. (1948). Clinical studies with pituitary adrenocorticotropin. *J. clin. Endocrin.* **8**, 15.

FOWLER, M. A. and CHESTER JONES, I. (1955). The adrenal cortex in the frog *Rana temporaria* and its relation to water and salt-electrolyte metabolism. *J. Endocrin.* **13**, vi.

FRASER, A. H. H. (1929). Lipin secretion in the Elasmobranch Interrenal. *Quart. J. micr. Sci.* **73**, 121.

FRASER, E. A. (1950). The development of the vertebrate excretory system. *Biol. Rev.* **25**, 159.

FREUDENBERGER, C. B. and CLAUSEN, F. W. (1937*a*). The effect of continued theelin injections on the body growth and organ weights of young female rats. *Anat. Rec.* **68**, 133.

FREUDENBERGER, C. B. and CLAUSEN, F. W. (1937*b*). Quantitative effects of theelin on body growth and endocrine glands in young albino rats. *Anat. Rec.* **69**, 171.

FREY, H. (1849). Suprarenal capsules. In: *The Cyclopaedia of Anatomy and Physiology.* Ed. R. B. Todd. Part **37**, 827.

FRIED, J. and SABO, E. F. (1953). Synthesis of 17α-hydroxycorticosterone and its 9α-halo derivatives from 11-epi-17α-hydroxycorticosterone. *J. Amer. chem. Soc.* **75**, 2273.

FRIED, J. and SABO, E. F. (1954). 9α-fluoro derivatives of cortisone and hydrocortisone. *J. Amer. chem. Soc.* **76**, 1455.

FRIEDBERG, F. and GREENBERG, D. M. (1947). Endocrine regulation of amino acid levels in blood and tissues. *J. biol. Chem.* **168**, 405.

FRIEDGOOD, H. B. (1946). *Endocrine function of the hypophysis.* N.Y: O.U.P.

FUHRMAN, F. A. and USSING, H. H. (1951). A characteristic response of the isolated frog skin potential to neurohypophysial principles and its relation to the transport of sodium and water. *J. cell. comp. Physiol.* **38**, 109.

FUSTINONI, O. (1938*a*). La astenia de la insuficiencia suprarrenal del Sapo. *Rev. Soc. argent. Biol.* **14**, 304.

The Adrenal Cortex

FUSTINONI, O. (1938b). L'asthénie des insuffisances hypophysaire et surrénale du crapaud. *C.R. Soc. Biol., Paris*, **129**, 1252.

FUSTINONI, O. and PORTO, J. (1938). Morfología das las glándulas adrenales del sapo *Bufo arenarum* (Hensel). *Rev. Soc. argent. Biol.* **14**, 315.

GANONG, W. F. and HUME, D. M. (1954). Absence of stress-induced and 'compensatory' adrenal hypertrophy in dogs with hypothalamic lesions. *Endocrinology*, **55**, 474.

GARDNER, L. I. and WALTON, R. L. (1954). Plasma 17-ketosteroids of the human fetus. *Helv. paediat. acta*, **4**, 311.

GARRETT, F. D. (1942). The development and phylogeny of the corpuscle of Stannius in ganoid and teleostean fishes. *J. Morph.* **70**, 41.

GASKELL, J. F. (1912). The distribution and physiological action of the suprarenal medullary tissue in *Petromyzon fluviatilis*. *J. Physiol.* **44**, 59.

GASSNER, F. X., NELSON, D. H., REICH, H., RAPALA, R. T. and SAMUELS, L. T. (1951). Isolation of an androgenic compound from the adrenal venous blood of cows. *Proc. Soc. exp. Biol., N.Y.*, **77**, 829.

GAUNT, R. (1944a). Water diuresis and water intoxication in relation to the adrenal cortex. *Endocrinology*, **34**, 400.

GAUNT, R. (1944b). Endocrine factors in water diuresis and water intoxication. *Trans. N.Y. Acad. Sci.* **6**, 179.

GAUNT, R. (1951). The adrenal cortex in salt and water metabolism. *Recent Progr. Hormone Res.* **6**, 247.

GAUNT, R. (1955). Biological studies with aldosterone (electrocortin). *Ciba Found. Coll. End.* **8**, 228.

GAUNT, R. and BIRNIE, J. H. (1951). *Hormones and body water*. Amer. Lect. Series no. 103. Springfield, Illinois: C. C. Thomas.

GAUNT, R., BIRNIE, J. H. and EVERSOLE, W. J. (1949). Adrenal cortex and water metabolism. *Physiol. Rev.* **29**, 281.

GAUNT, R., GORDON, A. S., RENZI, A. A., PADAWER, J., FRUHMAN, G. J. and GILMAN, M. (1954a). Studies of Electrocortin. *J. clin. Endocrin.* **14**, 783. Abst. 44.

GAUNT, R., GORDON, A. S., RENZI, A. A., PADAWER, J., FRUHMAN, G. J. and GILMAN, M. (1954b). Biological studies with electrocortin (aldosterone). *Endocrinology*, **55**, 236.

GAUNT, R. and HAYS, H. W. (1938). Role of progesterone and other hormones in survival of pseudopregnant adrenalectomized ferrets. *Amer. J. Physiol.* **124**, 767.

GEMZELL, C. A. (1952). Increase in the formation and the secretion of ACTH in rats following administration of oestradiol monobenzoate. *Acta endocr., Copenhagen*, **11**, 221.

GEMZELL, C. A. (1953). Blood levels of 17-hydroxycorticosteroids in normal pregnancy. *J. clin. Endocrin.* **13**, 898.

GÉRARD, G. (1913). Contribution à l'étude morphologique des artères des capsules surrénales de l'homme. *J. Anat., Paris*, **49**, 269.

GÉRARD, P. (1951). Sur la cortico-surrénale du Protoptère (*Protopterus Dolloi* Blgr.). *Arch. Biol., Paris*, **62**, 371.

References

GERSCHMAN, R. (1943). Variaciones estracionales o por hipofisectomía de los elementos minerales del plasma del sapo. *Rev. Soc. argent. Biol.* **19**, 172.

GERSH, I. and GROLLMAN, A. (1939). The relation of the adrenal cortex to the male reproductive system. *Amer. J. Physiol.* **126**, 368.

GERSH, I. and GROLLMAN, A. (1941). The vascular pattern of the adrenal gland of the mouse and rat and its physiological response to changes in glandular activity. *Carn. Inst. Wash. Publ.* no. **525**, 111.

GIACOMINI, E. (1897). Sopra la fina struttura delle capsule surrenali degli anfibi. *Atti Accad. Fisiocr. Siena.* 30 giugno 1897.

GIACOMINI, E. (1898). Brevi osservazioni interno alla minuta struttura del corpo interrenale e dei corpi soprarenali dei Selaci. *Atti Accad. Fisiocr. Siena*, Ser. 4, **10**, 835.

GIACOMINI, E. (1902*a*). Sulla esistenza della sostanza midollare nelle capsule surrenali dei Teleostei. *Monit. zool. ital.* **13**, 183.

GIACOMINI, E. (1902*b*). Contributo alla conoscenza delle capsule surrenali nei Ciclostomi. Sulle capsule surrenali dei Petromizonti. *Monit. zool. ital.* **13**, 143.

GIACOMINI, E. (1902*c*). Sopra la fina struttura delle capsule surrenali degli anfibi e sopra i nidi cellulari del simpatico di questi vertebrati. *Atti Accad. Fisiocr. Siena*, Ser. 4, **14**.

GIACOMINI, E. (1905). Contributo alla conoscenza del sistema delle capsule surrenali dei Teleostei. Sulla sostanza midollare (organi soprarenali o tessuto cromaffine) di *Amiurus catus* L. *R.C. Accad. Bologna (Classe di Scienze Fisiche, Nuova Serie)*, **9**, 183. 28 Maggio 1905.

GIACOMINI, E. (1906). Sulle capsule surrenali e sul simpatico dei Dipnoi. Ricerche in *Protopterus annectens. R.C. Accad. Lincei.* Ser. 5, **15**, 394.

GIACOMINI, E. (1908). Il sistema interrenale e il sistema cromaffine (sistema feocromo) nelle anguille adulte, nelle cieche e nei leptocefali. *Mem. R. Accad. Bologna (Sezioni delle Scienze Naturali)*, Serie VI, **5**, 113.

GIACOMINI, E. (1909*a*). Il sistema interrenale e il sistema cromaffine (sistema feocromo) in altre specie di Murenoidi. *R.C. Accad. Bologna (Classe di Scienze Fisiche, Nuova Serie)*, **13**, 87. 25 Aprile 1909.

GIACOMINI, E. (1909*b*). Il sistema interrenale e il sistema cromaffine (sistema feocromo) in altre specie di Murenoidi. Parte 1. *Mem. R. Accad. Bologna (Sezioni delle Scienze Naturali)*, Serie VI, **6**, 175.

GIACOMINI, E. (1910). Il sistema interrenale e il sistema cromaffine (sistema feocromo) in altre specie di Murenoidi. Parte 2. *Mem. R. Accad. Bologna (Sezioni delle Scienze Naturali)*, Serie VI, **7**, 113.

GIACOMINI, E. (1911*a*). Anatomia microscopica e sviluppo del sistema interrenale e del sistema cromaffine (sistema feocromo) dei Salmonidi. Parte 1: Anatomia microscopica. *Mem. R. Accad. Bologna (Sezioni delle Scienze Naturali)*, Serie VI, **8**, 67.

GIACOMINI, E. (1911*b*). Anatomia microscopica e sviluppo del sistema interrenale e del sistema cromaffine (sistema feocromo) dei Salmonidi. *R.C. Accad. Bologna (Classe di Scienze Fisiche, Nuova Serie)*, **15**, 107. 28 Maggio 1911.

GIACOMINI, E. (1911 c). Il sistema interrenale e il sistema cromaffine (sistema feocromo) dei Lofobranchi. *R.C. Accad. Bologna (Classe di Scienze Fisiche, Nuova Serie)*, **15**, 108. 28 Maggio 1911.

GIACOMINI, E. (1911 d). Il sistema interrenale e il sistema cromaffine (sistema feocromo) dei Ciprinidi. *R.C. Accad. Bologna (Classe di Scienze Fisiche, Nuova Serie)*, **15**, 109. 28 Maggio 1911.

GIACOMINI, E. (1912). Anatomia microscopica e sviluppo del sistema interrenale e del sistema cromaffine (sistema feocromo) dei Salmonidi. Parte 2. Sviluppo. *Mem. R. Accad. Bologna (Sezioni delle Scienze Naturali)*, Serie VI, **9**. 111.

GIACOMINI, E. (1920). Anatomia microscopica e sviluppo del sistema interrenale nei Lofobranchi. *R.C. Accad. Bologna (Classe di Scienze Fisiche, Nuova Serie)*, **24**, 129. 30 Maggio 1920.

GIACOMINI, E. (1921). Sul sistema interrenale e sul sistema cromaffine di alcuni Teleostei abissali (*Argyropelecus* e *Scopelus*). *R.C. Accad. Bologna (Classe di Scienze Fisiche, Nuova Serie)*, **25**, 130. 29 Maggio 1921.

GIACOMINI, E. (1922). Sull'anatomia microscopica e sullo sviluppo delle capsule surrenali dei Lofobranchi. *Arch. ital. Anat. Embriol.* **18**, 548.

GIACOMINI, E. (1928). Le capsule surrenali dei Teleostei. *Monit. zool. ital.* **39**, 48.

GIACOMINI, E. (1933). Il sistema interrenale e i corpuscoli di Stannius dei Ganoidi e dei Teleostei. *Boll. Soc. ital. Biol. sper.* **8**, 1215.

GINSBURG, M. (1954). The secretion of antidiuretic hormone in response to haemorrhage and the fate of vasopressin in adrenalectomized rats. *J. Endocrin.* **11**, 165.

GIROUD, A. and LEBLOND, C. P. (1934). Localisation histochimique de la vitamine C dans le cortex surrénal. *C.R. Soc. Biol., Paris,* **115**, 705.

GIROUD, A. and RATSIMAMANGA, A. R. (1942). Acide ascorbique—vitamine C. *Actualités sci. industr.* **921**, III, 1.

GIROUD, A., SANTA, N. and MARTINET, M. (1940). Microdosage biologique de l'hormone cortico-surrénale. *C.R. Soc. Biol., Paris,* **134**. 20.

GLAFKIDES, C. M., BENNETT, L. L. and GEORGE, R. (1952). Inhibition of the diabetes enhancing effect of ACTH by a diet high in potassium chloride. *Endocrinology,* **50**, 684.

GOMORI, G. (1942). Histochemical reactions for lipid aldehydes and ketones. *Proc. Soc. exp. Biol., N.Y.,* **51**, 133.

GORDON, A. S. (1954). Endocrine influences upon the formed elements of blood and blood-forming organs. *Recent Progr. Hormone Res.* **10**, 339.

GORDON, A. S., PILIERO, S. J. and LANDAU, D. (1951). The relation of the adrenal to blood formation in the rat. *Endocrinology,* **49**, 497.

GOTTSCHAU, M. (1883). Structur und embryonale Entwickelung der Nebennieren bei Säugethieren. *Arch. Anat. Physiol., Lpz. Anatom. Abt.* p. 412.

GOURFEIN, D. (1896). Recherches physiologiques sur la fonction des glandes surrénales. *Rev. méd. Suisse rom.* **16**, 113.

GOURFEIN, D. (1897). Le rôle de l'auto-intoxication dans le méchanisme de la mort des animaux décapsulés. *C.R. Acad. Sci., Paris,* **125**, 188.

References

GRAD, B., SCHER, O. and SYMCHOWICZ, S. (1953). Relative biological activity of some adrenal cortical steroids as determined by eosinopenic test. *Proc. Soc. exp. Biol., N.Y.,* **84,** 1.

GRAHAM, G. S. (1916). Toxic lesions of the adrenal gland and their repair. *J. med. Res.* **34,** 241.

GRATIOLET, P. (1853). Sur le système veineux des Reptiles. *L'Institut (Paris),* **21,** 60; and *Société Philomatique de Paris. Extraits des Procès Verbaux des Séances, pendant l'année* 1853, p. 7.

GRAY, H. (1852). On the development of the ductless glands in the chick. *Phil. Trans., Lond.,* **142,** 295.

GREENE, R. R., WELLS, J. A. and IVY, A. C. (1939). Progesterone will maintain adrenalectomized rats. *Proc. Soc. exp. Biol., N.Y.,* **40,** 83.

GREEP, R. O. and CHESTER JONES, I. (1950a). Steroids and pituitary hormones. In: *Symposium on steroid hormones.* Ed. E. S. Gordon. Univ. Wisconsin Press.

GREEP, R. O. and CHESTER JONES, I. (1950b). Steroid control of pituitary function. *Recent Progr. Hormone Res.* **5,** 147.

GREEP, R. O. and DEANE, H. W. (1947). Cytochemical evidence for the cessation of hormone production in the zona glomerulosa of the rat's adrenal cortex after prolonged treatment with desoxycorticosterone acetate. *Endocrinology,* **40,** 417.

GREEP, R. O. and DEANE, H. W. (1949a). Histological, cytochemical and physiological observations on the regeneration of the rat's adrenal gland following enucleation. *Endocrinology,* **45,** 42.

GREEP, R. O. and DEANE, H. W. (1949b). The cytology and cytochemistry of the adrenal cortex. *Ann. N.Y. Acad. Sci.* **50,** 596.

GREEP, R. O., KNOBIL, E., HOFMANN, F. G. and JONES, T. L. (1952). Adrenal cortical insufficiency in the Rhesus monkey. *Endocrinology,* **50,** 664.

GROLLMAN, A. (1936). *The Adrenals.* London: Baillière, Tindall and Cox.

GROLLMAN, A., FIROR, W. M. and GROLLMAN, E. (1934). The extraction of the adrenal cortical hormone from the interrenal body of fishes. *Amer. J. Physiol.* **108,** 237.

GROSGLIK, S. (1885). Zur Morphologie der Kopfniere der Fische. *Zool. Anz.* **8,** 605.

GROSGLIK, S. (1886). Zur Frage über die Persistenz der Kopfniere der Teleostier. *Zool. Anz.* **9,** 196.

GROSS, F. and GYSEL, H. (1954). The action of electrocortin in the adrenalectomised dog. *Acta endocr., Copenhagen,* **15,** 199.

GROSS, F., LOUSTALOT, P. and MEIER, R. (1955). Vergleichende Untersuchungen über die hypertensive Wirkung von Aldosteron und Desoxycorticosteron. *Experientia,* **11,** 67.

GRUBY, M. (1842). Recherches anatomiques sur le système veineux de la grenouille. *Ann. Sci. nat.,* sér. 2, Zool., **17,** 209.

GRUENWALD, P. and KONIKOV, W. M. (1944). Cell replacement and its relation to the zona glomerulosa in the adrenal cortex of mammals. *Anat. Rec.* **89,** 1.

GRUNDY, H. M., SIMPSON, S. A. and TAIT, J. F. (1952). Isolation of a highly active mineralocorticoid from beef adrenal extract. *Nature, Lond.,* **169**, 795.

GRYNFELLT, E. (1904*a*). Recherches anatomiques et histologiques sur les organes surrénaux des Plagiostomes. *Bull. sci. Fr. Belg.* **38**, 1.

GRYNFELLT, E. (1904*b*). Notes histologiques sur la capsule surrénale des amphibiens. *J. Anat., Paris,* **40**, 180.

GUIEYSSE, A. (1901). La capsule surrénale du cobaye—histologie et fonctionnement. *J. Anat., Paris,* **37**, 312.

GUILLEMIN, R. (1955). A re-evaluation of acetylcholine, adrenaline, noradrenaline and histamine as possible mediators of the pituitary adrenocorticotrophic activation by stress. *Endocrinology,* **56**, 248.

GUILLEMIN, R. and FORTIER, C. (1953). Role of histamine in the hypothalamo-hypophysial response to stress. *Trans. N.Y. Acad. Sci.* **15**, 138.

GULLIVER, G. (1840). A consideration of the function of the suprarenal capsules. *Dublin Medical Press,* **3**, 11.

GUTTMAN, P. H. (1930). A statistical analysis of 566 cases of Addison's disease. *Arch. Path.* **10**, 742.

HAAM, E. v., HAMMEL, M. A., RARDIN, T. E. and SCHOENE, R. H. (1941). Experimental studies on the activity and toxicity of stilbestrol. *Endocrinology,* **28**, 263.

HAINES, W. J. (1952). Studies on the biosynthesis of adrenal cortex hormones. *Recent Progr. Hormone Res.* **7**, 255.

HALES, W. M., HASLERUD, G. M. and INGLE, D. J. (1935). Time for development of incapacity to work in adrenalectomized rats. *Amer. J. Physiol.* **112**, 65.

HALLER, A. v. (1766). *Elementa physiologiae corporis humani.* Bernae.

HARLEY, G. (1858). The histology of the suprarenal capsules. *Lancet,* June 1858, p. 551.

HARMS, W. (1921). Morphologische und kausal-analytische Untersuchungen über das Internephridialorgan von *Physcosoma lanzarotae* nov. spec. *Arch. Entw.-Mech. Org.* **47**, 307.

HARRIS, G. W. (1948). Neural control of the pituitary gland. *Physiol. Rev.* **28**, 139.

HARRIS, G. W. (1952*a*). Hypothalamic control of the anterior pituitary gland. *Ciba Found. Coll. End.* **4**, 106.

HARRIS, G. W. (1952*b*). Functional hypophysial grafts. *Ciba Found. Coll. End.* **4**, 115.

HARRIS, G. W. (1955). The reciprocal relationship between the thyroid and adrenocortical responses to stress. *Ciba Found. Coll. End.* **8**, 531.

HARRIS, G. W., JACOBSOHN, D. and KAHLSON, G. (1952). The occurrence of histamine in cerebral regions related to the hypophysis. *Ciba Found. Coll. End.* **4**, 186.

HARRIS, J. E. (1941). The influence of the metabolism of human erythrocytes on their potassium content. *J. biol. Chem.* **141**, 579.

HARRISON, H. E. and DARROW, D. C. (1938). The distribution of body water and electrolytes in adrenal insufficiency. *J. clin. Invest.* **17**, 77.

References

HARRISON, R. G. (1951). A comparative study of the vascularization of the adrenal gland in the rabbit, rat and cat. *J. Anat., Lond.*, **85**, 12.

HARROP, G. A., SOFFER, L. J., ELLSWORTH, R. and TRESCHER, J. H. (1933). Studies on the suprarenal cortex. III. Plasma electrolytes and electrolyte excretion during suprarenal insufficiency in the dog. *J. exp. Med.* **58**, 17.

HARROP, G. A., SOFFER, L. J., NICHOLSON, W. M. and STRAUSS, M. (1934). Studies on the suprarenal cortex. IV. The effect of sodium salts in sustaining suprarenalectomised dogs. *J. exp. Med.* **61**, 839.

HARTMAN, F. A. (1946). Adrenal and thyroid weights in birds. *The Auk*, **63**, 42.

HARTMAN, F. A. and ALBERTIN, R. H. (1951). A preliminary study of the avian adrenal. *The Auk*, **68**, 202.

HARTMAN, F. A. and BROWNELL, K. A. (1934). Relation of adrenals to diabetes. *Proc. Soc. exp. Biol., N.Y.*, **31**, 834.

HARTMAN, F. A. and BROWNELL, K. A. (1949). *The adrenal gland*. London: Kimpton.

HARTMAN, F. A., KNOUFF, R. A., McNUTT, A. W. and CARVER, J. E. (1947). Chromaffin patterns in bird adrenals. *Anat. Rec.* **97**, 211.

HARTMAN, F. A., LEWIS, L. A., BROWNELL, K. A., ANGERER, C. A. and SHELDEN, F. F. (1944). Effect of interrenalectomy on some blood constituents in the skate. *Physiol. Zoöl.* **17**, 228.

HARTMAN, F. A., MacARTHUR, C. G., GUNN, F. D., HARTMAN, W. E. and MACDONALD, J. J. (1927). Kidney function in adrenal insufficiency. *Amer. J. Physiol.* **81**, 244.

HARTMAN, F. A., SHELDEN, F. F. and GREEN, E. L. (1943). Weights of interrenal glands of Elasmobranchs. *Anat. Rec.* **87**, 371.

HARTMAN, F. A., SMITH, D. E. and LEWIS, L. A. (1943). Adrenal functions in the opossum. *Endocrinology*, **32**, 340.

HARTMAN, F. A. and SPOOR, H. J. (1940). Cortin and the sodium factor of the adrenal. *Endocrinology*, **26**, 871.

HARTMAN, F. A., SPOOR, H. J. and LEWIS, L. A. (1939). The sodium factor of the adrenal. *Science*, **89**, 204.

HATAI, S. (1914). On the weight of some of the ductless glands of the Norway and of the albino rat according to sex and variety. *Anat. Rec.* **8**, 511.

HATEY, J. (1951 a). Influence de l'hypophysectomie sur la teneur en glycogène du foie de l'anguille (*Anguilla anguilla* L.). *C.R. Soc. Biol., Paris*, **145**, 172.

HATEY, J. (1951 b). La fonction glycogénique du foie de l'anguille (*Anguilla anguilla* L.) après hypophysectomie. *C.R. Soc. Biol., Paris*, **145**, 315.

HATEY, J. (1952). Interrénal et acide ascorbique des poissons Téléostéens. *C.R. Soc. Biol., Paris*, **146**, 566.

HATEY, J. (1954a). Recherches sur le contrôle hypophysaire de l'interrénal antérieur d'un Poisson Téléostéen: l'Anguille (*Anguilla anguilla* L.). Variation de l'acide ascorbique de l'interrénal antérieur consécutive à l'hypophysectomie. *C.R. Soc. Biol., Paris*, **148**, 231.

HATEY, J. (1954b). Recherches sur le contrôle hypophysaire de l'inter-rénal antérieur d'un Poisson Téléostéen: l'Anguille (*Anguilla anguilla* L.). Variation de l'acide ascorbique de l'interrénal antérieur après injection d'hormone corticotrope. *C.R. Soc. Biol., Paris*, **148**, 234.

HAYANO, M. and DORFMAN, R. I. (1952). The action of adrenal homo-genates on progesterone, 17-hydroxy-progesterone and 21-desoxycorti-sone. *Arch. Biochem.* **36**, 237.

HAYES, M. A. and BAKER, B. L. (1951). The effect of parenterally adminis-tered adrenocortical extract on the intradermal spreading action of hyaluronidase. *Endocrinology*, **49**, 379.

HAYNES, R., SAVARD, K. and DORFMAN, R. I. (1952). An action of ACTH on adrenal slices. *Science*, **116**, 690.

HAYS, E. E. and WHITE, W. F. (1954). The chemistry of the corticotrophins. *Recent Progr. Hormone Res.* **10**, 265.

HAYS, H. W. (1952). The influence of sodium chloride on water diuresis and its relation to the adrenal. *Ciba Found. Coll. End.* **4**, 481.

HAYS, V. J. (1914). The development of the adrenal glands of birds. *Anat. Rec.* **8**, 451.

HECHTER, O. (1949). Corticosteroid release from an isolated adrenal gland. *Fed. Proc.* **8**, 70.

HECHTER, O. (1950). Characterization of corticosteroids released from perfused cow adrenals. *Fed. Proc.* **9**, 58.

HECHTER, O. (1951). The biogenesis of adrenal cortical steroids. *Trans. 3rd Conf. on the adrenal cortex.* Ed. Ralli, p. 115. N.Y: Josiah Macy Jr. Foundation.

HECHTER, O. (1953). Biogenesis of adrenal cortical hormones. *Ciba Found. Coll. End.* **7**, 161.

HECHTER, O., JACOBSEN, R. P., JEANLOZ, R. W., LEVY, H., MARSHALL, C. W., PINCUS, G. and SCHENKER, V. (1949). The bio-oxygenation of 11-desoxycorticosterone at C_{11}. *J. Amer. chem. Soc.* **71**, 3261.

HECHTER, O., JACOBSEN, R. P., JEANLOZ, R. W., LEVY, H., MARSHALL, C. W., PINCUS, G. and SCHENKER, V. (1950). Bio-oxygenation of steroids at C_{11}. *Arch. Biochem.* **25**, 457.

HECHTER, O., JACOBSEN, R. P., JEANLOZ, R. W., LEVY, H., PINCUS, G. and SCHENKER, V. (1950). Pathways of corticosteroid synthesis. *J. clin. Endocrin.* **10**, 827.

HECHTER, O. and PINCUS, G. (1954). Genesis of adrenocortical secretion. *Physiol. Rev.* **34**, 459.

HECHTER, O., SOLOMON, M. M., ZAFFARONI, A. and PINCUS, G. (1953). Transformation of cholesterol and acetate to adrenal cortical hor-mones. *Arch. Biochem.* **46**, 201.

HECHTER, O., ZAFFARONI, A., JACOBSEN, R. P., LEVY, H., JEANLOZ, R. W., SCHENKER, V. and PINCUS, G. (1951). The nature and the biogenesis of the adrenal secretory product. *Recent Progr. Hormone Res.* **6**, 215.

HEGNAUER, A. H. and ROBINSON, E. J. (1936). The water and electrolyte distribution among plasma, red blood cells, and muscle after adrenal-ectomy. *J. biol. Chem.* **116**, 769.

References

HELLER, H. (1945). The effect of neurohypophysial extracts on the water balance of lower vertebrates. *Biol. Rev.* **20**, 147.

HELLER, H. (1950). The comparative physiology of the neurohypophysis. *Experientia*, **6**, 368.

HELLER, H. (1956). The hormonal control of water and salt electrolyte metabolism with special reference to the higher vertebrates. *Mem. Soc. Endocrin.* **5**, 25.

HENCH, P. S., KENDALL, E. C., SLOCUMB, C. H. and POLLEY, H. F. (1949). The effect of a hormone of the adrenal cortex (16-hydroxy-11-dehydro-corticosterone: compound E) and of pituitary adrenocorticotropic hormone on rheumatoid arthritis; preliminary report. *Proc. Mayo Clin.* **24**, 181.

HERNANDEZ, T. and COULSON, R. A. (1952). Hibernation in the alligator. *Proc. Soc. exp. Biol., N.Y.*, **79**, 145.

HERRICK, E. H. and FINERTY, J. C. (1940). The effect of adrenalectomy on the anterior pituitaries of fowls. *Endocrinology*, **27**, 279.

HERRICK, E. H. and TORSTVEIT, O. (1938). Some effects of adrenalectomy in fowls. *Endocrinology*, **22**, 469.

HESLOP, R. W., KROHN, P. L. and SPARROW, E. M. (1954). The effect of pregnancy on the survival of skin homografts in rabbits. *J. Endocrin.* **10**, 325.

HETT, J. (1928). Beobachtungen an der Nebenniere der Maus. II. Geschlechtsunterschiede im gegenseitigen Mengenverhältnis von Rinde und Mark bei wachsenden Tieren. *Z. mikrosk.-anat. Forsch.* **13**, 428.

HEUVERSWYN, J. VAN, FOLLEY, S. J. and GARDNER, W. U. (1939). Mammary growth in male mice receiving androgens, estrogens and desoxycorticosterone acetate. *Proc. Soc. exp. Biol., N.Y.*, **41**, 389.

HEWITT, W. F. (1947). The essential role of the adrenal cortex in the hypertrophy of the ovotestis following ovariectomy in the hen. *Anat. Rec.* **98**, 159.

HILL, R. T. (1937). Ovaries secrete male hormone. I. Restoration of the castrate type of seminal vesicle and prostate glands to normal by grafts of ovaries in mice. *Endocrinology*, **21**, 495.

HILL, R. T. and PARKES, A. S. (1934). Hypophysectomy of birds; effect on gonads, accessory organs and head furnishings. *Proc. roy. Soc. B*, **116**, 221.

HILLARP, N. A. and HÖKFELT, B. (1954). Evidence of adrenaline and noradrenaline in separate adrenal medullary cells. *Acta physiol. scand.* **30**, 55.

HIRASE, K. (1926). Der Einfluss der Nebennierenexstirpation auf die Arbeitsfähigkeit des Muskels. *Pflüg. Arch. ges. Physiol.* **212**, 582.

HOBERMAN, H. D. (1950). Endocrine regulation of N metabolism during fasting. *Yale J. Biol. Med.* **22**, 341.

HOERR, N. (1931). The cells of the suprarenal cortex in the guinea-pig. Their reaction to injury and their replacement. *Amer. J. Anat.* **48**, 139.

HOERR, N. (1936a). Histological studies on lipins. I. On osmic acid as a microchemical reagent with special reference to lipins. *Anat. Rec.* **66**, 149.

HOERR, N. (1936b). Histological studies on lipins. II. A cytological analysis of the liposomes in the adrenal cortex of the guinea-pig. *Anat. Rec.* **66**, 317.

HOFMANN, F. G. and DAVISON, C. (1953). Effect of ACTH and various substrates on *in vitro* adrenal cortical steroid synthesis. *J. clin. Endocrin.* **13**, 848.

HOFMANN, F. G. and DAVISON, C. (1954). Observations on *in vitro* adrenal steroid synthesis in the rat. *Endocrinology*, **54**, 580.

HOLLINSHEAD, W. H. (1936). The innervation of the adrenal glands. *J. comp. Neurol.* **64**, 449.

HOLMBERG, A. D. and SOLER, F. L. (1942). Some notes on the adrenals. Presence of a united adrenal in the marine tortoise. *Contributions from the laboratory of anatomy, comparative physiology, and pharmacodynamics. Univ. Buenos Aires,* **20**, 457 and 667.

HOLMES, W. (1950). The adrenal homologues in the lungfish *Protopterus. Proc. roy. Soc.* B, **137**, 549.

HOLMES, W. N. (1955). Histological variations in the adrenal cortex of the golden hamster, with special reference to the *X* zone. *Anat. Rec.* **122**, 271.

HOLZAPFEL, R. A. (1937). The cyclic character of hibernation in frogs. *Quart. Rev. Biol.* **12**, 65.

HOPPE, G. and VOGEL, G. (1951). Der Einfluss der Adrenalektomie auf den Zeitwert der muskelären Erregbarkeit. *Pflüg. Arch. ges. Physiol.* **253**, 518.

Hormones (1951). *A survey of their properties and uses.* London: Pharmaceutical Press.

HOUSSAY, B. A. (1945). The influence of adrenal insufficiency during pregnancy on the mother and on the offspring. *Rev. Soc. argent. Biol.* **21**, 316.

HOUSSAY, B. A. (1949). Hypophysial functions in the toad *Bufo arenarum* Hensel. *Quart. Rev. Biol.* **24**, 1.

HOUSSAY, B. A. and ARTUNDO, A. (1929). Métabolisme et thermorégulation des rats surrénaloprivés. *C.R. Soc. Biol., Paris,* **100**, 127.

HOUSSAY, B. A., BIASOTTI, A. and MAGDALENA, A. (1932). Hypophyse et thyroïde. Hypophyse et hypertrophie compensatrice de la thyroïde. *C.R. Soc. Biol., Paris,* **110**, 142.

HOUSSAY, B. A., GERSCHMAN, R. and RAPELA, C. E. (1950). Adrenalina y noradrenalina en la suprarenal del sapo normal o hipofisoprivo. *Rev. Soc. argent. Biol.* **26**, 29.

HOUSSAY, B. A. and MOLINELLI, E. A. (1926). L'irrigation et le poids des surrénales et leur relation avec le poids et la surface chez quelques animaux. *C.R. Soc. Biol., Paris,* **95**, 819.

HOUSSAY, B. A. and RODRIGUEZ, R. R. (1953). Diabetogenic action of different preparations of growth hormone. *Endocrinology*, **53**, 114.

HOUSSAY, B. A. and SAMMARTINO, R. (1933). Modifications histologiques de la surrénale chez les chiens hypophysoprivés ou à tuber lésé. *C.R. Soc. Biol., Paris,* **114**, 717.

HOWARD, E. (1927). A transitory zone in the adrenal cortex which shows age and sex relationships. *Amer. J. Anat.* **40**, 251.

References

HOWARD, E. (1930). The X zone of the suprarenal cortex in relation to gonadal maturation in monkeys and mice, and to epiphysial unions in monkeys. *Anat. Rec.* **46**, 93.

HOWARD, E. (1941). Effects of adrenalectomy and DOCA substitution therapy on the castrated male prostate; evidence for andromimetic function of the immature rat adrenal. *Endocrinology,* **29**, 746.

HOWARD, E. and BENUA, R. S. (1950). The effect of protein deficiency and other dietary factors on the X zone of the mouse adrenal. *J. Nutr.* **42**, 157.

HOWARD, E. and GENGRADOM, S. (1940). The effects of ovariectomy and administration of progesterone on the adrenal X zone and the uterus. *Endocrinology,* **26**, 1048.

HSIEH, K. M. (1950). Increase in weight of frogs after DOCA administration. *Fed. Proc.* **9**, 63.

HUME, D. M. (1952). The relationship of the hypothalamus to the pituitary secretion of ACTH. *Ciba Found. Coll. End.* **4**, 87.

HUME, D. M. (1954). In discussion on *Porter, 1954.*

HUME, D. M. and WITTENSTEIN, G. J. (1950). The relationship of the hypothalamus to pituitary-adrenocortical function. *Proc. 1st clin. ACTH conf.* p. 134. J. R. Mote, ed. Philadelphia: Blakiston and Co.

HUOT, E. (1897). Sur les capsules surrénales, les reins, le tissu lymphoïde des poissons lophobranches. *C.R. Acad. Sci., Paris,* **124**, 1462.

HUOT, E. (1898). Préliminaire sur l'origine des capsules surrénales des poissons lophobranches. *C.R. Acad. Sci., Paris,* **126**, 49.

HUOT, E. (1902). Recherches sur les poissons lophobranches. *Ann. Sci. nat.* Sér. 8, Zool. **14**, 197.

HUSCHKE, E. (1845). Capsules surrénales. In *Encyclopaedia of Anatomy*, transl. by A. J. L. Jourdan, p. 330. Paris.

HUXLEY, J. S. (1932). *Problems of relative growth.* London: Methuen.

HYMAN, C. and CHAMBERS, R. (1943). Effect of adrenal cortical compounds on edema formation of frogs' hind limbs. *Endocrinology,* **32**, 310.

IKEDA, M. (1932). Supplementary investigation of function of the hypophysis; effects of complete destruction of the hypophysis upon the general condition of the rabbit. *Jap. J. Obstet. Gynec.* **15**, 213.

INABA, M. (1891). Notes on the development of the suprarenal bodies in the mouse. *J. Coll. Sci. Tokyo,* **4**, 215.

INGLE, D. J. (1941). The production of glycosuria in the normal rat by means of 17-hydroxy-11-dehydrocorticosterone. *Endocrinology,* **29**, 649.

INGLE, D. J. (1944). The quantitative assay of adrenal cortical hormones by the muscle-work test in the adrenalectomized-nephrectomized rat. *Endocrinology,* **34**, 191.

INGLE, D. J. (1948). Effect of muscle work upon level of blood glucose in the eviscerated rat. *Proc. Soc. exp. Biol., N.Y.,* **67**, 299.

INGLE, D. J. (1951). The functional interrelationship of the anterior pituitary and the adrenal cortex. *Ann. int. Med.* **35**, 652.

INGLE, D. J. (1952a). The role of the adrenal cortex in homeostasis. *J. Endocrin.* **8**, xxiii.

INGLE, D. J. (1952 b). Some further studies on the relationship of adrenal cortical hormones to experimental diabetes. *Diabetes*, **1**, 345.

INGLE, D. J. and BAKER, B. L. (1951). The inhibition of hair growth by estrogens as related to adrenal cortical function in the male rat. *Endocrinology*, **48**, 764.

INGLE, D. J. and BAKER, B. L. (1953 a). *Physiological and therapeutic effects of corticotropin (ACTH) and cortisone.* Amer. Lect. Series no. 179. Springfield, Illinois: C. C. Thomas.

INGLE, D. J. and BAKER, B. L. (1953 b). A consideration of the relationship of experimentally produced and naturally occurring pathologic changes in the rat to the adaptation diseases. *Recent Progr. Hormone Res.* **8**, 143.

INGLE, D. J. and HIGGINS, G. M. (1938). Autotransplantation and regeneration of the adrenal gland. *Endocrinology*, **22**, 458.

INGLE, D. J., HIGGINS, G. M. and KENDALL, E. C. (1938). Atrophy of the adrenal cortex in the rat produced by administration of large amounts of cortin. *Anat. Rec.* **71**, 363.

INGLE, D. J. and LUKENS, F. J. (1941). Reversal of fatigue in the adrenal-ectomized rat by glucose and other agents. *Endocrinology*, **29**, 443.

INGLE, D. J., MEEKS, R. C. and HUMPHREY, L. M. (1953). Effects of exposure to cold upon urinary non-protein nitrogen and electrolytes in adrenalectomized and nonadrenalectomized rats. *Amer. J. Physiol.* **173**, 387.

INGLE, D. J. and NEZAMIS, J. E. (1948). Effect of adrenalectomy upon the tolerance of the eviscerated rat for intravenously administered glucose. *Amer. J. Physiol.* **152**, 598.

INGLE, D. J. and PRESTRUD, M. C. (1949). The effect of adrenal cortex extract upon urinary non-protein nitrogen and changes in weight in young adrenalectomized rats. *Endocrinology*, **45**, 143.

INGLE, D. J., PRESTRUD, M. C., LI, C. H. and EVANS, H. M. (1947). The relationship of diet to the effect of adrenocorticotrophic hormone upon urinary nitrogen, glucose and electrolytes. *Endocrinology*, **41**, 170.

INGLE, D. J., PRESTRUD, M. C. and NEZAMIS, J. E. (1948). Effect of adrenal-ectomy upon level of blood amino acids in the eviscerated rat. *Proc. Soc. exp. Biol., N.Y.*, **67**, 321.

INGLE, D. J., PRESTRUD, M. C. and RICE, K. L. (1950). The effect of cortisone acetate upon the growth of the Walker rat carcinoma and upon urinary non-protein nitrogen, sodium, chloride and potassium. *Endocrinology*, **46**, 510.

INGLE, D. J., SHEPPARD, R., EVANS, J. S. and KUIZENGA, M. H. (1945). A comparison of adrenal steroid diabetes and pancreatic diabetes in the rat. *Endocrinology*, **37**, 341.

INGLE, D. J., SHEPPARD, R., OBERLE, E. A. and KUIZENGA, M. H. (1946). A comparison of the acute effects of corticosterone and 17-hydroxy-corticosterone on body weight and the urinary excretion of sodium, chloride, potassium, nitrogen and glucose in the normal rat. *Endocrinology*, **39**, 52.

274

References

INGLE, D. J. and THORN, G. W. (1941). A comparison of effects of 11-desoxycorticosterone acetate and 17-hydroxy-11-dehydrocorticosterone in partially depancreatized rats. *Amer. J. Physiol.* **132**, 670.

INGLE, D. J., WINTER, H. A., LI, C. H. and EVANS, H. M. (1945). Production of glycosuria in normal rats by means of adrenocorticotropic hormone. *Science*, **101**, 671.

INGRAM, W. R. and WINTER, C. A. (1938). The effects of adrenalectomy upon the water exchange of cats with diabetes insipidus. *Amer. J. Physiol.* **122**, 143.

JACKSON, C. M. (1913). Postnatal growth and variability of the body and of the various organs in the albino rat. *Amer. J. Anat.* **15**, 1.

JACKSON, C. M. (1919). The postnatal development of the suprarenal gland—and the effects of inanition upon its growth and structure in the albino rat. *Amer. J. Anat.* **25**, 220.

JAILER, J. W. and BOAS, N. F. (1950). The inability of epinephrine or adrenocorticotropic hormone to deplete the ascorbic acid content of the chick adrenal. *Endocrinology*, **46**, 314.

JAILER, J. W. and KNOWLTON, A. I. (1950). Simulated adrenocortical activity during pregnancy in an addisonian patient. *J. clin. Invest.* **29**, 1430.

JANES, R. G. and NELSON, W. O. (1942). The influence of diethylstilboestrol on carbohydrate metabolism in normal and castrated rats. *Amer. J. Physiol.* **136**, 136.

JARKOWSKA, E. A. (1932). Sur la question de pathogénie de l'œdème de la grenouille après hypophysectomie. *J. med. Acad. Ukraine*, **4**, 291.

JENKIN, P. M. (1928). Note on the sympathetic nervous system of *Lepidosiren paradoxa*. *Proc. roy. Soc. Edinb.* **48**, 55.

JOHNSEN, V. K. and USSING, H. H. (1949). The influence of the corticotropic hormone from ox on the active salt uptake in the axolotl. *Acta physiol. scand.* **17**, 38.

JONES, M. E. and STEGGERDA, F. R. (1935). Water metabolism in normal and hypophysectomized frogs. *Amer. J. Physiol.* **112**, 397.

JORDAN, H. E. and SPIEDEL, C. C. (1931). Blood formation in the African lungfish, under normal conditions and under conditions of prolonged estivation and recovery. *J. Morph.* **51**, 319.

JØRGENSEN, C. B. (1947). Influence of adenohypophysectomy on the transfer of salt across the frog skin. *Nature, Lond.*, **160**, 872.

JØRGENSEN, C. B., LEVI, H. and USSING, H. H. (1946). On the influence of the neurohypophysial principles on the sodium metabolism in the axolotl (*A. mexicanum*). *Acta physiol. scand.* **12**, 350.

JOST, A. (1953). Problems of fetal endocrinology: The gonadal and hypophysial hormones. *Recent Progr. Hormone Res.* **8**, 379.

JUHN, M. and MITCHELL, J. B. (1929). On endocrine weights in brown leghorns. *Amer. J. Physiol.* **88**, 177.

JUSTIN-BESANÇON, L., LAMOTTE, M., LAMOTTE-BARRILLON, S. and BARBIER, P. (1951). Les hormones corticosurrénales et le métabolisme de l'eau et des électrolytes. *Acta clin. belg.* **6**, 639.

KAR, A. B. (1947a). The action of male and female sex hormones on the adrenals in the fowl. *Anat. Rec.* **97**, 551.

KAR, A. B. (1947b). The adrenal cortex testicular relations in the fowl: the effect of castration and replacement therapy on the adrenal cortex. *Anat. Rec.* **99**, 177.

KEENE, M. F. L. and HEWER, E. E. (1927). Observations on the development of the human suprarenal gland. *J. Anat., Lond.,* **61**, 302.

KEKWICK, A. and PAWAN, G. L. S. (1954). Oral aldosterone. Effect in a case of Addison's disease. *Lancet,* **ii**, 162.

KENDALL, E. C. (1948). The influence of the adrenal cortex on the metabolism of water and electrolytes. *Vitam. & Horm.* **6**, 277.

KEYES, P. H. (1949). Adrenocortical changes in Syrian hamsters following gonadectomy. *Endocrinology,* **44**, 274.

KEYS, A. and WILLMER, E. N. (1932). "Chloride-secreting cells" in the gills of fishes, with special reference to the common eel. *J. Physiol.* **76**, 368.

KINSELL, L. W., MICHAELS, G. D., MARGAN, S., PARTRIDGE, J. W., BOLING, L. and BALCH, H. E. (1954). The case for cortical steroid hormone acceleration of neoglucogenesis from fat in diabetic subjects. *J. clin. Endocrin.* **14**, 161.

KIRKALDY, J. W. (1894). On the head-kidney of Myxine. *Quart. J. micr. Sci.* **35**, 353.

KISCH, B. (1928). Untersuchungen über die Funktion des Interrenalorgans der Selachier. *Pflüg. Arch. ges. Physiol.* **219**, 426.

KISS, T. (1951). Experimentelle-morphologische Analyse der Nebennieren-innervation. *Acta anat., Basel,* **13**, 81.

KLEINHOLZ, L. H. (1938). The pituitary and adrenal glands in the regulation of the melanophores of *Anolis carolinensis. J. exp. Biol.* **15**, 474.

KLEMPERER, P. (1950). The concept of collagen diseases. *Amer. J. Path.* **26**, 505.

KLOSE, H. G. (1941). Über den Einfluss der Kastration auf Schilddrüsen, Hypophyse und Interrenalsystem der Urodelen. Zugleich ein Beitrag zur Morphologie und Histologie diesen Drüsen bei *Triton vulgaris vulgaris* L. und *T. cristatus cristatus* Laur. *Z. wiss. Zool.* **155**, 46.

KLYNE, W. (1953). Some aspects of the stereochemistry of C-20. *Ciba Found. Coll. End.* **7**, 127.

KNAUFF, R. E., NIELSON, E. D. and HAINES, W. J. (1953). Studies on a mineralocorticoid from hog adrenal extract. *J. Amer. chem Soc.* **75**, 4868.

KNIGGE, K. M. (1954a). The effect of acute starvation on the adrenal cortex of the hamster. *Anat. Rec.* **120**, 555.

KNIGGE, K. M. (1954b). The effect of hypophysectomy on the adrenal gland of the hamster (*Mesocricetus auratus*). *Amer. J. Anat.* **94**, 225.

KNOUFF, R. A. and HARTMAN, F. A. (1951). A microscopic study of the adrenal of the brown pelican. *Anat. Rec.* **109**, 161.

KOHN, A. (1902). Das chromaffine Gewebe. *Ergebn. Anat. Entw.-Gesch.* **12**, 253.

KOHNO, S. (1925). Zur vergleichenden Histologie und Embryologie der Nebenniere der Säuger und des Menschen. *Z. ges. Anat.* Abt. 1, **77**, 419.

References

KOJIMA, R. (1928). Qualitative und quantitative morphologische Reaktionen der Nebenniere (Meerschweinen) auf besondere Reize. *Beitr. path. Anat.* **81**, 264.

KÖLLIKER, A. v. (1854). *Mikroskopische Anatomie oder Gewebelehre des Menschen.* 2nd ed. **2**, 377. Leipzig: Hälfte.

KOLMER, W. (1918). Zur vergleichenden Histologie, Zytologie und Entwicklungsgeschichte der Säugernebenniere. *Arch. mikr. Anat.* Abt. I, **91**, 1.

KORENCHEVSKY, V. (1942). The natural relative hypoplasia of organs and the process of ageing. *J. Path. Bact.* **54**, 13.

KOSTITCH, A. and TELEBAKOVITCH, A. (1929). Sur un rythme vaginal chez les animaux ovariectomisés. *C.R. Soc. Biol., Paris,* **100**, 51.

KRAHL, M. E. (1951). The effect of insulin and pituitary hormones on glucose uptake in muscle. *Ann. N.Y. Acad. Sci.* **54**, 649.

KRAHL, M. E. and CORI, C. F. (1947). The uptake of glucose by the isolated diaphragm of normal, diabetic, and adrenalectomized rats. *J. biol. Chem.* **170**, 607.

KRAHL, M. E. and PARK, C. R. (1948). The uptake of glucose by the isolated diaphragm of normal and hypophysectomized rats. *J. biol. Chem.* **174**, 939.

KRAUTER, D. (1952). Über persistierende Vornieren bei Cyprinodonten. *Zool. Anz.* **148**, 23.

KREBS, H. A., EGGLESTON, L. V. and TERNER, C. (1951). In vitro measurements of the turnover rate of potassium in brain and retina. *Biochem. J.* **48**, 530.

KROGH, A. (1904). Some experiments on the cutaneous respiration of vertebrate animals. *Skand. Arch. Physiol.* **16**, 348.

KROHN, P. L. (1955). The effect of ACTH and cortisone on the survival of skin homografts and on the adrenal glands in monkeys (*Macaca mulatta*). *J. Endocrin.* **12**, 220.

KUCNEROWICZ, H. K. (1935). Sur les cellules d'été dans la glande surrénale de la grenouille, *R. esculenta. C.R. Soc. Biol., Paris,* **120**, 486.

KUIZENGA, M. H. and CARTLAND, G. F. (1939). Fractionation studies on adrenal cortex extract with notes on the distribution of biological activity among the crystalline and amorphous fractions. *Endocrinology,* **24**, 526.

KUIZENGA, M. H. and CARTLAND, G. F. (1940). Corticosterone and its esters. *Endocrinology,* **27**, 647.

KUMITA, M. (1909). Über die parenchymatösen Lymphbahnen der Nebenniere. *Arch. Anat. Physiol., Lpz. (Anat. Abt.),* 1909, p. 321.

KUNTZ, A. (1912). The development of the adrenals in the turtle. *Amer. J. Anat.* **13**, 71.

KURIYAMA, S. (1918). The adrenals in relation to carbohydrate metabolism. II. The influence of adrenalectomy upon the glycogenetic power of the liver. *J. biol. Chem.* **34**, 287.

LANCISIUS, J. M. (1722). *Tabulae anatomicae* of Bartholomaeus Eustachius. Amsterdam: R. and G. Wetstenios.

The Adrenal Cortex

LANDAUER, W. (1929). Thyrogenous dwarfism (myxoedema infantilis) in the domestic fowl. *Amer. J. Anat.* **43**, 1.

LANDAUER, W. and ABERLE, S. D. (1935). Studies on the endocrine glands of frizzle fowl. *Amer. J. Anat.* **57**, 99.

LANE, N. and DE BODO, R. C. (1952). Generalized adrenocortical atrophy in hypophysectomized dogs and correlated functional studies. *Amer. J. Physiol.* **168**, 1.

LANMAN, J. T. (1953). The fetal zone of the adrenal gland. Its developmental course, comparative anatomy, and possible physiologic functions. *Medicine, Baltimore,* **32**, 389.

LATIMER, H. B. (1924). Postnatal growth of the body, systems, and organs of the single-comb White Leghorn chicken. *J. agric. Res.* **29**, 363.

LATIMER, H. B. (1925). The relative postnatal growth of the systems and organs of the chicken. *Anat. Rec.* **29**, 367.

LATIMER, H. B. and LANDWER, M. F. (1925). The relative volumes and the arrangement of the cortical and the medullary cells of the suprarenal gland of the chicken. *Anat. Rec.* **29**, 389.

LAWTON, F. E. (1937). The adrenal-autonomic complex in *Alligator mississippiensis*. *J. Morph.* **60**, 361.

LAZAROW, A. and BERMAN, J. (1950). The production of diabetes in rats with cortisone and its relation to glutathione. *Anat. Rec.* **106**, 215.

LEBLOND, C. P. and NELSON, W. O. (1937). Modifications histologiques des organes de la souris après hypophysectomie. *C.R. Soc. Biol., Paris,* **124**, 9.

LEE, M. and AYRES, G. B. (1936). The composition of weight lost and the nitrogen partition of tissues in rats after hypophysectomy. *Endocrinology,* **20**, 489.

LEGALLOIS, J. J. C. (1801). Le sang est-il identique dans tous les vaisseaux qu'il parcourt? Thèse de Paris.

LEONARD, S. L. (1944). Effect of some androgenic steroids on the adrenal cortex of hypophysectomized rats. *Endocrinology,* **35**, 83.

LEONARD, S. L., MEYER, R. K. and HISAW, F. L. (1931). The effect of oestrin on development of the ovary in immature female rats. *Endocrinology,* **15**, 17.

LEONARD, S. L. and RICHTER, J. W. (1936). Endocrine weights of the Bantam fowl. *J. Hered.* **27**, 363.

LEROY, P. and BENOIT, J. (1954). Surrénalectomie bilatérale chez le canard mâle adulte: action sur le testicule. *J. Physiol. Path. gén.* **46**, 422.

LEVER, J. D. (1952). Observations on the adrenal blood vessels in the rat. *J. Anat., Lond.,* **86**, 459.

LEVER, J. D. (1954). Vascular zoning in the adrenal cortex of the normal and hypophysectomized rat with observations on the distribution of lipids. *J. Endocrin.* **10**, 133.

LEVER, J. D. (1955). Cellular-vascular relationships in the adrenal cortex, as studied in the rat. *Ciba Found. Coll. End.* **8**, 42.

LEVER, J. D., CATER, D. B. and STACK-DUNNE, M. P. (1953). Changes in the vascular and lipoid pattern of the adrenal cortex of the rat following hypophysectomy. *Nature, Lond.,* **172**, 33.

References

LEVINSKY, N. G. and SAWYER, W. H. (1952). Influence of the adenohypophysis on the frog water-balance response. *Endocrinology*, **51**, 110.

LEVINSKY, N. G. and SAWYER, W. H. (1953). Significance of the neurohypophysis in regulation of fluid balance in the frog. *Proc. Soc. exp. Biol.*, *N.Y.*, **82**, 272.

LEYDIG, F. (1851). Zur Anatomie und Histologie der *Chimaera monstrosa*. *Arch. Anat. Physiol.*, *Lpz.*, p. 241.

LEYDIG, F. (1852). *Beiträge zur mikroskopischen Anatomie und Entwicklungsgeschichte der Rochen und Haie*. Leipzig.

LEYDIG, F. (1853). *Anatomisch-histologische Untersuchungen über Fische und Reptilien*. Berlin: G. Reimer.

LI, C. H. (1952). On the bioassay of adrenocorticotropic hormone. *Ciba Found. Coll. End.* **4**, 327.

LI, C. H. (1953 a). Some aspects of the preparation and properties of adrenocorticotropic hormone. In: *The Suprarenal Cortex. Proc. 5th Symp. Colston Res. Soc.* p. 1. Ed. Yoffey. London: Butterworth.

LI, C. H. (1953 b). Discussion on paper by S. Zuckerman. In: *The Suprarenal Cortex. Proc. 5th Symp. Colston Res. Soc.* p. 91. London: Butterworth.

LI, C. H., EVANS, H. M. and SIMPSON, M. E. (1943). Adrenocorticotropic hormone. *J. biol. Chem.* **149**, 413.

LIDDLE, G. W., PECHET, M. M. and BARTTER, F. C. (1954). Enhancement of biological activities of corticosteroids by substitution of halogen atoms in 9α position. *Science*, **120**, 496.

LIEBERMAN, S. and TEICH, S. (1953). Recent trends in the biochemistry of the steroid hormones. *Pharmacol. Rev.* **5**, 285.

LINDERHOLM, H. (1952). Active transport of ions through frog skin with special reference to the action of certain diuretics. *Acta physiol. scand.* **27**, suppl. 97, 11.

LLOYD, C. W. (1952). Some clinical aspects of adrenal and fluid metabolism. *Recent Progr. Hormone Res.* **7**, 469.

LOBBAN, M. C. (1952). Structural variations in the adrenal cortex of the adult cat. *J. Physiol.* **118**, 565.

LOEB, R. F. (1932). Chemical changes in the blood in Addison's disease. *Science*, **76**, 420.

LOEB, R. F. (1933). Effect of sodium chloride treatment of a patient with Addison's disease. *Proc. Soc. exp. Biol.*, *N.Y.*, **30**, 808.

LOEB, R. F. (1941). Adrenal cortex insufficiency. *J. Amer. med. Ass.* **116**, 2495.

LOEB, R. F., ATCHLEY, D. W., BENEDICT, E. M. and LELAND, J. (1933). Electrolyte balance studies in adrenalectomized dogs with particular reference to the excretion of sodium. *J. exp. Med.* **57**, 775.

LOEWI, O. and GETTWERT, W. (1914). Über die Folgen der Nebennierenexstirpation. *Pflüg. Arch. ges. Physiol.* **158**, 29.

LONG, C. N. H. (1947). The relation of cholesterol and ascorbic acid to the secretion of the adrenal cortex. *Recent Progr. Hormone Res.* **1**, 99.

LONG, C. N. H. (1950). Factors regulating adrenal cortical secretion. *Symposium on Pituitary-Adrenal function. Amer. Ass. Adv. Sci.* p. 24. Baltimore: Horn-Shafer.

The Adrenal Cortex

LONG, C. N. H. (1952). The role of epinephrine in the secretion of the adrenal cortex. *Ciba Found. Coll. End.* **4**, 139.

LONG, C. N. H. (1953). Influence of the adrenal cortex on carbohydrate metabolism. *Ciba Found. Coll. End.* **6**, 136.

LONG, C. N. H., KATZIN, B. and FRY, E. G. (1940). The adrenal cortex and carbohydrate metabolism. *Endocrinology*, **26**, 309.

LONG, C. N. H. and LUKENS, F. D. W. (1936). The effects of adrenalectomy and hypophysectomy upon experimental diabetes in the cat. *J. exp. Med.* **63**, 465.

LONG, D. A. (1951). Some impressions of the symposium on the influence of the hypophysis and the adrenal cortex on biological reactions. Zürich: Sept. 30–Oct. 2. *Lancet*, **ii**, 733.

LOUSTALOT, P. and MEIER, R. (1954). Vergleichende histochemische Untersuchungen über die Wirkung von Aldosteron und anderen Nebennierenrindenhormonen (Desoxycorticosteron, Corticosteron, Cortison). *Schweiz. med. Wschr.* **84**, 1415.

LOWENSTEIN, B. E. and ZWEMER, R. L. (1946). The isolation of a new active steroid from the adrenal cortex. *Assoc. Study Intern. Secretions* 28th annual meeting.

LUCAS, G. H. W. (1926). Blood and urine findings in desuprarenalized dogs. *Amer. J. Physiol.* **77**, 114.

LUDFORD, R. J. (1933). Vital staining in relation to cell physiology and pathology. *Biol. Rev.* **8**, 357.

LUETSCHER, J. A., Jr., and JOHNSON, B. B. (1953). Chromatographic separation of a corticoid fraction containing sodium-retaining activity from the urine of patients with edema. *Amer. J. Med.* **15**, 417.

LUETSCHER, J. A., Jr., and JOHNSON, B. B. (1954). Chromatographic separation of the sodium-retaining corticoid from the urine of children with nephrosis, compared with observations on normal children. *J. clin. Invest.* **33**, 276.

LUFT, R., OLIVECRONA, H., SJÖGREN, B., IKKOS, D. and LJUNGGREN, H. (1955). Therapeutic results of hypophysectomy in metastatic carcinoma of the breast and in severe diabetes mellitus; adrenocortical function after hypophysectomy. *Ciba Found. Coll. End.* **8**, 438.

LUFT, R., SJÖGREN, B., IKKOS, D., LJUNGGREN, H. and TARUKOSKI, H. (1954). Clinical studies on electrolyte and fluid metabolism. Effect of ACTH, desoxycorticosterone acetate and cortisone; electrolyte and fluid changes in acromegaly. *Recent Progr. Hormone Res.* **10**, 425.

LUKENS, F. D. W. (1953). Hormonal influences in the synthesis of fat from carbohydrate. *Ciba Found. Coll. End.* **6**, 55.

MACCHI, I. A. and HECHTER, O. (1954a). Studies of ACTH action upon perfused bovine adrenals: corticosteroid biosynthesis in isolated glands maximally stimulated with ACTH. *Endocrinology*, **55**, 387.

MACCHI, I. A. and HECHTER, O. (1954b). Studies of ACTH action upon perfused bovine adrenals: minimal ACTH concentration requisite for maximal glandular response. *Endocrinology*, **55**, 426.

References

MACCHI, I. A. and HECHTER, O. (1954c). Studies of ACTH action upon perfused bovine adrenals: duration of ACTH action. *Endocrinology*, **55**, 434.

McEWEN, C., BUNION, J. J., BALDWIN, J. S., KUTTNER, A. J., APPEL, S. B. and KALTMAN, A. J. (1950). Effect of cortisone and ACTH on rheumatic fever. *Bull. N.Y. Acad. Med.* **26**, 212.

MACH, R. S., FABRE, J., DUCKERT, A., BORTH, R. and DUCOMMUN, P. (1954). Action clinique et métabolique de l'Aldostérone (électrocortine). *Schweiz. med. Wschr.* **84**, 407.

McKENZIE, T. and OWEN, R. (1919). *The glandular system in marsupials and monotremes.* Melbourne.

McMANUS, J. F. A. (1946). The histological demonstration of mucin after periodic acid. *Nature, Lond.*, **158**, 202.

McPHAIL, M. K. (1935). Hypophysectomy of the cat. *Proc. roy. Soc.* B, **117**, 45.

MAES, J. (1937). Étude du syndrome surrénoprivé de la grenouille. *Arch. int. Physiol.* **45**, 135.

MAGATH, T. B. (1915). The presence of acidophilous cells in the adrenals of certain amphibians. *Trans. Amer. micr. Soc.* **34**, 154.

MANDL, A. M. (1951). Cyclical changes in the vaginal smear of adult ovariectomized rats. *J. exp. Biol.* **28**, 585.

MANDL, A. M. (1954). The sensitivity of adrenalectomized rats to gonadotrophins. *J. Endocrin.* **11**, 359.

MANDL, A. M. and ZUCKERMAN, S. (1952). Factors influencing the onset of puberty in albino rats. *J. Endocrin.* **8**, 357.

MARENZI, A. D. (1936). Chemical changes in the muscle of the hypophysectomized toad. *Endocrinology*, **20**, 184.

MARENZI, A. D. and FUSTINONI, O. (1938). Potassium of the blood and tissues of the adrenalectomized toad. *Rev. Soc. argent. Biol.* **14**, 118.

MARINE, D. and BAUMANN, E. J. (1927). Duration of life after suprarenalectomy in cats and attempts to prolong it by injections of solutions containing sodium salts, glucose, and glycerol. *Amer. J. Physiol.* **81**, 86.

MARK, J. and BISKIND, G. R. (1941). The effect of long-term stimulation of male and female rats with estrone, estradiol benzoate, and testosterone propionate administered in pellet form. *Endocrinology*, **28**, 465.

MARKEE, J. E., SAWYER, C. H. and HOLLINSHEAD, W. H. (1948). Adrenergic control of the release of luteinizing hormone from the hypophysis of the rabbit. *Recent Progr. Hormone Res.* **2**, 117.

MARRIAN, G. F. (1951). The quantitative determination of the urinary adrenocortical steroids. *J. Endocrin.* **7**, lxix.

MARSHALL, E. K. (1932). Kidney secretion in reptiles. *Proc. Soc. exp. Biol., N.Y.*, **29**, 971.

MARX, W., SIMPSON, M. E., LI, C. H. and EVANS, H. M. (1943). Antagonism of pituitary adrenocorticotropic hormone to growth hormone in hypophysectomized rats. *Endocrinology*, **33**, 102.

MASON, H. L., HOEHN, W. M. and KENDALL, E. C. (1938). Chemical studies of the suprarenal cortex. IV. Structures of compounds C, D, E, F and G. *J. biol. Chem.* **124**, 459.

The Adrenal Cortex

MASON, H. L., HOEHN, W. M., McKENZIE, B. F. and KENDALL, E. C. (1937). Chemical studies of the suprarenal cortex. III. The structures of compounds A, B and H. *J. biol. Chem.* **120**, 719.

MASON, H. L., MYERS, C. S. and KENDALL, E. C. (1936). The chemistry of crystalline substances isolated from the suprarenal gland. *J. biol. Chem.* **114**, 613.

MASUI, K. and TAMURA, Y. (1926). The effect of gonadectomy on the structure of the suprarenal gland of mice, with reference to the functional relation between this gland and the sex gland of the female. *J. Coll. imp. Univ. Tokio,* **7**, 353.

MATTOX, V. R., MASON, H. L., ALBERT, A. and CODE, C. F. (1953). Properties of a sodium-retaining principle from beef adrenal extract. *J. Amer. chem. Soc.* **75**, 4869.

MECKEL, F. (1806). *Abhandlungen aus der menschlichen und vergleichenden Anatomie und Physiologie.* Halle, 1806.

MELLISH, C. H., BAER, A. J. and MACIAS, A. C. (1940). Experiments on the biological properties of stilbestrol and stilbestryl dipropionate. *Endocrinology,* **26**, 273.

MEYER, R. (1912). Nebennieren bei Anencephalen. *Virchows Arch.* **210**, 138.

MICHAEL, M. and WHORTON, C. M. (1951). Delay of the early inflammatory response by cortisone. *Proc. Soc. exp. Biol., N.Y.,* **76**, 754.

MILLER, M. R. (1952). The normal histology and experimental alteration of the adrenal of the viviparous lizard, *Xantusia vigilis. Anat. Rec.* **113**, 309.

MILLER, M. R. (1953). Experimental alteration of the adrenal histology of the urodele amphibian, *Triturus torosus. Anat. Rec.* **116**, 205.

MILLER, R. A. (1950). Cytological phenomena associated with experimental alterations of secretory activity in the adrenal cortex of mice. *Amer. J. Anat.* **86**, 405.

MILLER, R. A. and RIDDLE, O. (1938). Endocrine studies. *Carnegie Inst. Wash. Year Book* no. **37**, 52.

MILLER, R. A. and RIDDLE, O. (1939*a*). Rest, activity and repair in cortical cells of the pigeon adrenal. *Anat. Rec.* **75**, (suppl.) 103.

MILLER, R. A. and RIDDLE, O. (1939*b*). Stimulation of adrenal cortex of pigeons by anterior pituitary hormones and by their secondary products. *Proc. Soc. exp. Biol., N.Y.,* **41**, 518.

MILLER, R. A. and RIDDLE, O. (1941). Cellular response to insulin in suprarenals of pigeons. *Proc. Soc. exp. Biol., N.Y.,* **47**, 449.

MILLER, R. A. and RIDDLE, O. (1942). The cytology of the adrenal cortex of normal pigeons and in experimentally induced atrophy and hypertrophy. *Amer. J. Anat.* **71**, 311.

MILLER, R. A. and RIDDLE, O. (1943). Effects of prolactin and cortical hormones on body weight and food intake of adrenalectomized pigeons. *Proc. Soc. exp. Biol., N.Y.,* **52**, 231.

MILLER, Z. (1954). A study of the lymphocytolytic action of an adrenal cortical extract *in vitro. Endocrinology,* **54**, 431.

MILNE-EDWARDS, H. (1862). *Leçons sur la physiologie et l'anatomie comparée de l'homme et des animaux.* Masson, Paris, **7**, 215.

References

MINIBECK, H. (1939). The selective absorption of sugar in cold-blooded animals and the effect of adrenalectomy and hypophysectomy. *Pflüg. Arch. ges. Physiol.* **242**, 344.

MIRSKY, I. A., STEIN, M. and PAULISCH, G. (1954). The secretion of an antidiuretic substance into the circulation of adrenalectomized and hypophysectomized rats exposed to noxious stimuli. *Endocrinology*, **55**, 28.

MITCHELL, J. B. (1929). Experimental studies on bird hypophysis; effects of hypophysectomy in brown Leghorn fowl. *Physiol. Zool.* **2**, 411.

MITCHELL, R. M. (1948). Histological changes and mitotic activity in the rat adrenal during postnatal development. *Anat. Rec.* **101**, 161.

MIXNER, J. P., BERGMAN, A. J. and TURNER, C. W. (1943). Relation of certain endocrine glands to body weight in growing and mature guinea pigs. *Endocrinology*, **32**, 298.

MIXNER, J. P. and TURNER, C. W. (1942). Progesteron-like activity of some steroid compounds and of diethylstilbestrol in stimulating mammary lobule-alveolar growth. *Endocrinology*, **30**, 706.

MOON, H. D. (1937). Preparation and biological assay of adrenocorticotropic hormone. *Proc. Soc. exp. Biol., N.Y.*, **35**, 649.

MOORE, C. R. (1950). The role of the fetal endocrine glands in development. *J. clin. Endocrin.* **10**, 942.

MORGAGNI, J. B. (1763). *Opuscula miscellanea*, p. 62. De iis quae a Valsalva in Bononiensi Academia Instituti Scientiarum recitata fuerant, Epistola. Neapoli: Simoniana.

MORGAN, J. A. (1951). The influence of cortisone on the survival of homografts of skin in the rabbit. *Surgery*, **30**, 506.

MORRELL, J. A. and HART, G. W. (1941 a). Studies on stilbestrol. I. Some effects of continuous injections of stilbestrol in the adult female rat. *Endocrinology*, **29**, 796.

MORRELL, J. A. and HART, G. W. (1941 b). Studies on stilbestrol. II. The effect of massive doses on normal immature female rats. *Endocrinology*, **29**, 809.

MORRIS, C. J. O. R. and WILLIAMS, D. C. (1953). Estimation of individual adrenocortical hormones in human peripheral blood. *Ciba Found. Coll. End.* **7**, 261.

MOSCHINI, A. (1934). Le phosphogène musculaire après destruction totale des surrénales chez la grenouille. *C.R. Soc. Biol., Paris*, **115**, 215.

MÜLLER, J. (1929). Die Nebennieren von *Gallus domesticus* und *Columba livia domestica*. *Z. mikr.-anat. Forsch.* **17**, 303.

MULON, P. (1903). Divisions nucléaires et rôle germinatif de la couche glomérulaire des capsules surrénales du cobaye. *C.R. Soc. Biol., Paris*, **55**, 592.

MUNRO, S. S. and KOSIN, I. L. (1940). The relative potency of several estrogenic compounds tested on baby chicks of both sexes. *Endocrinology*, **27**, 687.

MUNTWYLER, E., MELLORS, R. C. and MAUTZ, F. R. (1940). Equilibria in the blood in adrenal insufficiency. *J. biol. Chem.* **134**, 345.

NACCARATI, S. (1922). On the relation between the weight of the internal secretory glands and the body weight and brain weight. *Anat. Rec.* **24**, 255.

NAGEL, M. (1836). Ueber die Structur der Nebennieren. *Arch. Anat. Physiol., Lpz.*, p. 365.

NELSON, D. H. and SAMUELS, L. T. (1952). A method for the determination of 17-hydroxycorticoids in blood, 17-hydroxycorticosterone in the peripheral circulation. *J. clin. Endocrin.* **12**, 519.

NELSON, D. H., SAMUELS, L. T. and REICH, H. (1951). The cortical steroid in mammalian blood after ACTH stimulation. *Proc. 2nd clin. ACTH conf.* **1**, 49. New York: Blakiston Co.

NELSON, W. O. (1941). Production of sex hormones in the adrenals. *Anat. Rec.* **81**, 97, suppl.

NELSON, W. O., GAUNT, R. and SCHWEIZER, M. (1943). Effects of adrenal cortical compounds on lactation. *Endocrinology*, **33**, 325.

NELSON, W. O. and MERCKEL, C. G. (1938). Maintenance of spermatogenesis in hypophysectomized mice with androgenic substance. *Proc. Soc. exp. Biol., N.Y.*, **38**, 737.

NICANDER, L. (1952). Histological and histochemical studies on the adrenal cortex of domestic and laboratory animals. *Acta anat.*, suppl. 16 to vol. **14**.

NIELSON, E. D., DRAKE, N. A. and HAINES, W. J. (1951). Incorporation of radioacetate into 17-hydroxycorticosterone by adrenal cortex slices. *Fed. Proc.* **10**, 228.

NOBLE, R. L. (1938). Effect of synthetic oestrogenic substances on the body-growth and endocrine organs of the rat. *Lancet*, **ii**, 192.

NOBLE, R. L. (1939). Effects of synthetic oestrogens and carcinogens when administered to rats by subcutaneous implantation of crystals or tablets. *J. Endocrin.* **1**, 216.

NOBLE, R. L. (1950). Physiology of the adrenal cortex. *The Hormones*, **2**, 65. New York: Academic Press Inc.

NOWELL, N. W. (1953). Some aspects of the adenohypophysial-adrenocortical relationships in the rat. *J. Endocrin.* **9**, xvii.

OCHOA, S. (1933). Über die Energetik der anaeroben Kontraktion von isolierten Muskeln nebennierenlosen Frösche. *Pflüg. Arch. ges. Physiol.* **231**, 222.

OLIVER, G. and SCHAEFER, E. A. (1894). The physiological effects of extracts of the suprarenal capsules. *J. Physiol.* **18**, 231.

OLIVEREAU, M. and FROMENTIN, H. (1954). Influence de l'hypophysectomie sur l'histologie de l'interrénal antérieur de l'anguille (*Anguilla anguilla* L.). *Ann. Endocr., Paris*, **15**, 805.

OWEN, R. (1847). Monotremata. *R. B. Todd's Cyclopaedia of Anatomy and Physiology*, **3**, 366. London.

PABST, M. L., SHEPPARD, R. and KUIZENGA, M. H. (1947). Comparison of liver glycogen deposition and work performance tests for the bioassay of adrenal cortex hormones. *Endocrinology*, **41**, 55.

PADAWER, J. and GORDON, A. S. (1952). A mechanism for the eosinopenia induced by cortisone and by epinephrine. *Endocrinology*, **51**, 52.

References

PADOA, E. (1938). La differenziazione del sesso invertita mediante la somministrazione di ormoni sessuali. Ricerche con follicolina in *Rana esculenta*. *Arch. ital. Anat. Embriol.* **40**, 122.

PANKRATZ, D. S. (1931). The development of the suprarenal gland in the albino rat, with a consideration of its possible relation to the origin of foetal movements. *Anat. Rec.* **49**, 31.

PARK, C. R. and DAUGHADAY, W. H. (1949). Effect of growth hormone on the glucose uptake and glycogen synthesis by the rat diaphragm. *Fed. Proc.* **9**, 212.

PARKER, W. N. (1892). On the anatomy and physiology of *Protopterus annectens*. *Trans. R. Irish Acad.* **30**, 109.

PARKES, A. S. (1945). The adrenal-gonad relationship. *Physiol. Rev.* **25**, 203.

PARKES, A. S. and SELYE, H. (1936). Adrenalectomy of birds. *J. Physiol.* **86**, 25.

PARKINS, W. M. (1931). An experimental study of bilateral adrenalectomy in the fowl. *Anat. Rec.* **51**, 39 (suppl.).

PASQUALINI, R. Q. (1938). Action des extraits pituitaires sur la diurèse du crapaud. *C.R. Soc. Biol., Paris*, **129**, 1240.

PATTEN, B. M. (1948). *Embryology of the pig*. 3rd ed. Philadelphia and Toronto: The Blakiston Company.

PATZELT, V. and KUBIK, J. (1913). Azidophile Zellen in der Nebenniere von *Rana esculenta*. *Arch. mikr.-anat.* **81**, 82.

PEARSE, A. G. E. (1953). *Histochemistry—thoeretical and applied*. London: J. and A. Churchill, Ltd.

PENDE, N. (1909). Sistema nervoso simpatico e glandole a secrezione interna. *Tommasi*, **4**, 732.

PERERA, G. A., PINES, K. L., HAMILTON, H. B. and VISLOCKY, K. (1949). Clinical and metabolic study of 11-dehydro-17-hydroxycorticosterone acetate (Kendall's compound E) in hypertension, Addison's disease, and diabetes mellitus. *Amer. J. Med.* **7**, 56.

PERLMUTTER, M. and GREEP, R. O. (1948). The uptake of glucose and the synthesis of glycogen by the isolated diaphragm of normal and pituitectomized rats. *J. biol. Chem.* **174**, 915.

PERRAULT, C. (1671). *Suite de Mémoires pour servir à l'histoire naturelle des animaux*, **1**, 155. Paris.

PETTIT, A. (1894). Sur les capsules surrénales de l'*Ornithorhynchus paradoxus*. *Bull. Soc. zool. France*, **19**, 158.

PETTIT, A. (1895). Sur les capsules surrénales et la circulation porte surrénale des reptiles. *Bull. Soc. zool. France*, **20**, 233.

PETTIT, A. (1896). Recherches sur les capsules surrénales. *J. Anat., Paris*, **32**, 301.

PICK, J. W. and ANSON, B. J. (1940). The inferior phrenic artery: origin and suprarenal branches. *Anat. Rec.* **78**, 413.

PICKFORD, G. E. (1953*a*). Disturbance of mineral metabolism and osmo-regulation in hypophysectomized *Fundulus*. *Anat. Rec.* **115**, 409.

PICKFORD, G. E. (1953*b*). A study of the hypophysectomized male killifish, *Fundulus heteroclitus* (Linn.). *Bull. Bingham oceanogr. Coll.* **14** (2), 5.

The Adrenal Cortex

PINCUS, G., HECHTER, O. and ZAFFARONI, A. (1951). The effect of ACTH upon steroidogenesis by the isolated perfused adrenal gland. *Proc. 2nd clin. ACTH conf.* **1**, 40. New York: Blakiston Co.

PITOTTI, M. (1938). Interrene, surreni, maturita sessuale e gestazione nei Selaci. *Pubbl. Staz. zool. Napoli*, **17**, 22.

PLAGER, J. E. and SAMUELS, L. T. (1952). Enzyme system involved in the oxidation of carbon-21 in steroids. *Fed. Proc.* **11**, 383.

POLL, H. (1904a). Allgemeines zur Entwickelungsgeschichte der Zwischenniere. *Anat. Anz.* **25**, 16.

POLL, H. (1904b). Die Anlage der Zwischenniere bei der Europäischen Sumpfschildkröte (*Emys europea*) nebst allgemeinen Bemerkungen über die Stamme und Entwickelungsgeschichte des Interrenalsystems der Nebennieren. *Int. Mscht. Anat. Physiol.* **21**.

POLL, H. (1906). Die vergleichende Entwicklungsgeschichte der Nebennierensysteme der Wirbeltiere. In: *Hertwig's Handbuch der vergleichenden und experimentellen Entwicklungslehre der Wirbeltiere*, **3**, 443. Jena: Fischer.

PORGES, O. (1909). Hypoglykaemie bei Morbus Addisonii sowie bei nebennierenlosen Hunden. *Z. klin. Med.* **69**. 341.

PORGES, O. (1910). Ueber Glykogenschwund nach doppelseitigen Nebennierenextirpation. *Z. klin. Med.* **70**, 14.

PORTER, R. W. (1954). The central nervous system and stress-induced eosinopenia. *Recent Progr. Hormone Res.* **10**, 1.

PORTO, J. (1940). Relaciones entre la hipofises y la adrenal en el sapo. *Rev. Soc. argent. Biol.* **16**, 389.

PRICE, D. (1936). Normal development of the prostate and seminal vesicles of the rat with a study of experimental postnatal modifications. *Amer. J. Anat.* **60**, 79.

PRICE, T. D. and RITTENBERG, D. (1950). The metabolism of acetone. I. Gross aspects of catabolism and excretion. *J. biol. Chem.* **185**, 449.

PRICE, W. H., CORI, C. F. and COLOWICK, S. P. (1945). Effect of anterior pituitary extract and of insulin on hexokinase reaction. *J. biol. Chem.* **160**, 633.

PRICE, W. H., SLEIN, M. W., COLOWICK, S. P. and CORI, G. T. (1946). Effect of adrenal cortex extract on the hexokinase reaction. *Fed. Proc.* **5**, 150.

PRUNTY, F. T. G., CLAYTON, B. E., MCSWINEY, R. R. and MILLS, I. H. (1955). Studies of the inter-relationship between the adrenal cortex and ascorbic acid metabolism. *Ciba Found. Coll. End.* **8**, 324.

PRUNTY, F. T. G., MCSWINEY, R. R. and MILLS, I. H. (1955). Biological effects of aldosterone with especial reference to man. *Proc. roy. Soc. Med.* **48**, 629.

QUINAN, C. and BERGER, A. A. (1933). Observations on human adrenals with especial reference to the relative weight of the normal medulla. *Ann. int. Med.* **6**, 1180.

QUIRING, D. P. (1941). The scale of being according to the power formula. *Growth*, **5**, 301.

RABEN, M. S. and WESTERMEYER, V. W. (1952). Differentiation of growth

References

hormone from the pituitary factor which produces diabetes. *Proc. Soc. exp. Biol.*, *N.Y.*, **80**, 83.

RADU, V. (1931). Étude cytologique de la glande surrénale des amphibiens anoures. (Note préliminaire.) *Bull. Histol. Tech. micr.* **8**, 249.

RAGAN, C., HOWES, E. L., PLOTZ, C. M., MEYER, K. and BLUNT, J. W. (1949). Effect of cortisone on production of granulation tissue in the rabbit. *Proc. Soc. exp. Biol.*, *N.Y.*, **72**, 718.

RAMALHO, A. M. (1917). Sur les corps birefringents de l'organe interrénal de la Torpille. *Bull. Soc. portug. Sci. nat.* **8**, 23.

RAMALHO, A. M. (1921). Sur l'appareil surrénal des Téléostéens. *C.R. Soc. Biol.*, *Paris*, **84**, 589.

RAMALHO, A. M. (1923). Sur la morphologie de l'organe interrénal antérieur des Téléostéens. *C.R. Ass. Anat.* **18**, 435.

RASQUIN, P. (1951). Effects of carp pituitary and mammalian ACTH on the endocrine and lymphoid systems of the teleost *Astyanax mexicanus*. *J. exp. Zool.* **117**, 317.

RASQUIN, P. and ATZ, E. H. (1952). Effects of ACTH and cortisone on the pituitary, thyroid and gonads of the teleost *Astyanax mexicanus*. *Zoologica*, *N.Y.*, **37**, 77.

RASQUIN, P. and ROSENBLOOM, L. (1954). Endocrine imbalance and tissue hyperplasia in teleosts maintained in darkness. *Bull. Amer. Mus. nat. Hist.* **104**, 365.

RATHKE, H. (1825). Ueber die Entwicklung der Geschlechtstheile bei den Amphibien. *Beitr. Gesch. Tierheilk.* **3**, 34.

RATHKE, H. (1827). Bemerkungen über den inneren Bau des Querders (*Ammocoetes branchialis*) und des kleinen Neunauges (*Petromyzon planeri*). *Schr. naturf. Ges. Danzig.* **2**, 3.

RATHKE, H. (1828). In: Burdach's '*Physiologie als Erfahrungswissenschaft*', **2**, 600.

RATHKE, H. (1839). *Entwicklungsgeschichte der Natter*. Königsberg.

REEDER, C. F. and LEONARD, S. L. (1944). Alterations in mammary structure following adrenalectomy in the immature male rat. *Proc. Soc. exp. Biol.*, *N.Y.*, **55**, 61.

REESE, A. M. (1931). The ductless glands of *Alligator mississippiensis*. *Smithson misc. Coll.* **82**, no. 16, 1.

REESE, J. D. and MOON, H. D. (1938). The Golgi apparatus of the cells of the adrenal cortex after hypophysectomy and on the administration of the adrenocorticotropic hormone. *Anat. Rec.* **70**, 543.

REICHSTEIN, T. (1936). Über Cortin, das Hormon der Nebennierenrinde. *Helv. chim. acta*, **19**, 29.

REICHSTEIN, T. (1937). Über Bestandteile der Nebennierenrinde. X. Zur Kenntnis des Cortico-sterons. *Helv. chim. acta*, **20**, 953.

REICHSTEIN, T. (1953). In discussion on paper by Bush, I. E. *Ciba Found. Coll. End.* **7**, 228.

REICHSTEIN, T. and EUW, J. v. (1938). Über Bestandteile der Nebennierenrinde. XX. Isolierung der Substanzen Q (Desoxy-corticosteron) und R sowie weitere Stoffe. *Helv. chim. acta*, **21**, 1197.

The Adrenal Cortex

REICHSTEIN, T. and EUW, J. v. (1939). Über Bestandteile der Nebennieren-rinde und verwandte Stoffe. Substanz T. *Helv. chim. acta*, **22**, 1109.

REICHSTEIN, T. and SHOPPEE, C. W. (1943). The hormones of the adrenal cortex. *Vitam. & Horm.* **1**, 345.

REID, E. (1951). Assay of diabetogenic pituitary preparations. *J. Endocrin.* **7**, 120.

REID, E. (1953). Relationship of the diabetogenic activity of ox pituitary extracts to their growth hormone content. *J. Endocrin.* **9**, 210.

REISS, M., BALINT, J., OESTREICHER, F. and ARONSON, V. (1936). Morpho-genetic action and biological standardization of corticotropic sub-stance from anterior lobe. *Endokrinologie*, **18**, 1.

RENNELS, E. G., HESS, M. and FINERTY, J. C. (1953). Response of pre-putial and adrenal glands of the rat to sex hormones. *Proc. Soc. exp. Biol., N.Y.*, **82**, 304.

RENZI, A. A., GILMAN, M. and GAUNT, R. (1954). ACTH-suppressing action of aldosterone. *Proc. Soc. exp. Biol., N.Y.*, **87**, 144.

RETZIUS, A. J. (1819). *Observationes in anatomiam chondropterygiorum*. Lund.

RETZIUS, A. J. (1830). *Anatomisk undersökning öfver några delar Python bivit-tatus jemte comparativa an markningar*. Stockholm.

RETZLAFF, E. W. (1949). The histology of the adrenal gland in the alli-gator lizard, *Gerrhonotus multicarinatus*. *Anat. Rec.* **105**, 19.

RICHTER, C. P. (1936). The spontaneous activity of adrenalectomized rats treated with replacement and other therapy. *Endocrinology*, **20**, 657.

RIDDLE, O. (1923). Studies on the physiology of reproduction in birds. XIV. Suprarenal hypertrophy coincident with ovulation. *Amer. J. Physiol.* **66**, 322.

RIDDLE, O., SMITH, G. C. and MILLER, R. A. (1944). The effect of adrenal-ectomy on heat production in young pigeons. *Amer. J. Physiol.* **141**, 151.

RIOLAN, J. (1629). *Œuvres anatomiques*. Paris.

ROAF, R. (1935). A study of the adrenal cortex of the rabbit. *J. Anat., Lond.*, **70**, 126.

ROBERTS, K. E. and PITTS, R. F. (1952). The influence of cortisone on renal function and electrolyte excretion in the adrenalectomized dog. *Endocrinology*, **50**, 51.

ROBERTSON, J. D. (1954). The chemical composition of the blood of some aquatic chordates, including members of the Tunicata, Cyclostomata and Osteichthyes. *J. exp. Biol.* **31**, 424.

ROBINSON, F. J., POWER, M. H. and KEPLER, E. J. (1941). Two new pro-cedures to assist in the recognition and exclusion of Addison's disease. *Proc. Mayo Clin.* **16**, 577.

ROE, J. H. and KUETHER, C. A. (1943). The determination of ascorbic acid in whole blood and urine through the 2, 4-dinitrophenylhydra-zine derivative of dehydroascorbic acid. *J. biol. Chem.* **147**, 399.

ROEMMELT, J. C., SARTORIUS, O. W. and PITTS, R. F. (1949). Excretion and reabsorption of sodium and water in the adrenalectomized dog. *Amer. J. Physiol.* **159**, 124.

References

ROESEL VON ROSENHOF, A. J. (1758). *Die natürliche Historie der Frösche hiesigen Landes etc.* Nürnberg.

ROGERS, P. V. and RICHTER, C. P. (1948). Anatomical comparison between the adrenal glands of wild Norway, wild Alexandrine and domestic Norway rats. *Endocrinology*, **42**, 46.

ROGOFF, J. M. and STEWART, G. N. (1926a). Studies on adrenal insufficiency in dogs. I. Control animals not subjected to any treatment. *Amer. J. Physiol.* **78**, 683.

ROGOFF, J. M. and STEWART, G. N. (1926b). Studies on adrenal insufficiency in dogs. II. Blood studies in control animals not subjected to treatment. *Amer. J. Physiol.* **78**, 711.

ROGOFF, J. M. and STEWART, G. N. (1928). Studies on adrenal insufficiency. IV. Influence of intravenous injections of Ringer's solution upon the survival period in adrenalectomised dogs. *Amer. J. Physiol.* **84**, 649.

ROGOFF, J. M. and STEWART, G. N. (1929). The survival period of untreated adrenalectomised cats. *Amer. J. Physiol.* **88**, 162.

ROLLESTON, Sir H. D. (1936). *The endocrine organs in health and disease.* London: Oxford Medical Publications, O.U.P.

ROSEMBERG, E., CORNFIELD, J., BATES, R. W. and ANDERSON, E. (1954). Bioassay of adrenal steroids in blood and urine based on eosinophil response: a statistical analysis. *Endocrinology*, **54**, 363.

ROTTER, W. (1949a). Das Wachstum der fötalen und kindlichen Nebennierenrinde. *Z. Zellforsch.* **34**, 547.

ROTTER, W. (1949b). Die Entwicklung der fötalen und kindlichen Nebennierenrinde. *Virchows Arch.* **316**, 590.

ROWNTREE, L. G. and SNELL, A. M. (1931). *A clinical study of Addison's disease.* Philadelphia.

RUPP, J. J., PASCHKIS, K. E. and CANTAROW, A. (1955). Role of potassium in the protein-catabolic effect of cortisone and ACTH. *Endocrinology*, **56**, 21.

RUSSELL, J. A. (1939). Effects of anterior pituitary and adrenal cortical extracts on metabolism of adrenalectomised rats fed glucose. *Proc. Soc. exp. Biol., N.Y.*, **41**, 626.

RUSSELL, J. A. and BENNETT, L. C. (1936). Maintenance of carbohydrate levels in fasted hypophysectomised rats treated with anterior pituitary extracts. *Proc. Soc. exp. Biol., N.Y.*, **34**, 406.

SACARRÃO, G. F. (1943). Contribution à l'étude de tissu conjonctif des capsules surrénales des vertébrés. *Bull. Soc. portug. Sci. nat.* **14**, 167.

SAFFRON, M., GRAD, B. and BAYLISS, M. J. (1952). Production of corticoids by rat adrenal *in vitro. Endocrinology*, **50**, 639.

SAKAMI, W. and LAFAYE, J. M. (1951). The metabolism of acetone in the intact rat. *J. biol. Chem.* **193**, 199.

SALMON, T. N. and ZWEMER, R. L. (1941). A study of the life history of corticoadrenal gland cells of the rat by means of trypan blue injections. *Anat. Rec.* **80**, 421.

SAMUELS, L. T. (1953). Studies of the enzymes involved in the synthesis and degradation of the hormones of the adrenal cortex. *Ciba Found. Coll. End.* **7**, 176.

SAMUELS, L. T., EVANS, G. T. and McKELVEY, J. L. (1943). Ovarian and placental function in Addison's disease. *Endocrinology*, **32**, 422.

SANTA, N. (1940). Valeur endocrine des corps interrénaux des Sélaciens. Présence de l'hormone corticale, type corticostérone. *C.R. Soc. Biol., Paris*, **133**, 417.

SARASON, E. L. (1943). Morphologic changes in the rat's adrenal cortex under various experimental conditions. *Arch. Path.* **35**, 373.

SARETT, L. H. (1948). A new method for the preparation of 17(α)-hydroxy-20-ketopregnanes. *J. Amer. chem. Soc.* **70**, 1454.

SARTORIUS, O. W., CALHOON, D. and PITTS, R. F. (1953). Studies on the interrelationships of the adrenal cortex and renal ammonia excretion by the rat. *Endocrinology*, **52**, 256.

SARTORIUS, O. W. and ROBERTS, K. (1949). The effects of pitressin and desoxycorticosterone in low dosage on the excretion of sodium, potassium and water by the normal dog. *Endocrinology*, **45**, 273.

SAWYER, W. H. (1951 a). Effect of posterior pituitary extract on permeability of frog skin to water. *Amer. J. Physiol.* **164**, 44.

SAWYER, W. H. (1951 b). Effect of posterior pituitary extracts on urine formation and glomerular circulation in the frog. *Amer. J. Physiol.* **164**, 457.

SAWYER, W. H. (1956). The hormonal control of water and salt electrolyte metabolism, with special reference to the Amphibia. *Mem. Soc. Endocrin.* **5**, 44.

SAWYER, W. H. and SAWYER, M. K. (1952). Adaptive responses to neurohypophysial fractions in vertebrates. *Physiol. Zool.* **25**, 84.

SAYERS, G. (1950). The adrenal cortex and homeostasis. *Physiol. Rev.* **30**, 241.

SAYERS, G. and SAYERS, M. A. (1948). The pituitary-adrenal system. *Recent Progr. Hormone Res.* **2**, 81.

SAYERS, G., SAYERS, M. A., LIANG, T.-Y. and LONG, C. N. H. (1945). The cholesterol and ascorbic acid content of the adrenal, liver, brain and plasma following hemorrhage. *Endocrinology*, **37**, 96.

SAYERS, G., SAYERS, M. A., LIANG, T.-Y. and LONG, C. N. H. (1946). The effect of adrenotropic hormone on the cholesterol and ascorbic acid content of the adrenal of the rat and the guinea pig. *Endocrinology*, **38**, 1.

SAYERS, G., WHITE, A. and LONG, C. N. H. (1943). Preparation and properties of pituitary adrenotropic hormone. *J. biol. Chem.* **149**, 425.

SCHAEFER, W. H. (1933). Hypophysectomy and thyroidectomy of snakes. *Proc. Soc. exp. Biol., N.Y.*, **30**, 1363.

SCHARRER, E. and SCHARRER, B. (1954). *Neurosekretion. Handb. mikrosk. Anat. Mensch.* **6**, 953. Berlin: Springer-Verlag.

SCHILLER, E. (1944). Über Kernsekretion in der Nebennierenrinde. *Z. mikr.-anat. Forsch.* **54**, 598.

SCHMIDT (1785). *Dissertatio de glandulis surprarenalibus.* Traji ad viadicum.

References

SCHOOLEY, J. P. (1939). Technique for hypophysectomy of pigeons. *Endocrinology*, **25**, 373.

SCHULER, W., DESAULLES, P. and MEIER, R. (1954). Elektrocortin-Wirkung im Glykogentest. *Experientia*, **10**, 142.

SCHWABE, E. L. and EMERY, F. E. (1939). Progesterone in adrenalectomized rats. *Proc. Soc. exp. Biol., N.Y.*, **40**, 383.

SCHWEIZER, M. and LONG, M. E. (1950). Partial maintenance of the adrenal cortex by anterior pituitary grafts in fed and starved guinea pigs. *Endocrinology*, **46**, 191.

SECKEL, H. P. G. (1940). The influence of various physiological substances on the glycogenolysis of surviving rat liver. *Endocrinology*, **26**, 97.

SEGAL, S. J. (1953). Morphogenesis of the estrogen induced hyperplasia of the adrenals in larval frogs. *Anat. Rec.* **115**, 205.

SELKURT, E. E. (1954). Sodium excretion by the mammalian kidney. *Physiol. Rev.* **34**, 287.

SELYE, H. (1940). Compensatory atrophy of the adrenals. *J. Amer. med. Ass.* **115**, 2246.

SELYE, H. (1941). Effect of dosage on the morphogenetic actions of testosterone. *Proc. Soc. exp. Biol., N.Y.*, **46**, 142.

SELYE, H. (1947). *Textbook of Endocrinology*. Acta Endocrinologica. Canada: Univ. Montreal.

SELYE, H. (1953). The diseases of adaptation: introductory remarks. *Recent Progr. Hormone Res.* **8**, 117.

SELYE, H., COLLIP, J. D. and THOMSON, D. L. (1935). Effect of oestrin on ovaries and adrenals. *Proc. Soc. exp. Biol., N.Y.*, **32**, 1377.

SELYE, H. and STONE, H. (1950). *On the experimental morphology of the adrenal cortex*. American Lecture Series. Springfield, Illinois: C. C. Thomas.

SEMON, R. (1891). Studien über den Bauplan des Urogenitalsystems der Wirbeltiere. Dargelegt an der Entwicklung dieses Organsystems bei *Ichthyophis glutinosus. Jena Z. Med. Naturw.* **26**, 89.

SHAW, J. H. and GREEP, R. O. (1949). Relationships of diet to the duration of survival, body weight and composition of hypophysectomized rats. *Endocrinology*, **44**, 520.

SHEEHAN, H. L. (1939). Simmond's disease due to postpartum necrosis of anterior pituitary. *Quart. J. Med.* **8**, 277.

SHEPHERD, D. M. and WEST, G. B. (1951). Noradrenaline and the suprarenal medulla. *Brit. J. Pharmacol.* **6**, 665.

SHERWOOD JONES, E. (1955). Cellular electrolytes and adrenal steroids. *Nature, Lond.*, **176**, 269.

SHOPPEE, C. W. (1946). Steroids and related compounds. *Annu. Rep. Progr. Chem.* **43**, 200.

SILVETTE, H. and BRITTON, S. W. (1936). Carbohydrate and electrolyte changes in the opossum and marmot following adrenalectomy. *Amer. J. Physiol.* **115**, 618.

SILVETTE, H. and BRITTON, S. W. (1938a). Renal function in normal and adrenalectomized opossums and effects of post-pituitary and cortico adrenal extracts. *Amer. J. Physiol.* **121**, 528.

SILVETTE, H. and BRITTON, S. W. (1938b). Renal function in the opossum and the mechanism of cortico-adrenal and post-pituitary action. *Amer. J. Physiol.* **123**, 630.

SIMPSON, M. E., EVANS, H. M. and LI, C. H. (1943). Bioassay of adrenocorticotropic hormone. *Endocrinology,* **33**, 261.

SIMPSON, S. A. and TAIT, J. F. (1953). Physico-chemical methods of detection of a previously unidentified adrenal hormone. *Mem. Soc. Endocrin.* **2**, 9.

SIMPSON, S. A. and TAIT, J. F. (1955). The possible role of electrocortin in normal human metabolism. *Ciba Found. Coll. End.* **8**, 204.

SIMPSON, S. A., TAIT, J. F. and BUSH, I. E. (1952). Secretion of a salt-retaining hormone by the mammalian adrenal cortex. *Lancet,* **ii**, 226.

SIMPSON, S. A., TAIT, J. F., WETTSTEIN, A., NEHER, R., EUW, J. v. and REICHSTEIN, T. (1953). Isolierung eines neuen kristallisierten Hormons aus Nebennieren mit besonders hoher Wirksamkeit auf den Mineralstoffwechsel. *Experientia,* **9**, 333.

SIMPSON, S. A., TAIT, J. F., WETTSTEIN, A., NEHER, R., EUW, J. v., SCHINDLER, O. and REICHSTEIN, T. (1954a). Konstitution des Aldosterons, des neuen Mineralocorticoids. *Experientia,* **10**, 132.

SIMPSON, S. A., TAIT, J. F., WETTSTEIN, A., NEHER, R., EUW, J. v., SCHINDLER, O. and REICHSTEIN, T. (1954b). Aldosteron. Isolierung und Eigenschaften. Über Bestandteile der Nebennierenrinde und verwandte Stoffe. *Helv. chim. acta,* **37**, 1163.

SIMPSON, S. L. (1948). *Major endocrine disorders.* 2nd ed. London: Oxford University Press.

SINGER, B. and STACK-DUNNE, M. P. (1954). Secretion of aldosterone and corticosterone by the rat adrenal. *Nature, Lond.,* **174**, 790.

SINGER, B. and STACK-DUNNE, M. P. (1955). The secretion of aldosterone and corticosterone by the rat adrenal. *J. Endocrin.* **12**, 130.

SINGER, E. and ZWEMER, R. L. (1934). Microscopic observations of structural changes in the adrenal gland of the living frog under experimental conditions. *Anat. Rec.* **60**, 183.

SLUITER, J. W., MIGHORST, J. C. A. and VAN OORDT, J. T. (1949). The changes in the cytology of the adrenals of *Rana esculenta* following hypophysectomy. *Proc. Kon. Akad. Wetensch. Amsterdam,* **52**, 1214.

SLUSHER, M. A. and ROBERTS, S. (1954). Fractionation of hypothalamic tissue for pituitary-stimulating activity. *Endocrinology,* **55**, 245.

SMITH, C. L. (1950). Seasonal changes in blood sugar, fat body, liver glycogen and gonads in the common frog, *Rana temporaria. J. exp. Biol.* **26**, 412.

SMITH, D. E., LEWIS, L. A. and HARTMAN, F. A. (1943). Sodium retention in the opossum. *Endocrinology,* **32**, 437.

SMITH, H. W. (1932). Water regulation and its evolution in the fishes. *Quart. Rev. Biol.* **7**, 1.

SMITH, H. W. (1951). *The Kidney. Structure and function in health and disease.* New York: Oxford University Press.

SMITH, P. E. (1920). The pigmentary, growth, and endocrine disturbances

References

induced in the anuran tadpole by the early ablation of the pars buccalis of the hypophysis. *Amer. anat. Mem.* **11.**

SMITH, P. E. (1927). The disabilities caused by hypophysectomy and their repair. *J. Amer. med. Ass.* **88,** 158.

SMITH, P. E. (1930). Hypophysectomy and a replacement therapy in the rat. *Amer. J. Anat.* **45,** 205.

SMITH, P. E. and SMITH, I. P. (1923). The function of the lobes of the hypophysis as indicated by replacement therapy with different portions of the ox gland. *Endocrinology,* **7,** 579.

SODDU, L. (1898). Intorno agli effetti dello estirpazione delle capsule surrenali nel cane. *Sperimentale,* **52,** 87.

SOFFER, L. J., EISENBERG, J., IANNACCONE, A. and GABRILOVE, J. L. (1955). Cushing's syndrome. *Ciba Found. Coll. End.* **8,** 487.

SOSKIN, S. and LEVINE, R. (1946). *Carbohydrate Metabolism.* Chicago: University of Chicago Press.

SOULIÉ, A. H. (1903). Développement des capsules surrénales chez les vertébrés supérieurs. *J. Anat., Paris,* **39,** 197.

SPANNER, R. (1929). Über die Wurzelgebiete der Nieren-, Nebennieren-, und Leberpfortader bei Reptilien. *Morph. Jb.* **63,** 314.

SPARROW, E. M. (1953). The behaviour of skin autografts and skin homografts in the guinea-pig, with special reference to the effect of cortisone acetate and ascorbic acid on the homograft reaction. *J. Endocrin.* **9,** 101.

SPEERT, H. (1940). Gynecogenic action of desoxycorticosterone in the rhesus monkey. *Johns Hopk. Hosp. Bull.* **67,** 189.

SPEIRS, R. S. (1953). Eosinopenic activity of epinephrine in adrenalectomised mice. *Amer. J. Physiol.* **172,** 520.

SPEIRS, R. S. and MEYER, R. K. (1949). The effects of stress, adrenal and adrenocorticotrophic hormones on the circulating eosinophils of mice. *Endocrinology,* **45,** 403.

SPEIRS, R. S. and MEYER, R. K. (1950a). A new method of assaying cortical hormones based on a decrease in the number of circulating eosinophils in the mouse. *Anat. Rec.* **106,** 249.

SPEIRS, R. S. and MEYER, R. K. (1950b). The effects of epinephrine on the eosinophils of adrenalectomised mice. *Anat. Rec.* **106,** 289.

SPEIRS, R. S. and MEYER, R. K. (1951). A method of assaying adrenal cortical hormones based on a decrease in the circulating eosinophil cells of adrenalectomised mice. *Endocrinology,* **48,** 316.

SPEIRS, R. S., SIMPSON, S. A. and TAIT, J. F. (1954). Certain biological activities of crystalline electrocortin. *Endocrinology,* **55,** 233.

SPIGELIUS, A. (1627). *De humani corporis fabrica.* Venice.

SPRAGUE, R. G., MASON, H. L. and POWER, M. H. (1951). Physiologic effects of cortisone and ACTH in man. *Recent. Progr. Hormone Res.* **6,** 315.

SPRAGUE, R. G., POWER, M. H., MASON, H. L., ALBERT, A., MATHIESON, D. R., HENCH, P. S., KENDALL, E. C., SLOCUMB, C. H. and POLLEY, H. F. (1950). Observations on physiologic effects of cortisone and ACTH in man. *Arch. int. Med.* **85,** 199.

SRDINKO, O. V. (1898). Über Bau und Entwicklung der Nebennieren des Frösches. *Sitzgsber. böhm. Kaiser Franz Josephs Akad. Prag*, 2 Kl., nr. 12.

SRDINKO, O. V. (1900a). Beiträge zur Kenntnis der Entwicklung der Nebennieren bei den Amphibien. *Sitzgsber. böhm. Kaiser Franz Josephs Akad. Prag*, 2 Kl., nr. 32.

SRDINKO, O. V. (1900b). Bau und Entwicklung der Nebennieren bei Anuren. *Anat. Anz.* **18**, 500.

STACK-DUNNE, M. P. (1953). The action of pituitary preparations on the adrenal cortex. *Ciba Found. Coll. End.* **5**, 133.

STACK-DUNNE, M. P. and YOUNG, F. G. (1951). The properties of ACTH. *J. Endocrin.* **7**, lxvi.

STACK-DUNNE, M. P. and YOUNG, F. G. (1954). Biochemistry of hormones. *Annu. Rev. Biochem.* **23**, 405.

STANNIUS, H. (1839). Die Nebennieren bei Knochenfischen. *Arch. Anat. Physiol., Lpz.*, p. 97.

STANNIUS, H. (1846). *Lehrbuch der vergleichenden Anatomie der Wirbelthiere*, p. 118. Berlin.

STANNIUS, H. (1854). *Zootomie der Fische und Amphibien*, 1854. Berlin.

STARLING, E. H. (1905). The chemical correlation of the functions of the body. *Lancet*, **ii**, 339.

STEIGER, M. and REICHSTEIN, T. (1937). Desoxy-cortico-steron (21-oxy-progesteron) aus Δ⁵-3-oxy-atio-cholensaure. *Helv. chim. acta*, **20**, 1164.

STENGER, A. H. and CHARIPPER, H. A. (1946). A study of adrenal cortical tissue in *Rana pipiens* with special reference to metamorphosis. *J. Morph.* **78**, 27.

STILLING, H. (1887). Zur Anatomie der Nebennieren. *Virchows Arch.* **109**, 324.

STILLING, H. (1898). Zur Anatomie der Nebennieren. *Arch. mikr. Anat.* **52**, 176.

STIPPICH, K. (1935). Die Wirkung des Nebennierenrindenhormons B auf das Herz normaler und nebennierenloser Frösche. *Z. Biol.* **96**, 522.

STOERK, O. and HABERER, H. v. (1908a). Beitrag zur Morphologie des Nebennierenmarkes. *Arch. mikr. Anat.* **72**, 481.

STOERK, O. and HABERER, H. v. (1908b). Über das anatomische Verhalten interrenal eingepflanzten Nebennierengewebes. *Arch. klin. Chir.* **87**, 893.

STÖHR, P. (1935). Zur innervation der menschlichen Nebennieren. *Z. ges. Anat., Abt. I. Z. Anat. EntwGesch.* **104**, 475.

STOLZ, F. (1904). Über Adrenalin und Alkylaminoacetobrenzkatechin. *Ber. dtsch. chem. Ges.* **37**, 4149.

STREETER, D. H. P. and SOLOMON, A. K. (1954). The effect of ACTH and adrenal steroids on K transport in human erythrocytes. *J. gen. Physiol.* **37**, 643.

STREHL, H. and WEISS, O. (1901). Beiträge zur Physiologie der Nebenniere. *Pflüg. Arch. ges. Physiol.* **86**, 107.

SUN, T. P. (1932). Histo-physiogenesis of the glands of internal secretion—thyroid, adrenal, parathyroid and thymus—of the chicken embryo. *Physiol. Zool.* **5**, 384.

References

SUTHERLAND, E. W. (1950). The effect of the hyperglycemic factor of the pancreas and of epinephrine on glycogenolysis. *Recent Progr. Hormone Res.* **5**, 441.

SUTHERLAND, E. W. (1951). The effect of the hyperglycemic factor and epinephrine on enzyme systems of liver and muscle. *Ann. N.Y. Acad. Sci.* **54**, 693.

SWAMMERDAM, J. (1758). *The book of nature or the history of insects.* Translated from the Dutch and Latin by Thomas F. Lloyd. London: C. G. Seyffert.

SWANN, H. G. (1940). The pituitary-adrenocortical relationship. *Physiol. Rev.* **20**, 493.

SWEAT, M. L. (1951). Enzymatic synthesis of 17-hydroxycorticosterone. *J. Amer. chem. Soc.* **73**, 4056.

SWEAT, M. L., ABBOTT, W. E., JEFFRIES, W. M. and BLISS, E. L. (1953). Adrenocortical steroids in human peripheral and adrenal venous blood as determined by fluorescence. *Fed. Proc.* **12**, 141.

SWEAT, M. L. and FARRELL, G. L. (1954). Decline of corticosteroid secretion following hypophysectomy. *Proc. Soc. exp. Biol., N.Y.*, **87**, 615.

SWINGLE, W. W. (1927). Studies on the functional significance of the suprarenal cortex. I. Blood changes following bilateral epinephrectomy in cats. *Amer. J. Physiol.* **79**, 666.

SWINGLE, W. W., FEDOR, E. J., BARLOW, G., COLLINS, E. J. and PERLMUTT, J. (1951). Induction of pseudopregnancy in rat following adrenal removal. *Amer. J. Physiol.* **167**, 593.

SWINGLE, W. W., MAXWELL, R., BEN, M., BAKER, C., LeBRIE, S. J. and EISLER, M. (1954). A comparative study of aldosterone and other adrenal steroids in adrenalectomised dogs. *Endocrinology*, **55**, 813.

SWINGLE, W. W., PARKINS, W. M., TAYLOR, A. R., HAYS, H. W. and MORRELL, J. A. (1937). Effect of oestrus (pseudopregnancy) and certain pituitary hormones on the life-span of adrenalectomised animals. *Amer. J. Physiol.* **119**, 675.

SWINGLE, W. W., PFIFFNER, J. J., VARS, H. M., BOTT, P. A. and PARKINS, W. M. (1933). The function of the adrenal cortical hormone and the cause of death from adrenal insufficiency. *Science*, **77**, 58.

SWINGLE, W. W. and REMINGTON, J. W. (1944). The role of the adrenal cortex in physiological processes. *Physiol. Rev.* **24**, 89.

SWINYARD, C. A. (1937). The innervation of the suprarenal glands. *Anat. Rec.* **68**, 417.

SWINYARD, C. A. (1940). Volume and cortico-medullary ratio of the adult human suprarenal gland. *Anat. Rec.* **76**, 69.

SYDNOR, K. L., KELLEY, V. C., RAILE, R. B., ELY, R. S. and SAYERS, G. (1953). Blood adrenocorticotrophin in children with congenital adrenal hyperplasia. *Proc. Soc. exp. Biol., N.Y.*, **82**, 695.

TÄHKÄ, H. (1951). On the weight and structure of the adrenal glands and the factors affecting them, in children of 0–2 years. *Acta paediatr. Stockh.* Suppl. 81, 1.

TAKAMINE, J. (1901). Adrenalin the active principle of the suprarenal glands, and its mode of preparation. *Amer. J. Pharm.* **73**, 523.

The Adrenal Cortex

TAMURA, Y. (1926). Structural changes in the suprarenal gland of the mouse during pregnancy. *J. exp. Biol.* **4**, 81.

TENG, C. T., SINEX, F. M., DEANE, H. W. and HASTINGS, A. B. (1952). Factors influencing glycogen synthesis by rat liver slices *in vitro*. *J. cell. comp. Physiol.* **39**, 73.

THOMSON, J. S. (1932). The anatomy of the tortoise. *Sci. Proc. roy. Dublin Soc.* **20**, 359.

THORN, G. W., ENGEL, L. L. and LEWIS, R. A. (1941). The effect of 17-hydroxycorticosterone and related adrenal cortical steroids on sodium and chloride excretion. *Science*, **94**, 348.

THORN, G. W. and FORSHAM, P. H. (1950). Adrenal cortical insufficiency. In: *Textbook of Endocrinology*, p. 248. Ed. R. H. Williams. London: W. B. Saunders Co.

THORN, G. W. and FORSHAM, P. H. (1951). In: Harrison's *Principles of Internal Medicine*. London: H. K. Lewis.

THORN, G. W., FORSHAM, P. H. and EMERSON, K. (1949). *The diagnosis and treatment of adrenal insufficiency*. American Lecture Series, no. 29. Springfield, Illinois: C. C. Thomas.

THORN, G. W., FORSHAM, P. H., PRUNTY, F. T. G. and HILLS, A. G. (1948). A test for adrenal cortical insufficiency. The response to pituitary adrenocorticotropic hormone. *J. Amer. med. Ass.* **137**, 1005.

THORN, G. W., HOWARD, R. P., EMERSON, K. and FIROR, W. M. (1939). Treatment of Addison's disease with pellets of crystalline adrenal cortical hormone implanted subcutaneously. *Johns Hopk. Hosp. Bull.* **64**, 339.

THORN, G. W., KOEPF, G. F., LEWIS, R. A. and OLSEN, E. F. (1940). Carbohydrate metabolism in Addison's disease. *J. clin. Invest.* **19**, 813.

TIZZONI, G. (1889). Über die Wirkungen der Exstirpation der Nebennieren auf Kaninchen. Experimentaluntersuchungen. *Beitr. path. Anat.* **5**, 3.

TOBIAN, L. (1949). Cortical steroid excretion in edema of pregnancy, preeclampsia and essential hypertension. *J. clin. Endocrin.* **9**, 319.

TONUTTI, E. (1941). Histochemische Befunde an der Diphtherienebenniere mittels der Plasmalreaktion. *Klin. Wschr.* **20**, 1196.

TONUTTI, E. (1942a). Zur Histophysiologie der Nebennierenrinde: Bau und Histochemie bei der Atrophie des Organs nach Hypophysektomie. *Z. mikr.-anat. Forsch.* **51**, 346.

TONUTTI, E. (1942b). Zur Histophysiologie der Nebennierenrinde: Bau und Histochemie bei der Atrophie des Organs nach Hypophysektomie. *Klin. Wschr.* **21**, 739.

TONUTTI, E. (1942c). Die Umbauvorgänge in den Transformationsfeldern der Nebennierenrinde als Grundlage der Beurteilung der Nebennierenrindenarbeit. *Z. mikr.-anat. Forsch.* **52**, 32.

TONUTTI, E. (1953). Experimentelle Untersuchungen zur Pathophysiologie der Nebennierenrinde. *Verh. dtsch. path. Ges.* **36**, 123.

TONUTTI, E., BAHNER, F. and MUSCHKE, E. (1954). Die Veränderungen der Nebennierenrinde der Maus nach Hypophysektomie und nach ACTH-Behandlung, quantitativ betrachtet am Verhalten der Zellkernvolumina. *Endokrinologie*, **31**, 266.

References

TUCHMANN-DUPLESSIS, H. (1945). Correlations hypophyso-endocrines chez le Triton. Déterminisme hormonal des caractères sexuels secondaires. *Actualités sci. industr. Paris*, **987**, ix. (Histophysiologie.)

TURNER, C. D. (1939). Homotransplantation of suprarenal glands from prepuberal rats into the eyes of adult hosts. *Anat. Rec.* **73**, 145.

TYLER, F. H., MIGEON, C., FLORENTIN, A. A. and SAMUELS, L. T. (1954). The diurnal variation of 17-hydroxycorticosteroid levels in plasma. *J. clin. Endocrin.* **14**, 774. Abst. no. 25.

UCHOA JUNQUEIRA, L. C. (1944). Nota sobre a morfologia das adrenais dos ofidios. *Rev. brasil. Biol.* **4**, 63.

UOTILA, U. U. (1939a). The masculinising effect of some gonadotropic hormones on pullets compared with spontaneous ovariogenic virilism in hens. *Anat. Rec.* **74**, 165.

UOTILA, U. U. (1939b). On the fuchsinophile and pale cells in the adrenal cortex tissue of the fowl. *Anat. Rec.* **75**, 439.

UOTILA, U. U. (1940). The early embryological development of the fetal and permanent adrenal cortex in man. *Anat. Rec.* **76**, 183.

URANO, F. (1908). Neue Versuche über die Salze des Muskels. *Z. Biol.* **52**, 72.

USSING, H. H. (1949). The active ion transport through the isolated frog skin in the light of tracer studies. *Acta physiol. scand.* **17**, 1.

USSING, H. H. (1952). Ion transport across living membranes. In: *Renal Function. Trans. 4th conf. Josiah Macy Jr. Foundation*, p. 88. New York: Corlies, Macy and Co. Inc.

USSING, H. H. (1954). Active transport of inorganic ions. *Symp. Soc. exp. Biol.* **8**, 407.

VACCAREZZA, A. J. (1945). Histofisiología de la corticosuprarrenal. *Medicina, B. Aires*, **5**, 425, and **6**, 46.

VACCAREZZA, A. J. (1946). Histofisiología de la corticoadrenal. *Rev. Asoc. méd. argent.* **60**, 9.

VALLE, J. R. DO (1945). Survival of snakes following adrenalectomy. *Mem. Inst. Butantan.* **18**, 237.

VALLE, J. R. DO and SOUZA, P. R. (1942). Observacões sobre o sistema endocrino dos ofidios. *Rev. brasil. Biol.* **2**, 81.

VAN OORDT, G. J., SLUITER, J. W. and VAN OORDT, P. G. W. J. (1951). Spermatogenesis in normal and hypophysectomised frogs (*Rana temporaria*), following gonadotrophin administration. *Acta endocr., Copenhagen*, **7**, 257.

VAZQUEZ-LOPEZ, E. and WILLIAMS, P. C. (1952). Nerve fibres in the rat adenohypophysis under normal and experimental conditions. *Ciba Found. Coll. End.* **4**, 54.

VENNING, E. H. (1946). Adrenal function in pregnancy. *Endocrinology*, **39**, 203.

VENNING, E. H., KAZMIN, V. E. and BELL, J. C. (1946). Biological assay of adrenal corticoids. *Endocrinology*, **38**, 79.

VENZKE, W. G. (1943). Endocrine gland weights of chick embryos. *Growth*, **7**, 265.

VERNEY, E. B. (1947). The antidiuretic hormone and the factors which determine its release. *Proc. roy. Soc. B*, **135**, 25.

VERZÁR, F. (1952). The influence of corticoids on enzymes of carbohydrate metabolism. *Vitam. & Horm.* **10**, 297.

VERZÁR, F. and WENNER, V. (1948). The action of steroids on glycogen breakdown in surviving muscle. *Biochem. J.* **42**, 48.

VILLEE, C. A. (1943). Effect of adrenocorticotrophic hormone on the interrenals of *Triturus torosus*. *J. Elisha Mitchell sci. Soc.* **59**, 23.

VILLEE, C. A. and HASTINGS, A. B. (1949). The metabolism of C14-labelled glucose by the rat diaphragm *in vitro*. *J. biol. Chem.* **179**, 673.

VILLEE, C. A., WHITE, V. K. and HASTINGS, A. B. (1952). Metabolism of C14-labelled glucose and pyruvate by rat diaphragm muscle *in vitro*. *J. biol. Chem.* **195**, 287.

VINCENT, S. (1896). The suprarenal capsules in the lower vertebrates. *Proc. Birm. nat. Hist. phil. Soc.* **10**, 1.

VINCENT, S. (1897a). The suprarenal bodies in fishes and their relation to the so-called head-kidney. *Trans. zool. Soc. Lond.* **14**, (3) 41.

VINCENT, S. (1897b). On the morphology and physiology of the suprarenal capsules in fishes. *Anat. Anz.* **13**, 39.

VINCENT, S. (1898a). The effects of extirpation of the suprarenal bodies of the eel (*Anguilla anguilla*). *Proc. roy. Soc.* **62**, 354.

VINCENT, S. (1898b). Further observations upon the comparative physiology of the suprarenal capsules. *Proc. roy. Soc.* **62**, 176.

VINCENT, S. (1924). *Internal secretion and the ductless glands.* 3rd ed. London: Edward Arnold and Co.

VINCENT, S. and CURTIS, F. R. (1927). A note on the teleostean adrenal bodies. *J. Anat., Lond.,* **62**, 110.

VOGT, M. (1945). The effect of chronic administration of adrenaline on the adrenal cortex and the comparison of this effect with that of hexestrol. *J. Physiol.* **104**, 60.

VOGT, M. (1955). Medullary-cortical relationships in the adrenal. *Ciba Found. Coll. End.* **8**, 241.

WACHHOLDER, K. and MORGENSTERN, V. (1933). Der Einfluss von Wirksubstanzen der Nebennierenrinde und von Adrenalin auf die Leistungen der Muskeln normaler und nebennierenloser Frösche. *Pflüg. Arch. ges. Physiol.* **232**, 444.

WALAAS, E. and WALAAS, O. (1944). Studies on the compensatory hypertrophy of the fetal adrenal glands in the albino rat, produced by adrenalectomy during pregnancy. *Acta path. microbiol. scand.* **21**, 640.

WALSH, E. L., CUYLER, W. K. and McCULLAGH, D. R. (1934). The physiologic maintenance of the male sex glands. *Amer. J. Physiol.* **107**, 508.

WARING, H. (1935). The development of the adrenal gland of the mouse. *Quart. J. micr. Sci.* **78**, 329.

WARING, H. (1942). Effect of hormones on degeneration of the X zone in the mouse adrenal. *J. Endocrin.* **3**, 123.

WARING, H. and SCOTT, E. (1937). Some abnormalities of the adrenal gland of the mouse and a discussion on cortical homology. *J. Anat., Lond.,* **71**, 299.

References

WATSON, C. (1907). A note on the adrenal gland in the rat. *J. Physiol.* **35**, 230.

WEAVER, H. M. and NELSON, W. O. (1943). Changes in the birefringent material in the adrenal cortex of the rat following administration of adrenotropic hormone. *Anat. Rec.* **85**, 51.

WEBER, A. F., McNUTT, S. H. and MORGAN, B. B. (1950). Structure and arrangement of zona glomerulosa cells in the bovine adrenal. *J. Morph.* **87**, 393.

WELDON, W. F. R. (1884). On the head kidney of Bdellostoma, with a suggestion as to the origin of the suprarenal bodies. *Quart. J. micr. Sci.* **24**, 171.

WELDON, W. F. R. (1885). On the suprarenal bodies of vertebrates. *Quart. J. micr. Sci.* **25**, 137.

WELLS, B. B. and KENDALL, E. C. (1940a). A qualitative difference in the effect of compounds separated from the adrenal cortex on distribution of electrolytes and on atrophy of the adrenal and thymus glands of rats. *Proc. Mayo Clin.* **15**, 133.

WELLS, B. B. and KENDALL, E. C. (1940b). The influence of corticosterone and C_{17}-hydroxydehydrocorticosterone (compound E) on somatic growth. *Proc. Mayo Clin.* **15**, 324.

WELT, I. D., STETTEN, D., INGLE, D. J. and MORLEY, E. H. (1952). Effect of cortisone upon rates of glucose production and oxidation in the rat. *J. biol. Chem.* **197**, 57.

WELT, I. D. and WILHELMI, A. E. (1950). The effect of adrenalectomy and of the adrenocorticotrophic and growth hormones on the synthesis of fatty acids. *Yale J. Biol. Med.* **23**, 99.

WEST, G. B. (1955). The comparative pharmacology of the suprarenal medulla. *Quart. Rev. Biol.* **30**, 116.

WETTSTEIN, A., KAHNT, F. W. and NEHER, R. (1955). The biosynthesis of aldosterone (electrocortin) in the adrenal. *Ciba Found. Coll. End.* **8**, 170.

WEXLER, B. C. (1952). Adrenal ascorbic acid changes in the gonadectomised golden hamster following a single injection of diethylstilbestrol. *Endocrinology*, **50**, 531.

WEXLER, B. C., RINFRET, A. P., GRIFFIN, A. C. and RICHARDSON, H. L. (1955). Evidence for pituitary control of the lipid content of the zona glomerulosa of the rat adrenal cortex. *Endocrinology*, **56**, 120.

WHARTON, T. (1656). *Adenographia.* London.

WHITEHEAD, R. (1933). The involution of the transitory cortex of the mouse suprarenal. *J. Anat., Lond.*, **67**, 387.

WHITNEY, J. E. and BENNETT, L. L. (1952). Inhibition of the catabolic effect of adrenocorticotropic hormone (ACTH) in rats by a diet high in potassium chloride. *Endocrinology*, **50**, 657.

WHITNEY, J. E., BENNETT, L. L. and LI, C. H. (1952). Reduction of urinary sodium and potassium produced by hypophysial growth hormone in normal female rats. *Proc. Soc. exp. Biol., N.Y.*, **79**, 584.

WHITTAM, R. and DAVIES, R. E. (1954). Relations between metabolism and the rate of turnover of sodium and potassium in guinea-pig kidney-cortex slices. *Biochem. J.* **56**, 445.

WILDER, R. M., KENDALL, E. C., SNELL, A. M., KEPLER, E. J., RYNEARSON, E. H. and ADAMS, M. (1937). Intake of potassium, an important consideration in Addison's disease. A metabolic study. *Arch. int. Med.* **59**, 367.

WILKINS, L., BONGIOVANNI, A. M., CLAYTON, G. W., GRUMBACH, M. M. and VAN WYK, J. (1955). Virilizing adrenal hyperplasia: its treatment with cortisone and the nature of the steroid abnormalities. *Ciba Found. Coll. End.* **8**, 460.

WILLIAMS, R. G. (1947). Studies of adrenal cortex: regeneration of the transplanted gland and the vital quality of autogenous grafts. *Amer. J. Anat.* **81**, 199.

WINDHAUS, A. (1910). Über die quantitative Bestimmung des Cholesterins und der Cholesterinester in einigen normalen und pathologischen Nieren. *Hoppe-Seyl. Z.* **65**, 110.

WINTER, C. A., GROSS, E. G. and INGRAM, W. R. (1938). Serum sodium, potassium and chloride after suprarenalectomy in cats with diabetes insipidus. *J. exp. Med.* **67**, 251.

WINTER, C. A., INGRAM, W. R., GROSS, E. G. and SATTLER, D. G. (1941). Sodium and chloride balance in cats as affected by diabetes insipidus, adrenal insufficiency, and pitressin injections. *Endocrinology*, **28**, 535.

WINTERSTEINER, O. and PFIFFNER, J. J. (1935). Chemical studies on the adrenal cortex. II. Isolation of several physiologically inactive crystalline compounds from active extracts. *J. biol. Chem.* **111**, 599.

WITSCHI, E. (1951). Embryogenesis of the adrenal and the reproductive glands. *Recent Progr. Hormone Res.* **6**, 1.

WITSCHI, E. (1953). The experimental adrenogenital syndrome in the frog. *J. clin. Endocrin.* **13**, 316.

WITSCHI, E., BRUNER, J. A. and SEGAL, S. J. (1953). The pluripotentiality of the mesonephric blastema. *Anat. Rec.* **115**, 381.

WOODBURY, D. M., CHENG, C. P., SAYERS, G. and GOODMAN, L. S. (1950). Antagonism of adrenocorticotrophic hormone and adrenal cortical extract to desoxycorticosterone: electrolytes and electroshock threshold. *Amer. J. Physiol.* **160**, 217.

WOOLLEY, G. W. (1950). Experimental endocrine tumours with special reference to the adrenal cortex. *Recent Progr. Hormone Res.* **5**, 383.

WOTTON, R. M. and ZWEMER, R. L. (1943). A study of the cytogenesis of cortico-adrenal cells in the cat. *Anat. Rec.* **86**, 409.

WRIGHT, A. and CHESTER JONES, I. (1955). Chromaffin tissue in the lizard adrenal gland. *Nature, Lond.*, **175**, 1001.

YEAKEL, E. H. (1946). Changes with age in adrenal glands of Wistar albino and gray Norway rats. *Anat. Rec.* **96**, Abst. 59.

YOFFEY, J. M. (1950). The mammalian lymphocyte. *Biol. Rev.* **25**, 314.

YOFFEY, J. M. (1953). The suprarenal cortex: the structural background. In: *The Suprarenal Cortex. Proc. 5th Symp. Colston Res. Soc.* p. 31. Ed. Yoffey. London: Butterworth.

References

YOFFEY, J. M. (1955). Some observations on the problem of cortical zoning. *Ciba Found. Coll. End.* **8**, 18.

YOFFEY, J. M. and BAXTER, J. S. (1947). The formation of birefringent crystals in the suprarenal cortex. *J. Anat., Lond.*, **81**, 335.

YOFFEY, J. M. and BAXTER, J. S. (1949). Histochemical changes in the suprarenal gland of the adult male rat. *J. Anat., Lond.*, **83**, 89.

YOUNG, F. G. (1952). The growth hormone and carbohydrate metabolism. *Ciba Found. Coll. End.* **4**, 255.

YOUNG, J. Z. (1950). *The life of vertebrates*. Oxford: Clarendon Press.

ZAFFARONI, A. and BURTON, R. B. (1953). Corticosteroids present in adrenal vein blood of dogs. *Arch. Biochem.* **42**, 1.

ZAFFARONI, A., HECHTER, O. and PINCUS, G. (1951 a). Adrenal conversion of C^{14}-labelled cholesterol and acetate to adrenal cortical hormones. *J. Amer. chem. Soc.* **73**, 1390.

ZAFFARONI, A., HECHTER, O. and PINCUS, G. (1951 b). Conversion of C^{14}-labelled acetate and cholesterol to adrenocortical hormones by perfused adrenal glands. *Fed. Proc.* **10**, 150.

ZALESKY, M., WELLS, L. J., OVERHOLSER, M. D. and GOMEZ, E. T. (1941). Effects of hypophysectomy and replacement therapy on the thyroid and adrenal glands of the male ground squirrel. *Endocrinology*, **28**, 521.

ZARROW, M. X. (1942). Protective action of desoxycorticosterone acetate and progesterone in adrenalectomized mice exposed to low temperatures. *Proc. Soc. exp. Biol., N.Y.*, **50**, 135.

ZARROW, M. X. and MONEY, W. L. (1949). Involution of the adrenal cortex of rats treated with thiouracil. *Endocrinology*, **44**, 345.

ZECKWER, I. T. (1938). The adrenals and gonads of rats following thyroidectomy considered in relation to pituitary histology. *Amer. J. Physiol.* **121**, 224.

ZIZINE, L., SIMPSON, M. E. and EVANS, H. M. (1950). Direct action of the male sex hormone on the adrenal cortex. *Endocrinology*, **47**, 97.

ZONDEK, B. (1936). Impairment of anterior pituitary functions by follicular hormone. *Lancet*, **ii**, 842.

ZUCKERMAN, S. (1941). The effect of desoxycorticosterone on the endometrium of monkeys. *J. Endocrin.* **2**, 311.

ZUCKERMAN, S. (1952). The influence of environmental changes on the pituitary. *Ciba Found. Coll. End.* **4**, 213.

ZUCKERMAN, S. (1953). The adreno-genital relationship. In: *The suprarenal cortex. Proc. 5th Symp. Colston Res. Soc.* p. 69. Ed. J. M. Yoffey. London: Butterworth.

ZUCKERMAN, S. (1955). The possible functional significance of the pituitary portal vessels. *Ciba Found. Coll. End.* **8**, 551.

ZWARENSTEIN, H. (1933). Metabolic changes associated with endocrine activity and the reproductive cycle in *Xenopus laevis*. Part 2. The effect of hypophysectomy on the potassium content of the serum. *J. exp. Biol.* **10**, 201.

ZWEMER, R. L., WOTTON, R. M. and NORKUS, M. G. (1938). A study of cortico-adrenal cells. *Anat. Rec.* **72**, 249.

INDEX

(*Italic figures indicate main references in the text.*)

Index

Index

cortisone (*cont.*)
  reptiles, 181, 183; blood glucose, 182; water and salt-electrolyte metabolism, 180; teleosts, 138; corpuscles of Stannius, 140
cortone, *see* cortisone
Cushing's syndrome, 70, 93, 94, 102, *118*
cyclopentenophenanthrene system, 29

dark cells (zona reticularis), *see* cells
DCA, *see* deoxycorticosterone acetate
deamination, 68
decalcification, 93
deciduoma, 108
dehydration, 75, 116, 119
dehydrocorticosterone
  Amphibia, 162
  chemistry and properties, 31, 33, 34, 36, *39*, 40, 41, 42
  Eutheria, carbohydrate metabolism, 64
dehydroepiandrosterone, 45, 120
11-dehydro-17-hydroxycorticosterone, *see* cortisone
dehydrogenase, 91, 120
deoxycorticosterone
  Amphibia, 154, 160, 162, 164
  biosynthesis, 45, 46, 54
  birds, 200, 201
  chemistry and properties, 31, 34, *35-7*, 38, 41
  elasmobranchs, 130
  Eutheria, Addison's disease, 117; adrenal-gonad relationships, 101; carbohydrate metabolism, 64; lactation, 99; mammary tissue, 99; mineral metabolism, 83, 87, 88, 89, 91; relation to cortical zones, 235
  marsupials, 216
  reptiles, 181
  teleosts, 235
deoxycorticosterone acetate, *see* deoxycorticosterone
deoxycorticosterone glycoside, 90; *and see* deoxycorticosterone
deoxycortone, *see* deoxycorticosterone
dermis, 93
deuterium, 70
diabetes, adrenal steroid, 64

diabetes insipidus, 80, 81, 82
diabetes mellitus, 62, 63
  Cushing's syndrome, 118
  formation of glycogen, 67
  mineral metabolism, 84
diabetogenic hormone, 62
diarrhoea, 58, 116
diethylstilboestrol, *see* stilboestrol
digitonin reaction, 20
diphosphopyridine nucleotide (DPN), 44
diuresis, after adrenalectomy, 75
DOC, *see* deoxycorticosterone

elastic tissue, 12, 176, 194
electrocortin, *see* aldosterone
embryology
  adrenal, Amphibia, 225, 226; birds, 226, 227; cyclostomes, 223; elasmobranchs, 223, 224; mammals, 227, 228; man, 228, 229; reptiles, 226; teleosts, 224
  corpuscle of Stannius, teleosts, 139, 223, 224, 225
enucleation, adrenal, 236
enzyme systems
  general, 53, 60, 79, 82, 89, 91, 120
  hydroxylation, 44, 45, 46
eosinopaenia, 95
eosinophil cells, *see* cells
eosinophil test, 36, 49, 94, 95
epidermis, 93
epinephrine, *see* adrenaline
equation
  allometric, 6
  surface-area/weight relation, 6
equilenin, 156
ergosterol, 29
erythrocytes, *see* cells
erythroid cells, *see* cells
erythropoiesis, 96
escalator theory, *see* cell migration theory
esterase, 89
ethinyl-oestradiol, 156
exhaustion, stage of, 121

fasciculata, *see* zona fasciculata
fat, 18
  brown, 4, 142
  metabolism, *69-70*
  white, 4; *and see* lipid

307

20-2

Index

Index

mineral metabolism (*cont.*)
  birds, 197
  elasmobranchs, 129, 130
  Eutheria, 72–91; Addison's disease,
    116; adrenogenital syndrome, 119;
    Cushing's syndrome, 118; Sim-
    mond's disease, 118
  marsupials, 216, 217
  reptiles, 180, 183, 184
  teleosts, 138, 139
mineralocorticoid, 37, 38, 230, 235
mitochondria
  cortical cells, birds, 193, 194, 196,
    197, 198, 201, 204; Eutheria, 16,
    17, 18, 25; reptiles 177
  kidney, 82, 90
mitotic division, 91, 94
  in adrenal, 198, 231–4, 237, 242, 243
monocytes, *see* cells
muscle
  asthenia, Amphibia, 157; Eutheria, 58
  chloride, Eutheria, 75
  glycogen, Eutheria, 58, 61
  potassium, Amphibia, 159, 160, 161,
    162; Eutheria, 58, 73, 75, 82; rep-
    tiles, 180, 183; teleosts, 138
  sodium, Amphibia, 159, 160, 162,
    163, 166; Eutheria, 58, 73, 75, 81;
    marsupials, 217; reptiles, 180, 183;
    teleosts, 138
  water, Amphibia, 159, 160, 161;
    Eutheria, 58, 75, 76; marsupials,
    217; reptiles, 180, 183
  work, Amphibia, 158; Eutheria, 71–2
myeloid elements, *see* cells

nausea, 58, 116
neoplasm, cortical, 119
nephrectomy, 68
nephrogenic blastema, *see* blastema
nephrogenic tissue, 219, 221, 222, 224
nerve fibres
  adenohypophysis, 55
  adrenal, cortex, 22; medulla, 22
  hypothalamus, 56
neurohumour, 56
neurohypophysectomy, 23
neurohypophysial extracts, 163
neurohypophysis, 57; removal, 80, 81
  Amphibia, 164, 165; *and see* pituitary,
    posterior lobe
neutrophilic cells, *see* cells

nitrogen excretion, 63–5
non-protein nitrogen, blood
  Addison's disease, 116
  adrenalectomy, 74
Noradrenaline, 54, 56, 62
  distribution in vertebrate chromaffin
    tissue, 243, 244
  methylation, 243
  production by chromaffin tissue, 14
  reptiles, 175, 178
norepinephrine, *see* noradrenaline

obesity, Cushing's syndrome, 118
oedema
  Amphibia, 157, 159
  Eutheria, 97
oestradiol, birds, 204
oestrogens
  adrenal origin, 106
  Amphibia, 156
  birds, 204, 205
  Eutheria, 93, 100–5, 110, 113
oestrone, 29, 30, 156, 204
oestrous cycle, 7, 11, 106, 107
opisthotonus, 128
organ of Giacomini, *see* interrenal,
    teleost, anterior
osmic acid, 18
osmophilia
  Amphibia, 149, 152
  birds, 192
  elasmobranchs, 126
  Eutheria, 18, 19, 21
  reptiles, 176, 177
osmoreceptors
  Amphibia, 167
  Eutheria, 80
osmoregulation, 244, 245
  teleosts, 138; *and see* mineral metabolism
osteoblasts, *see* cells
osteogenesis, 91
osteoporosis, 92, 93
  Cushing's syndrome, 118
ovariectomy, 106; *and see* adrenal-gonad
    relationships
ovary, text-fig. 14, 106, 107, 113
oxycellulose adsorption, preparation of
    ACTH, 47

packed cell volume, *see* haematocrit
pancreas, 3
  in fat metabolism 70

311

Index

Index

Index

Printed in the United States
By Bookmasters